Investigating
Equipment Failures Through
Root Cause Failure Analysis

9th Discipline on World Class Maintenance Management

(Volume 9)

AF440721

By Rolly Angeles

Investigating Equipment Failures through Root Cause Failure Analysis

9th Discipline on World Class Maintenance Management (Volume 9)

Copyright @2021 Rolly Angeles

Original Printing in the Philippines. All rights reserved.

No part of this publication and document may be reproduced, stored in a retrieval system, or transmitted in any form or by any means through electronic, mechanical, photocopying, recording, or otherwise without the prior written notice from the author of this book.

10 9 8 7 6 5 4 3 2 1

First Edition: Printed in the Philippines by **Central Books**
Head Office: Phoenix Bldg. 927 Quezon Avenue, Quezon City, Philippines 1101

The National Library of Philippine Catalogue

Kindle Asin Amazon: B09H7LNB79
Paperback Amazon: 9798485193621
Hardcover Amazon: 9798485195083
Paperback Ingramspark: 9798885260015
Hardcover Ingramspark: 979-8885260022

Published by:

Published by Amazon KDP for Kindle and Paperback 2021
Published by Ingram Spark for Hardcover Format 2021

This book was designed and produced by:

RSA Reliability and Maintenance Consultancy Firm, 2021
Sta. Rosa, Laguna, Philippines 4026
Website: http://www.rsareliability.com
Email: rollyangeles@rsareliability.com

First Printing: October 2021

All Rights Reserved ® RSA Reliability and Maintenance Consultancy Firm

Disclaimer: The author and publisher of this book provide no guarantees concerning the level of success you may experience by following the advice and strategies contained in this book, and you accept the risk that results will differ for each industry. The testimonials and case studies stated in this book show exceptional results, which may or may not apply by just reading this book, and are not intended to represent or guarantee that the reader will achieve the same or similar results without the proper consultation and guidance of a qualified RCFA third party practitioner from RSA Reliability and Maintenance Consultancy Firm.

<u>Original Concept of World Class Maintenance – The 12 Disciplines</u>

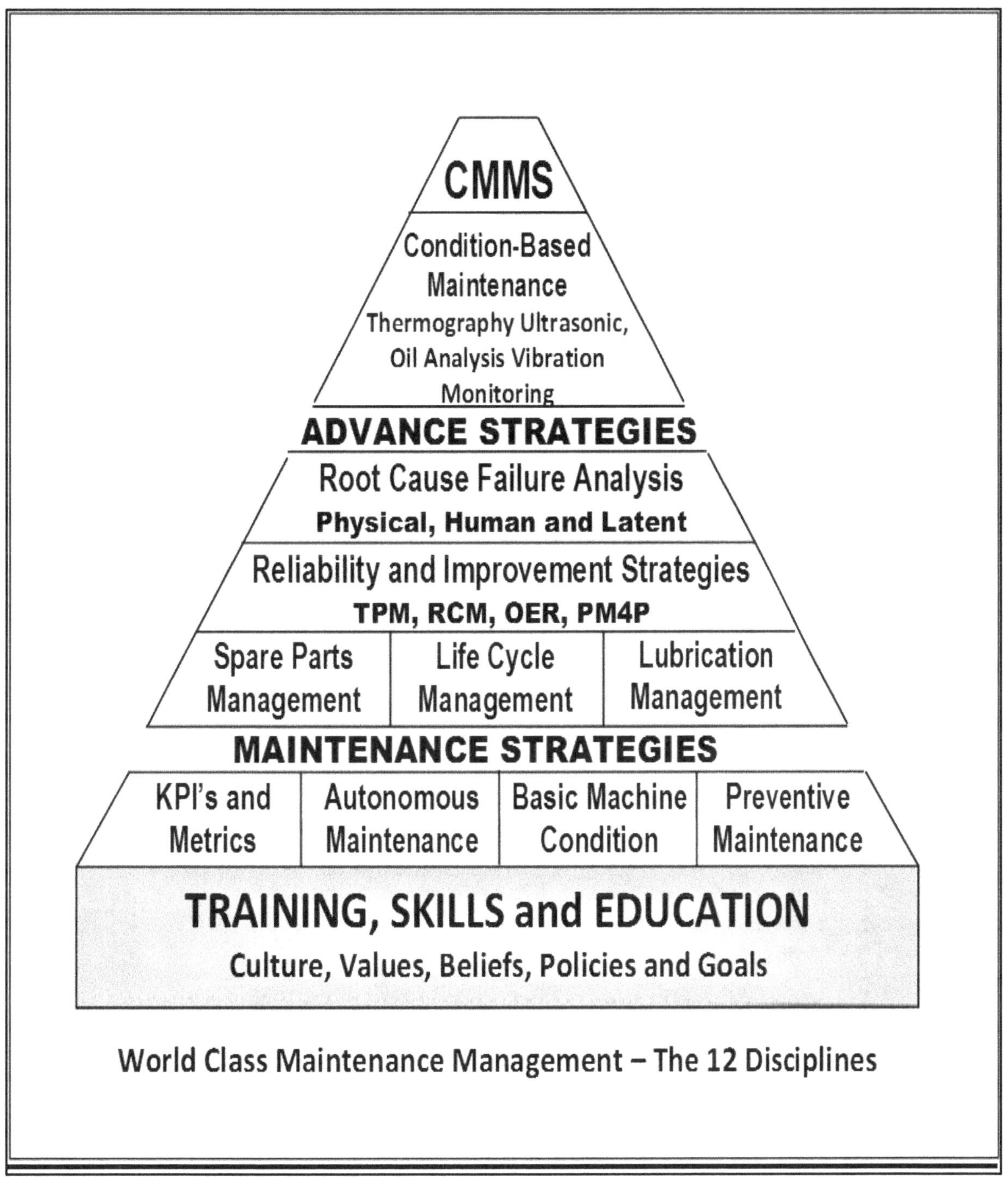

Figure A: Original Concept of World Class Maintenance Management – The 12 Disciplines

Figures

Note: For simplification purposes, all pictures, graphs, charts, tables, drawings in this book will be referred to as figures consecutively below.

<u>**List of Figures**</u>

Chapter 4: The Life-blood of Root Cause Failure Analysis

5: Understanding the Physical Cause of Failures

Chapter 6: The Study of Human Errors

Chapter 7: Understanding the Latent Cause of the Problem

Chapter 8: Guidelines for Conducting an RCFA Investigation

Chapter 9: RCFA Corrective Actions and Countermeasures

Serving Maintenance Mankind Worldwide

RSA Maintenance Books Collection in Series

Table of Contents

In Loving Memory

This book is dedicated to my Mom Cecilia Santiago Angeles who passed away last July 28, 2021, at the age of 88 years old. She audited my first book on World Class Maintenance Management – The 12 Disciplines and my other two books on Maintenance – Roadmap to Reliability as well as Reliability – Shared Responsibility for Operators and Maintenance. May you be with Dad and Rest in Peace.

Figure B: In Loving Memory of my Mom

<u>About the Author</u>

Rolly is a seasoned international maintenance and reliability consultant with over 30 years of solid experience in the field. He has been invited to different countries and has conducted reliability and maintenance training in United Arab Emirates, Qatar, India, Malaysia, Indonesia, Brunei, Thailand, Nigeria, Bangladesh, South Africa, China, and Botswana. His maintenance training portfolio includes maintenance and reliability courses on TPM, Lubrication, Tribology, Condition-Based Maintenance, RCM, RCFA, TPM Planned Maintenance, Autonomous Maintenance 7 Steps, World Class Maintenance Management, The 12 Disciplines, Oil Contamination Control, Maintenance Indices, and KPI's, Maintenance and Reliability Management Strategies and much more. Rolly previously worked with Amkor Technology Philippines as a TPM Senior Engineer, an industry engaged in manufacturing Integrated Circuit products and spearheaded their Planned Maintenance organization, composed of maintenance managers and engineers. He was also responsible for dramatically reducing unplanned breakdowns in their TPM Journey and RCM implementation on their Facilities AHU units and substation equipment. Rolly is currently working as an independent reliability and maintenance consultant.

Rolly is a graduate of Mechanical Engineering from Mapua Institute of Technology in the Philippines, batch 1985, and passed the licensure board examination the following year in 1986. With 30 years of solid experience, he had worked in various industries from shipping, woodworking, foundry, cast-iron machining, assembly lines, semiconductor manufacturing, and the mining industry. Here, he gained hands-on experience and understanding of TPM and RCM, respectively, a strategy from both the west and the east. His last corporate employment was in 2002, where he worked as a technical training specialist at Lepanto Consolidated Mining Industry. In 2005, Rolly retired early from the industry and decided to establish his own consulting business, RSA Reliability and Maintenance Consultancy Firm, where he dedicates his time and passion for working as an independent reliability and maintenance consultant. He provides in-house training, consultation, and facilitation to different maintenance and reliability best practices. You can reach him through his email at rollyangeles@rsareliability.com, or you can visit his website at http://www.rsareliability.com.

- The RCFA training is a perfect combination of technical and inspirational training. Mr. Angeles used very practical and real-life situations to explain his points in videos regarding Michael Jackson which creates a relaxing atmosphere. Mr. Angeles is very patient in answering all our questions. We see his passion and expertise very much with the subject matter. He has a broad knowledge in other fields like history when he discussed JFK's life story. I personally enjoyed and learned a lot from this training. I will truly keep all my notes for reference in the future. Mr. Angeles is indeed a world-class trainer. *From Lisaer Pineda, Process Engineer, Petron Corporation*

<u>Acknowledgment</u>

First, I would like to thank all my students who have attended my training on Root Cause Failure Analysis in the past. I hope that the knowledge gained from the RCFA training provides these people with the groundwork for conducting a Root Cause Failure Analysis investigation in their industries.

This is to acknowledge a dear friend of mine, Mr. C. Robert Nelms, or Bob, as we call him. Bob is the president of Failsafe Network and teaches The Latent Cause Experience to industries. His knowledge and mentorship allowed me to open my eyes to understand the real meaning of what Root Cause Failure Analysis actually is all about and what it is not.

I would also like to thank my family, most especially my three kids: Marie Vic, Kathleen Kay, and Christian Joseph, my wife, Marites, and my wonderful granddaughter Kalie.

I would like to thank my mom who recently passed away this July 2021 for the values, moral support, and wisdom she has provided me throughout my life. I would want to dedicate this book to my mom. She also edited my first three books on World Class Maintenance Management, The 12 Disciplines, Maintenance – Roadmap to Reliability and Reliability – A Shared Responsibility for both operators and maintenance.

Trust in the Lord with all your heart, and lean not on your own understanding, in all your ways, acknowledge Him, and He will make your path straight – Proverbs 3.5

Take Quiz on RCFA Part 1

1. Root Cause Failure Analysis is designed to;
   a) Sharpen our response to impending failure
   b) Eliminate the risks of recurrence of the same cause of failure
   c) Help quickly recover from an unexpected failure
   d) Apply only to high visibility failures

2. Currently, Root Cause Analysis is not recommended to;
   a) Safety Incidents
   b) Fires and Explosions
   c) Punish the culprit
   d) Chronic Events

3. Ideally, who should lead a Root Cause Failure Analysis?
   a) Training department representative
   b) Principal Investigator
   c) Cross selection of engineers
   d) Technical expert in the failure

4. According to Roger Boisjoly, from Morton Thiokol, most RCFA efforts fail because of;
   a) Lack of interest in the workforce
   b) Poor or no RCFA training
   c) Management Support and Commitment
   d) Lack of expertise available in the facility

5. Root Cause Failure Analysis concludes when;
   a) The failed component is found
   b) Latencies are uncovered
   c) Sensitive information is uncovered
   d) The guilty person is punished

6. The science behind any successful RCFA methodology typically is associated with;
   a) Cause and Effect relationship
   b) Software automation
   c) Categorization
   d) Witch-Hunting

7. In RCFA, a hypothesis that is proven to be true is called a;
   a) Failure Modes
   b) Probable Causes
   c) Evidence and Fact
   d) Casual Factors

8. A true Root Cause Failure Analysis means uncovering of;
   a) Physical roots
   b) Human roots
   c) Latent roots
   d) All of the above

9. An example of a Latent Root Cause is;
   a) Communication problem
   b) Bearing fatigue
   c) Obsolete procedure
   d) Inspection not completed

10. For the greatest return, RCFA should be conducted on;
   a) All failures
   b) Events with greatest annualize impact
   c) High-cost single events
   d) None of the above

11. In conducting a failure analysis the investigation will end on;
   a) Latent Cause of the Problem
   b) Physical Cause of the Problem
   c) Human Cause of the Problem
   d) System Cause of the Problem

12. In conducting failure analysis, we are specifically interested in;
   a) How the component failed?
   b) The person who is responsible
   c) Effect of lubrication in the failure
   d) Why the failure had occurred?

13. Being proactive means;
   a) Addressing the root cause of the problem
   b) Preventing or Predicting the problem
   c) Performing a Root Cause Investigation and implementing Corrective Actions
   d) Being 1 step ahead of the failure

14. In effectively pinpointing the cause of the problem means;
   a) Separate the facts from the fiction
   b) Address all the causes of the problem
   c) Provide countermeasures on all causes
   d) Restoring the equipment

15. The lifeblood on any Root Cause Failure Analysis is;
   a) Data
   b) People
   c) Evidence
   d) The Output of Brainstorming

16. Flawed management decision caused the lives of the seven astronauts of the Challenger Disaster in 1986. This is a classic example of;
   a) Physical Cause
   b) Human Cause
   c) System Cause
   d) Latent Cause

17. Most abrasive wear starts off as what type of wear;
    a) Adhesive wear
    b) Fatigue wear
    c) Abrasive wear
    d) Erosive wear

18. Continuous bombardment of particles, either fluid or gas, will definitely result to;
    a) Adhesive wear
    b) Fatigue wear
    c) Abrasive wear
    d) Erosive wear

19. What kind of stress can be experienced on shafts?
    a) Tensile Stress
    b) Compressive Stress
    c) Shear Stress
    d) Torsional Stress

20. This type of wear leads to a fracture under repeated or fluctuating stress.
    a) Adhesive wear
    b) Fatigue wear
    c) Abrasive wear
    d) Erosive wear

21. This type of wear is caused by a micro or small crack or fracture and propagates to a point of rupture.
    a) Adhesive wear
    b) Fatigue wear
    c) Abrasive wear
    d) Erosive wear

22. According to the author, which evidence is the easiest to evaporate?
    a) People Evidence
    b) Paper Evidence
    c) Physical Evidence
    d) Positional Evidence

23. A fractographer can perform an investigation as to what causes the fracture on the mechanical part. Their investigation will conclude on the;
    a) Physical Cause
    b) Human Cause
    c) System Cause
    d) Latent Cause

24. The driver of a company's ambulance disregard the sign in the plant that says maximum speed limit of 5 kilometers per hour since it is a matter of life and death. This situation is a classic example of;
    a) Mistake
    b) Human Error

c) Routine Violation
d) Exceptional Violation

25. According to the Author of this book, the best song to explain the Latencies is.
a) I Can't Get No Satisfaction by The Rolling Stones
b) Man in the Mirror by Michael Jackson
c) Let it Be by The Beatles
d) Beast of Burden by The Rolling Stones

Take Quiz on RCFA Part 2

1. Root cause can only be performed if there is data.
a) True
b) False

2. Root Cause Failure Analysis is reactive.
a) True
b) False

3. Most Root Cause Failure Analyses can be completed in less than 24 hours, especially if this is a maxi event.
a) True
b) False

4. Pareto Analysis, Brainstorming, and Ishikawa Diagram are tools design to address the root cause of the problem.
a) True
b) False

5. In Root Cause Failure Analysis, it is highly recommended that the person who is found guilty should be disciplined and punish to prevent a recurrence.
a) True
b) False

6. Root Cause Failure Analysis will conclude when the physical cause of the problem had been found and determined.
a) True
b) False

7. The root cause of the Challenger Explosion is the O-Ring erosion problem on the right solid rocket booster.
a) True
b) False

8. Performing Root Cause Failure Analysis and these problem-solving tools are just one and the same.
a) True
b) False

9. The depth of probe in performing Root Cause Failure Analysis and Failure Analysis on a specific failure is the same.
    a) True
    b) False

10. Root Cause Failure Analysis can be done by a single person with extensive knowledge on the subject matter for a medium and large-scale event.
    a) True
    b) False

11. One of the human errors on the RMS Titanic that caused it to sink was that they place the engines in full reverse when turning the ship to the port (left side) to avoid collision with the iceberg.
    a) True
    b) False

12. Lack of communication between the maintenance and operations people is a classic example of human cause.
    a) True
    b) False

13. People are the company's greatest asset.
    a) True
    b) False

14. Fatigue fracture on the raceway of a bearing is an example of a root cause.
    a) True
    b) False

15. The Golden Rule on Root Cause Failure Analysis is Amnesty. This means that no one should be punished unless the failure is a clear case of sabotage.
    a) True
    b) False

16. A Root Cause Failure Analysis will require a cross-selection of members to address the problem.
    a) True
    b) False

17. A system and latent cause are one and the same.
    a) True
    b) False

18. Performing 5-Why will address the root cause of the problem.
    a) True
    b) False

19. Among the three types of evidence, the most sensitive type of evidence is people's evidence.
    a) True
    b) False

20. Chronic failures tend to be more expensive than catastrophic failures in the long run.
    a) True
    b) False

Refer to Appendix A for the answers

Preface

This book may be different from other books on root cause as the thoughts I have written on this book defies the conventional methods and tools for conducting a root cause. I only learned the true meaning of what root cause is when I attended a training on a 4-day Latent Cause Experience with my friend Bob Nelms from Failsafe Network. Reminiscing the times when I was still employed, we were already conducting what I thought was a Root Cause Analysis on defects and equipment-related failures. I was very wrong on this. My only regret was how I wish that I knew this subject when I was still employed so that we can learn from the lessons of failure. Almost all industries have their own way of performing a root cause for related safety, defects, and failures. My only question is, have we really understand the root cause of our failures, or do the very same causes keep on recurring back? This book will answer what does it take to perform a Root Cause Failure Analysis investigation on failures and incidents in our industry?

There are many versions, techniques, books, training, and consultants worldwide that explain and teach about how to conduct a Root Cause Analysis or Root Cause Failure Analysis. Some of these people who conduct these RCA or RCFA training are either my connections on LinkedIn or even personal friends. In fact, there is no standard or, what we can say, a universal approach on how root cause is and should be done, including this book. This book may also contradict other books on how to conduct a root cause failure investigation. I am not stating that my method or their method is right or wrong, but rather my point is that the context of this book is based on what I teach and what I believed. I have been employed for many years in different industries, and the word root cause is perhaps a very saturated word. We always discussed the root cause of this when the boss asks about it but only to realized after several years that what we have been doing is not really meant to address the root cause of the problem. Thanks to my good and dear friend Bob Nelms for enlightening me on what a true root cause is all about.

The author of this book based his interpretation on his personal knowledge, experience, and former affiliation with one of the leading providers of Root Cause. I believe that uncovering the root cause is based on 100 % facts and pieces of evidence unfolded. Any investigation that is not based on evidence will not generate the root cause but only the probable or the most likely cause of the failure.

The words Root Cause Analysis (RCA), and Root Cause Failure Analysis (RCFA) will be used respectively in this book. The author would prefer to use the word Root Cause Failure Analysis for equipment-related problems, while Root Cause Analysis will be used for non-equipment-related problems such as accident or safety investigation, quality defects investigation, and administrative incident investigation. This will be explained more in Chapter 2 of this book.

Chapter 1: Why Failures are Important? This chapter explains the importance of understanding failures whether from the lessons on life or from industries problems. This chapter briefly discussed four successful and inspiring people and their struggles to achieve their goals. They are Thomas Edison, Mr. Sirivat Voravetvuthikun also known as Thailand's Sandwich Man, Jackie Chan, and Usain Bolt. Failures are important because it teaches us something. In fact, failure can be our greatest teacher if we can just learn from it. I do not know of any person on this planet that achieves success without ever failing whether from the lessons of life or from industries. Another topic covered in this chapter is about the slow and the fast train which is a metaphor for explaining what is going on typically in industries and why the majority of industries prefer maintenance who can repair fast rather than those who can analyze equipment-related failures. Lastly, we discussed why MTTR and RCFA are just the opposite of both worlds.

Chapter 2: Understanding Root Cause Failure Analysis explains what root cause is all about. The author clarifies the difference between conducting a Failure Analysis, Root Cause Analysis, and Root Cause Failure Analysis. Although both RCA and RCFA processes will be conducted the same way, which is addressing the physical, human, system, and latent cause of the problem. The main difference between RCA and RCFA is where you apply it. It is also important for industries to understand the reasons why conducting root cause is important and why many industries fail in their root cause initiative. Industries must also understand the golden rule on conducting a root cause investigation Also discussed in this chapter is whether conducting root cause is reactive or proactive.

Chapter 3: Different Problem-Solving Tools and RCFA reveals the difference between conducting a Root Cause Failure Analysis investigation and an Analytical Problem Solving tool. The most common problem-solving tools such as Fishbone or Ishikawa Diagram, FMEA/FMECA, 5-Why, P-M Analysis, Pareto's 80/20 Rule, Fault Tree Analysis, 8-Disciplines, and Kepner Tregoe are briefly explained in this chapter. The author provides his own thoughts on whether these problem-solving tools are designed to address the root cause of the problem or only the most probable cause. The basic steps on conducting an improvement process are also covered in this chapter. What is important for the reader is to understand that there is a big difference between a probable or most likely cause and a root cause itself.

Chapter 4: The Life-blood of Root Cause Failure Analysis disclosed the most important part of conducting a Root Cause Failure Analysis investigation which is collecting pieces of evidence. The three types of evidence which include physical evidence, people evidence, and paper evidence are likewise defined as well as how to proceed with the gathering of evidence. This chapter also discussed how to construct an RCFA Logic Tree diagram based on the evidence uncovered to determine the sequence of events that lead to the physical cause. The RCFA logic tree diagram will then be used to derived the physical, human, and system cause of the problem. A quiz is provided at the end of this chapter.

Chapter 5: Understanding The Physical Cause of Failures converse about the process of wear and the most common types of wear that can occur on mechanical components. This is the first of a series of causes that should be uncovered in any root cause investigation. This chapter also explains the main reason why we can no longer use our equipment. The physical cause of failure also refers to conducting failure analysis on the failed part, component, or item. Wear can either be natural which means that the lifespan has been reached or premature in which the mechanical component or item failed prematurely. This chapter also explains the most common types of wear as well as the different types of mechanical stress. Other topics discussed in this chapter include failures attributed to lubrication, bearing failures, and a brief discussion on common electronic failures. The physical cause of failures will not only be limited to mechanical components and parts but to anything that can fail on the equipment which includes electronic and electrical parts as well.

Chapter 6: The Study of Human Errors, this chapter provides the readers an understanding of human errors and will answer if it is possible to totally eliminate them, or not. It also discussed the circadian rhythm which is related in a way to human fatigue. A classic case of human error is about the RMS Titanic sinking. There are many cases of human error committed by different people that led to the sinking of the most luxurious ship during that period in 1912. This chapter also explains the blame game happening in industries whenever someone is accused of something which is part of our human nature. Another topic of interest to the reader is how can we reduce human error in our industries. Finally, this chapter ends with a simple exercise about the E-Experiment to gauge whether the reader will commit an error while reading this chapter or not.

Chapter 7: Understanding the Latent Cause of the Problem, discussed the subject of latent cause and why it is important in any RCA or RCFA investigation. There are two questions on the latent cause that needs to be answered which is, what is it about the way we are as an organization that contributed to the problem? Another question that should be answered is about our personal latencies or what is it about the way I am that contributed to the problem? The song Man in the Mirror by Michael Jackson reflects about latencies, the author also interprets the meaning of the song for industries. Three classic cases on latencies are included in this chapter which is about the latent cause of the Challenger Disaster in January of 1986, the Union Carbine Bhopal Tragedy in December of 1984 that killed more than 6000 people, and the latencies on why the RMS Titanic sank on April 15, 1912.

Chapter 8: Guidelines for Conducting an RCFA Investigation, this chapter starts with the basic requirements on RCFA or RCA which is about training. It is also important to set up an independent RCFA Core Team. The function and roles of the RCFA core team are explained in this chapter. It is also important for the organization to have a list of qualified principal investigators (PI). The traits and qualities of a principal investigator are provided in this chapter. Preparatory or preliminary steps are provided before proceeding with the root cause investigation. A detailed step-by-step guideline and flow chart on how to conduct a Root Cause Failure Analysis investigation is provided and explained. Another important topic discussed is how to conduct the Stakeholder meeting, who are the RCFA stakeholders and what is their role in the RCFA investigation are also explained in this chapter.

Chapter 9: RCFA Corrective Actions and Countermeasures unravel the different ways and means to do a corrective action to eliminate or mitigate the chances of the same cause of the failure to recur. Doing corrective actions will only be limited to the physical, human, and system causes of the problem. Latencies will not require corrective action but will require changes in the organization and oneself. This chapter also explains the concept of horizontal replication or fan-out to similar equipment with similar symptoms of the problem. Lastly, corrective actions should be verified 3 to 6 months after implementation to determine if the problem is recurring or not.

Chapter 10: Case Study on RCFA provides two classic cases on RCFA. One case study is a small-scale personal problem I have had in my home for several years that provided me a life-long lesson. The other is a large-scale incident that happened with one of my clients I have assisted and trained on RCFA. The important part of the RCFA investigation is constructing the logic tree diagram based on the evidence provided by the evidence gathering team. It also explains the role of the stakeholder in the RCFA investigation process.

Chapter 11: FAQs, Tips, and Don'ts on RCFA discussed the most frequently asked questions on RCFA that have been asked during my class, as well as the answers to it. This chapter also covers some key points and tips about RCFA. Lastly, this chapter discusses the don't's and the most common reason why RCFA fails in industries.

Chapter 12: The Conclusion reveals why it is important for industries to take action and perform root cause investigations for their equipment-related failures. Another topic of interest is how we can integrate RCFA into the RCM process. Failures must not only be prevented nor predicted but what is important is challenging failures and learning from them. Finally, this book ends on what the real message of Root Cause Failure Analysis is all about.

Chapter 1

Why Failures are Important?

> *We can learn from failures if we take the time to analyze them. It is an opportunity for us to begin more intelligently. Like the lessons of life, failures tell us something: to take things slowly and find their causes, not just the symptoms. If we just take care of the symptoms, then failures will just repeat themselves again. Fixing failures will not stop them from recurring; what is important is to learn from the failure.*

1.1: Failure Defined

If we look for the meaning of the word failure, it generally refers to the state or condition of not meeting a desirable or intended objective. It may be viewed as the opposite of success. **Webster Online Dictionary** states that failure is the act or fact of failing short, losing strength, breaking down, going bankrupt, not doing or succeeding, while **dictionary.com** defines failure as the inability of a system or system component to perform a required function within specified limits. A failure may be produced when a fault is encountered. Failure of a component indicates it had become completely or partially unusable or has deteriorated to the point that it is undependable or is unsafe for normal sustained service.

Before we discuss about Root Cause Failure Analysis, let us talk about life itself. Many say that failure is the opposite of success. I think otherwise; failure is not the opposite of success, but it is part of being successful; we need to fail first before achieving success. In fact, the most successful people we know failed miserably not once but several times in their life. We cannot achieve success without failing. The problem is that most people give up easily and never understand their own failures and adversities. Failure is just part of the long road to success. There is no one on this planet I know of that achieves success without failing. Failing is a painful part of life that we need to accept and learn about. The truth is, nobody can achieve success without failing.

All people have their high and low moments in their life. When we talked about the human side of failure, people suffer pain, grief, tribulations, and adversities. Many give up easily. Pain and sufferings are an unpleasant but necessary part of life. It is the price we have to pay for being alive. Only dead people don't feel pain, sufferings and have no problems. According to Doctor Norman Vincent Peal, a religious leader, and speaker, said that problems are a sign of life. The more problems you have, the more alive you are. In fact, if we take a problem and analyze it carefully, it offers us an opportunity to solve them. If we take a problem and look at it, it becomes a situation. If we were able to analyze this situation, it becomes a challenge, and when we think of our ability to solve and overcome the problem, then it becomes an opportunity. This means that pain, sufferings, and failures are just temporary and can actually be good if we can learn from them. It allows us to become a better person.

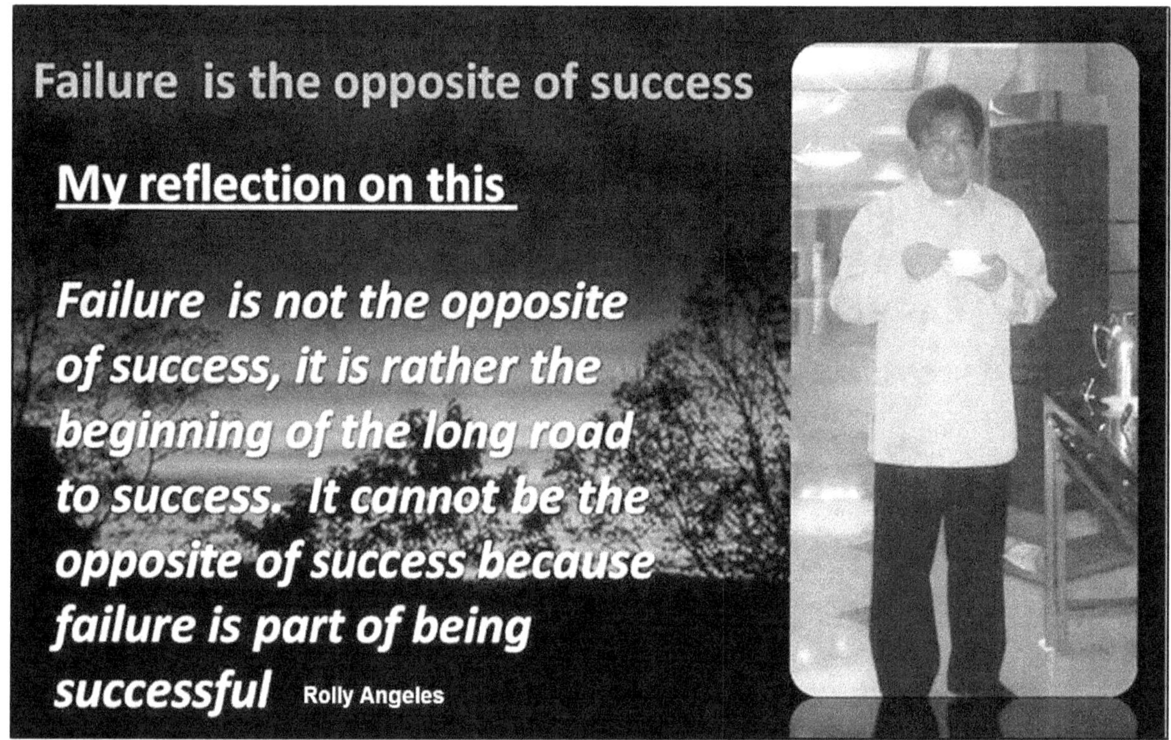

Figure 1.1: Failure is not the Opposite of Success

During my Root Cause Failure Analysis classes, I usually asked my students, what if God came down from the heavens and talked to you and asked you if you want him to remove all the pain and sufferings you feel physically? Would you like that to happen? Most of my students would agree and say yes for as long as they are not dead.

You see, approximately 1 out of every 400,000 babies born every year is fated to live a short life that none of us would envy. A life in which the child will frequently hurt himself physically, sometimes seriously, and without knowing or even feeling it. If you cut the child's skin or pour boiling water at that child, they will not feel any pain. That child has a rare genetic disease known as familial dysautonomia. Familial dysautonomia is a genetic disorder that affects the development and survival of certain nerve cells. The disorder disturbs cells in the autonomic nervous system, which controls involuntary actions such as digestion, breathing, production of tears, the regulation of blood pressure and body temperature. It also affects the sensory nervous system, which controls activities related to the senses, such as taste and the perception of pain, heat, and cold. This means that the person may not feel any pain. A child will cut himself, burn himself, break his bones, and never know that something is wrong since they cannot feel any pain.

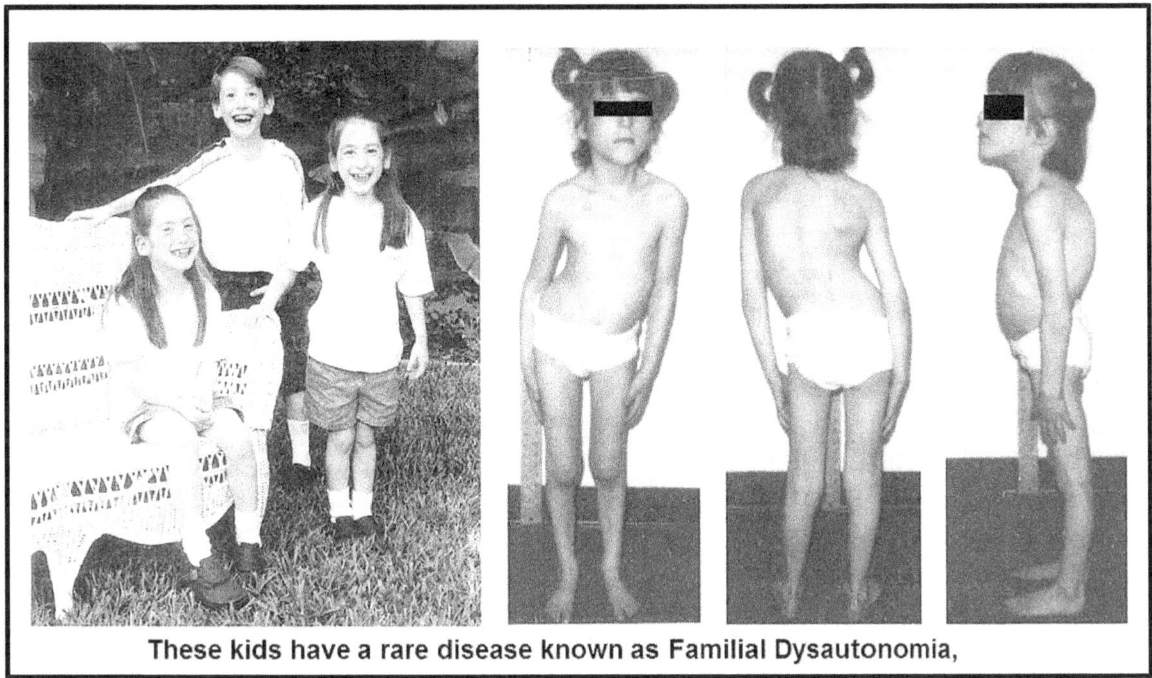

Figure 1.2: Rare Case of Disease Called Familial Dysautonomia

Not being able to feel pain is dangerous, most especially for children. We need pain to tell us what you should and shouldn't do. We need pain to tell us when to move away from something because it is causing us harm. Most people try to avoid causing pain to themselves, but it is because they have felt pain that they could avoid those things that initially caused it. Let me share some examples of people that failed miserably in their lives before they reached their pinnacle of success.

1.2: A Vision of Light: The Thomas Alba Edison Story

Perhaps only a few people really know the struggles and difficulties Thomas Edison went through before lighting up the first incandescent bulb and perfecting it. As a young student, his teacher used to call him "stupid." Although Edison was not a bright boy, he wandered a lot, walking on his way home at night with only the light of the moon; after many years, he still wanders; what if this light can be in his home and in everyone's home? He had a vision that he wanted to see this light in every house and imagined the benefit that each family can have. So he started to make his vision a reality and started working on it. In his experimental research, he knew that light can be produced by heating a wire with electricity. Still, the difficult part was to do this without burning the wire into two. Still, every element conceivable in the periodic table was tried, which burned his wire in half. People passing by his home often have mistaken him as crazy and laughed at him. People close to him advised him to stop, which only motivates Edison to continue. Perhaps in his mind, what he was thinking was there must be a way to do it and he need to find it.

He laid down his periodic table and experimented on every sort of element but failed miserably. He tried one element after another and combined elements. Again, every attempt he made failed. Edison spends at least 2 to 4 hours of sleep each day, again waking up in the morning trying a different combination of elements but ending up in vain. People close to him often tell him to give up. Still, Thomas never gave up on his vision. He knew that his invention was half completed. Still, it was of no practical use since the wire always burns and breaks in half until he can find the missing link; finally, after many experiments, Thomas found the missing link. In the study of physics, there can be no combustion without the presence of oxygen. The difficult part of his invention was to control the heat, so he placed his electric light apparatus inside a glass globe, shut off all the oxygen, and lo, the mighty incandescent light was a reality. Some say that it took Thomas Edison close to 10,000 attempts before lighting a single incandescent bulb. He said, I have not failed; I've just found 10,000 ways that don't work.

They say that the innovation of today will be the standard of tomorrow. Every night when we say goodnight to our children and close the light, we never glimpse that it was this man's vision that made it all possible. If Thomas had given up, maybe we will be spending our night with candles or lamps. Perhaps other people would have invented it which I cannot say, but what I know is that this average man that they called "stupid" had surpassed his vision of having a light in every home, buildings, billboards, and everywhere, from the first incandescent electric bulb, came to its diverse applications such as the car's taillight, headlight, flashlight, Christmas lights, strobe lights, Halogen lamps, neon light, and many more kinds of lights, you name it, all boiled out to this single invention of what Thomas Alba Edison did. In the book of Napoleon Hill's "Law of Success," Thomas Alba Edison failed more than 10,000 times before he can make his Incandescent Electric Bulb worked.

1.3: Thailand's Sandwich Man: From Riches to Rags and Bouncing Back

Mr. Sirivat Voravetvuthikun is a former millionaire, tycoon, stockbroker, and investor in Thailand. When the stock market crashed in 1997, his money was taken away by the banks. Mr. Sirivat was left with no money and a large amount of debt to the banks. With nothing, Sirivat asked his wife what to do, and his wife told him to sell sandwiches. He followed his wife's advice, removed his tie, swallowed all his pride and ego, wore an ordinary shirt, and walked down the streets as a peddler and sell sandwiches. Mr. Sirivat and his wife started making and selling the first 20 sandwiches on April 20, 1997. It took them 6 1/2 hours to sell them. Today, Sirivat owns several coffee shops and still sells sandwiches in Thailand. He had been a motivational speaker and an inspiration to Thailand. He is better known as Thailand's Sandwich Man. The story of the "Sandwich Man" is a symbol of optimism and tenacity for the people of Thailand.

The transition from being a former millionaire stockbroker to being a peddler and selling sandwiches in the streets is not something every millionaire will do. Although Sirivat made his fortune in stocks before the market collapse in 1997, all his investments disappear instantly, leaving him with $30.4 million in debt. In an interview, he said that his life was totally changed from living a luxurious lifestyle to just a common one. When asked what made him decide to sell sandwiches on the streets, his reply was I would rather go bankrupt than be dead.

The former millionaire said that it was a difficult time for his family. While the banks and his former creditors sued him, he swallowed his pride and remove his expensive suit and tie, and started selling sandwiches in the streets of Bangkok. On his very first day, he earned $14. His wife, Vilailuk, says she often felt sad for his husband, but she realized that if both of them give up, nobody will help us. He said to his husband that they need to be together, help each other, and fight to survive for their family's sake no matter the odds. Today, Mr. Sirivat is well known in Thailand as the "Sandwich Man." He is once again on the road to success. Sirivat's assets include his two coffee shops and a sushi catering service, which employs 14 people. He plans to expand his business in the future. Sirivat said in an interview, today, as you can see, I can smile. Although I was formerly rich, I was a bankrupt person. Today, I run a very, very small business. And I still sell sandwiches on the streets of Bangkok, although I'm going to have shops now one after the other, and I feel happy about that.

Figure 1.3: Mr. Sirivat Selling Sandwiches on the Streets of Thailand

1.4: The Jackie Chan Story: I Failed Miserably as Bruce Lee

When I was a child, I loved watching martial arts movies on Bruce Lee way back in the 1970s. In fact, my parents and I would go to a movie in Manila on a Saturday or Sunday just to watch a full-packed and standing room Bruce Lee movie. In fact, I have watched all his movies. Everyone knew who Bruce Lee was during the 70s until his untimely death on July 20, 1973, at the young age of 32. Bruce Lee was the talk of the town, and little or nothing is known about one of Bruce Lee's stuntmen by the name of Jackie Chan. In fact, this is how the career of Jackie Chan started out.

Jackie Chan worked as a stuntman on two of Bruce Lee's films, Fist of Fury and Enter the Dragon. In one of the fight scenes, Jackie Chan was accidentally hit with a wooden stick on his face by Bruce Lee, and when the director shouted cut, only then Bruce Lee saw Jackie Chan still on the floor and walk towards him and asked him if he was okay. Jackie Chan pretended that he was still hurt since he wanted to be noticed by Bruce Lee. In a recent interview with Jackie Chan, he said that it felt good to be hurt by Bruce Lee; that was the finest moment for Jackie Chan since he was finally recognized by the great actor himself. With the untimely death of Bruce Lee and his unfinished movie Game of Death, they need someone to fill in the vacuum. This movie also featured the great NBA basketball player Kareem Abdul Jabbar, who is 7 feet and 2 inches (2.18 meters). Kareem is also one of Bruce Lee's students on Jeet Kun Do in real life. Jackie Chan tried to replace Bruce Lee and realized that he cannot mimic the way Bruce Lee does his martial arts. Again in an interview, Jackie Chan said that it is very difficult to be Bruce Lee. We have a different style; I love action, but I hate violence, so I added comedy to my action films. He said that instead of being him (Bruce Lee), I decided to be myself. And the rest was history.

Figure 1.4: Jackie Chan and Bruce Lee

1.5: Usain Bolt: The Fastest Man on Earth

Perhaps one of the most inspiring success stories is non-other than the story of the fastest sprinter of all times, Usain Bolt. He has broken many world records and holds several Guinness Book of World Records, but perhaps his greatest record of all time is topping a speed of 9.58 seconds in the 100 meters in 2009, breaking his own record of 9.69 seconds. Bolt started his first competition in 2001 at the age of 14, where he won the high school championship and earned the silver medal in the 200-meter race. At 15 years of age, Bolt competed at the 2003 World Junior Championship in Kingston, Jamaica, and won the 200-meter race to become the youngest junior gold medalist. This is where he earned the name "Lighting Bolt."

Usain Bolt has earned a total of eight gold medals which he earned in three consecutive Olympic events in the year 2008, 2012, and 2016 in the 100-meters, 200-meters, and 400-meters events. He also earned his 9th Gold Medal in the Olympics, but it was vacated when his fellow runner was tested positive for drugs. Note that the Olympics is held every 4 years. He was the only athlete in his field to earn a gold medal in 3 Olympic Events behind Carl Lewis

Perhaps only a few knew that Usain Bolt has Scoliosis, which is a condition where there is an abnormal lateral curvature of the spine, but this condition does not deter him from competing and aiming for the gold in the Olympics. In an interview with Bolt, he quote, one failure doesn't outweigh past successes. In fact, after losing a 100-meter race, someone reminded me how even boxer Muhammad Ali lost his last fight. For me, I have proven myself year in year out and throughout my whole career. I don't think one championship or one race or what I did in my last race will change the fact of what I have done in the sport.

Figure 1.5: Usain Bolt Breaking the World Record for 100 meters

Usain Bolt retired in track and field sports in 2017 and devoted his time to his business foundation and endorsements. He also owns a restaurant in Jamaica called Tracks and Records. In addition, Bolt has a sponsorship deal with Puma, which pays Bolt more than $10 million annually. . He also has endorsements with products such as Gatorade, Visa, Virgin Media, and Nissan Motors.

According to a 2016 Forbes Magazine, Bolt earned around 33 million US Dollars a year, which is a Whopping 10 times more than any track and field athlete can make, making him the 23rd highest-paid athlete in the world. Perhaps as Forrest Gump said, I just feel like running. To us, it might seem easy as anyone can run, but whatever time he dedicated to training and conditioning his mind and body, throughout the years, molded him to be the fastest man who ever run in the 100, 200, and 400-meters. He is considered to be the highest-paid athlete in the history of sprinting. The Lighting Bolt may have hung up his sprinting career last 2017, but Usain Bolt's record of 9.58 seconds for the 100 meters will be the challenge and benchmark of the younger and future generations to challenge.

Usain Bolt's had run for 114 seconds in the Olympics, and have won 8 gold medals. He has set the world record holder of 100 meters, 200 meters, and the 400 meters relay and has beaten every competition there is. But what truly inspires us is the amount of time he spent training since he won his first gold medal at the age of 15. If ever there is one competitor he needs to beat, that would be himself in which he did, which is what makes him truly a legend.

1.6: Why Failures are Important

Life is not as easy as we might think. We failed many times and it is painful mentally and emotionally. I failed several times too and there are times that I was way too close to giving up. I was rejected more than 10 x as a maintenance trainer. When I was starting my career in training, I received many calls from large international training organizations. They asked me if I can deliver this particular training on maintenance which I confirmed and said yes. After submitting my personal CV and resume as part of their requirement, I never heard from them again. Later on, after checking their website, I saw photos of the training which I was supposed to be conducting being delivered by a white skin person. Perhaps that's because, I'm from the Philippines and my skin is brown and every time different invitations come in, the same thing happens where I was denied and discriminated against that opportunity. My family even advised me to go back to the factory and work once more. I never let go of wanting to teach industries. What I believe is that one day I will be teaching maintenance to different races which I did and still doing as of this point of writing.

When I graduated from Mechanical Engineering in 1985, I applied to more than 200 companies in the Philippines and all 200 companies rejected me. I still keep their letters as a souvenir. Employment during that time was so hard. There may be more than 500 to 1000 new graduates applying for the same position. I can even recall that since there were so many applicants, we wait outside the gate of the plant under the burning sun holding our yellow envelope that contains my resume. I remembered taking a qualifying examination with Shell Pilipinas in 1987. After a couple of weeks later, they sent me a letter that my qualifying exams failed so I did not get the job but several years later they hired me to teach them.

We experienced failures in our life. We have our low points, but through this, pain, agony, suffering, tribulations, and adversities, it allow us to reflect on what went wrong with our lives. Successful people have something in common. They learned from their mistakes and failures. In fact, failure can be our greatest teacher if we can just learn from them. Success may mean different things to different people. In our four cases above:

• Thomas Edison's goal is to provide light to every home with his invention, the incandescent bulb.

- Mr. Sirivat, Thailand's Sandwich Man who has lost his fortune but never accepted defeat and bounced back once more. He serves as an inspiration to the people of Thailand.
- Jackie Chan stopped pretending to be Bruce Lee and decided to be himself. It took some time for the world to accept him. Despite the lack of formal education, he was awarded the honorary doctor of Social Science at Hong Kong Baptist University, and an honorary fellow of the Hong Kong Academy of Performing Arts. Although, despite the odds, Jackie Chan was able to achieve what he is due to his hard work and dedication in his chosen field of martial arts. His style of martial arts with a combination of comedy proved to be a great success.
- Usain Bolt perhaps just run a few minutes but what we actually don't know is the training, pain, difficulties, and time he devoted to being marked as the fastest man on the planet. His 100-meter record of 9.58 seconds still remains unbroken up to this point of writing. Nobody has ever run that fast.

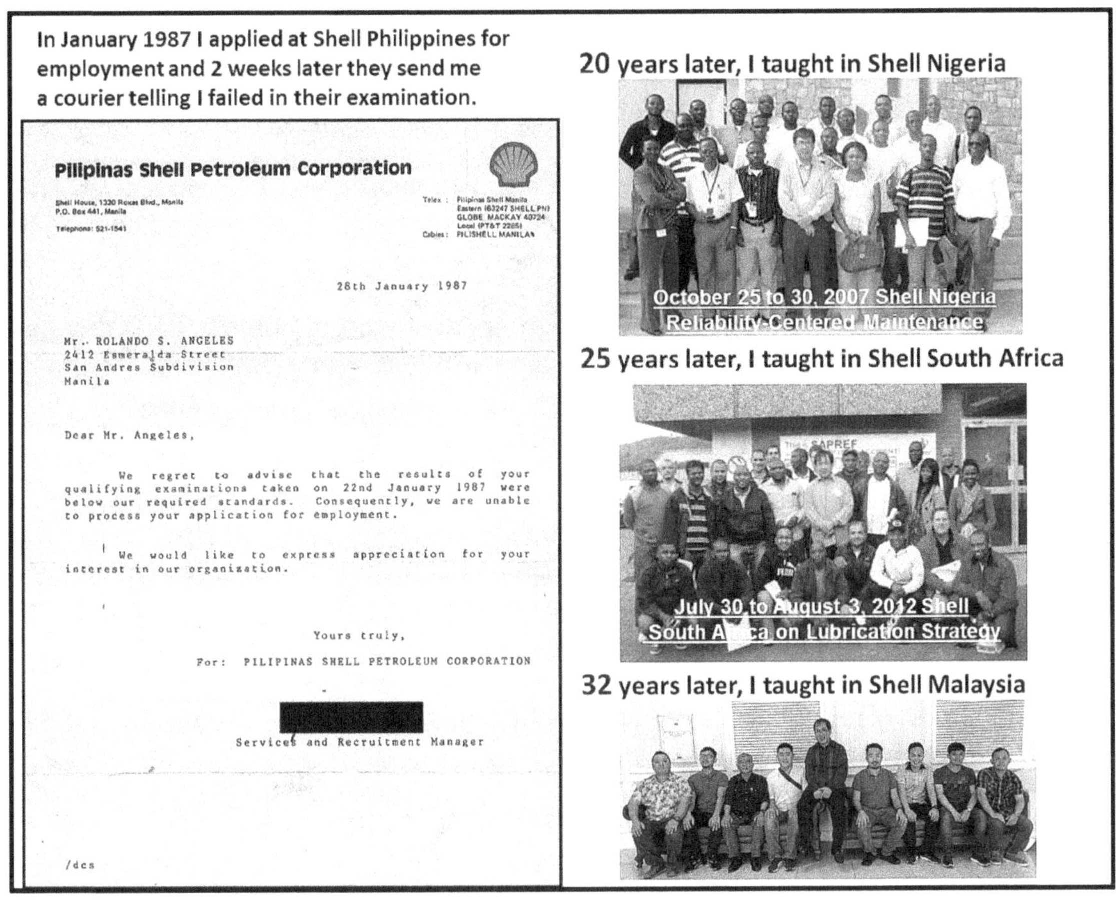

Figure 1.6: I Failed in Shell Philippines Qualifying Exams

If you asked, what makes these people stand out from the rest? My response is simple, they face their fears and challenge them. Just like equipment failures in industries, much can be learned from the failure if we can take the time to investigate and analyze them. But in reality, that will not be hardly the case. Many say that failure is the opposite of success; what I think is that failure is part of being successful. We need to fail first to achieve success. In fact, the most successful people we know today failed miserably in their life before they achieve success. There is always a lesson we can absorb from failures, and when we learn from our failures, we can be on the road to success. When someone does something wrong to us and apologizes sincerely, we tell them to forget it. The only difference between equipment failures is that in industries, we forgive, but we will never forget as we do not want these failures to recur once again in the future. If we study these people's lives, there is always a common trend: they learned from their mistakes and failures. Success is not about having lots of money but being able to achieve what you want in life. Remember that the happiest people may not possess everything; they just make the most of everything they got. Another important thing these people possess is persistence. To them, failure is not an option; it is a way to succeed. I learned from Bob Nelms of Failsafe Network that the root cause is not only limited to industries' problems but can also be used in our personal lives if we really want to improve our way of life.

1.7: Famous Quotes on Failure

Here are just some quotes by famous people who experienced failure before truly becoming successful in their lives. If you study their lives, there is actually a pattern. They learned from their failures by challenging them and becoming persistent. While most people quit, these people used their creative minds to turn their failure into an opportunity.

- I've missed more than 9000 shots in my career. I've lost almost 300 games. 26 times, I've been trusted to take the game-winning shot and missed. I've failed over and over again in my life. And that is why I succeed. *From Michael Jordan, NBA Goat*

- If it turns out that my best wasn't good enough, at least I won't look back and say that I was afraid to try; failure makes me work even harder. *From Michael Jordan, NBA Goat*

- Just because you fail once doesn't mean you're gonna fail at everything. Keep trying, hold on, and always, always, always believe in yourself, because if you don't, then who will, sweetie? *From American Actress Marilyn Monroe*

- The real test is not whether you avoid this failure because you won't. It's whether you let it harden or shame you into inaction, or whether you learn from it, whether you choose to persevere. *From Former US President Barack Obama*

- Do not judge me by my successes; judge me by how many times I fell down and got back up again. *From Nelson Mandela*

- Most great people have attained their greatest success just one step beyond their greatest failure. *From Napoleon Hill, Author of Laws of Success*

- You, me, or nobody is gonna hit as hard as life. But it ain't about how hard you hit. It's about how hard you can get hit and keep moving forward. How much you can take and keep moving forward. That's how winning is done. *From the Movie Rocky 4 by Sylvester Stallone*

- What people see of my success is only one percent, but what they don't see is ninety-nine percent, which is my failures. *From Soichiro Honda, Founder of Honda Motors Corporation*

- Only a man who knows what it is like to be defeated can reach down to the bottom of his soul and come up with the extra ounce of power it takes to win when the match is even. *From Muhammad Ali, The Greatest of All Times*

Chapter 1: Why Failures are Important

- You build on failure. You use it as a stepping stone. Close the door on the past. You don't try to forget the mistakes, but you don't dwell on them. You don't let it have any of your energy, or any of your time, or any of your space. *From Johnny Cash, Singer, and Musician*

- Failure is so important. We speak about success all the time. It is the ability to resist failure or use failure that often leads to greater success. I've met people who don't want to try for fear of failing. *From J.K. Rowling, Author of Harry Potter*

- It's fine to celebrate success, but it is more important to heed the lessons of failure. *From Bill Gates*

- Just like in bodybuilding, failure is also a necessary experience for growth in our own lives, for if we're never tested to our limits, how will we know how strong we really are? How will we ever grow? *From Arnold Schwarzenegger, Movie Actor and Politician*

- I remember that I ate one meal a day and sometimes slept in the street as a little boy. I will never forget that, and it inspires me to fight hard, stay strong and remember all the people of my country trying to achieve better for themselves. *Manny Pacquiao 8 time World Boxing Title Division*

- In the most difficult time in your life, you only have three allies, you, your mind, and most especially God. Take it from someone who had been there. *Book Author and Maintenance Teacher*

- There is a tendency to walk away from failure and leave it buried. An enormous amount of institutional learning gets buried because failures don't get analyzed. So the real learning is what's learned from failure. *From Andrew Grove, Chairman, and Co-founder of Intel Corporation*

- You don't learn about yourself through your success. You only learn through your failures and mistakes. *From Wynonna (The Judds)*

- Most people make a common mistake by thinking of failure as the enemy of success. You've got to put failure to work for you. Go ahead and make a mistake. Make all you can. Remember, that's where you'll find success on the far side of failure. *From Thomas Watson of IBM*

- Every adversity, every failure, every heartache carries with it the seed of an equal or greater benefit. *From Napoleon Hill, Author Laws of Success*

- The only real mistake is the one from which we learn nothing. *From Henry Ford*

- Remembering that I'll be dead soon is the most important tool I've ever encountered to help me make the big choices in life. Because almost everything, all external expectations, all pride, all fear of embarrassment or failure–these things just fall away in the face of death, leaving only what is truly important. *From Steve Jobs, Founder of Apple*

- Part of being a man is learning to take responsibility for your successes and for your failures. You can't go blaming others or being jealous. Seeing somebody else's success as your failure is a cancerous way to live. *From Kevin Bacon, Movie Actor*

- Failure should be our teacher, not our undertaker. Failure is a delay, not a defeat. It is a temporary detour, not a dead end. Failure is something we can avoid only by saying nothing, doing nothing, and being nothing. *From Denis Waitley*

• Winners are not afraid of losing but losers are. Failure is part of the process of success. People who avoid failure also avoid success. *From Robert T. Kiyosaki (Author of Rich Dad, Poor Dad)*

It is OK to Fail as long as we learn from it

Thomas Alva Edison

Thomas Edison boyhood teacher told him that he was too stupid to learn anything. He was known today as one of the worlds greatest inventors

Elvis Presley

King of Rock and Roll was fired after just one show at the Grand Ole Opry and was told "that you ain't going nowhere son"

Michael Jordan

Before joining the NBA Jordan was just an ordinary person that he was cut from high school basketball team because of his lack of skills

Albert Einstein

His parents thought he was mentally retarded & his grades at school were poor that his teacher ask him to quit and said that he will never amount to anything

Soichiro Honda

Honda was turned down by Toyota during a job interview as an engineer during WWII and again rejected by Toyota as a supplier of piston rings

The Wright Brothers

They say only birds can fly. heavier than air flying machines are impossible to fly according to Lord Kelvin 1895

Bill Gates

Before starting Microsoft he was a dropout at Harvard University and started his software by purchasing it from someone at USD 50.00

Jackie Chan

When Bruce Lee died, he along others were picked up to fill the vacuum and failed miserably, so instead of being him (Bruce Lee), he decided to be himself

Chester Carlson

In 1930 his idea was turned down by 20 companies including the biggest in the country, after 7 years his idea was accepted it was the XEROX machine

The Beatles

In one of their early auditions, we don't like their sound and guitar music is on the way out Decca Recording Company

Clint Eastwood

You have a chip in your tooth, your Adam's apple sticks out too far and you talk too slow by a universal pictures executive 1959

Henry Ford

The horse is here to stay but the automobile is only a novelty, a fad, President of Michigan Savings Bank telling Henry Ford.

Figure 1.7: People That Fail Before They Become Successful

1.8: Why Not Just Copy the Success of Others?

If you were asked to choose, which would you prefer? Would you just copy the success of someone or learn from your own failures? Many people want to talk about successful people but nobody wants to know about their failures. As Soichiro Honda quote, many people talked about my success which is just 1%, what they do not know is the 99 %, which are my failures. For industries, there is a saying "Don't Reinvent the Wheel," so they just want to copy the other's industries' success which will just end up in a massive failure. And so we asked, what did we do wrong? How come it did not work with our industry and so on?

Today, only around 52 US companies have made it to the Fortune 500 companies since 1955. This means that only 10.4% of these companies have remained existing until today after 64 years since 2019. More than 89% of these companies have either gone bankrupt, merged, or acquired by other industries or have fallen out from the original Top 500 Fortune Companies in the United States. Many of these companies were big shots during their time in 1955. Still, today, they are now unrecognizable, forgotten, or merely gone. Although these industries have different reasons for their failure, one thing they have in common is that they did not learn from the things that went wrong with their industries until it was finally too late for them to realize.

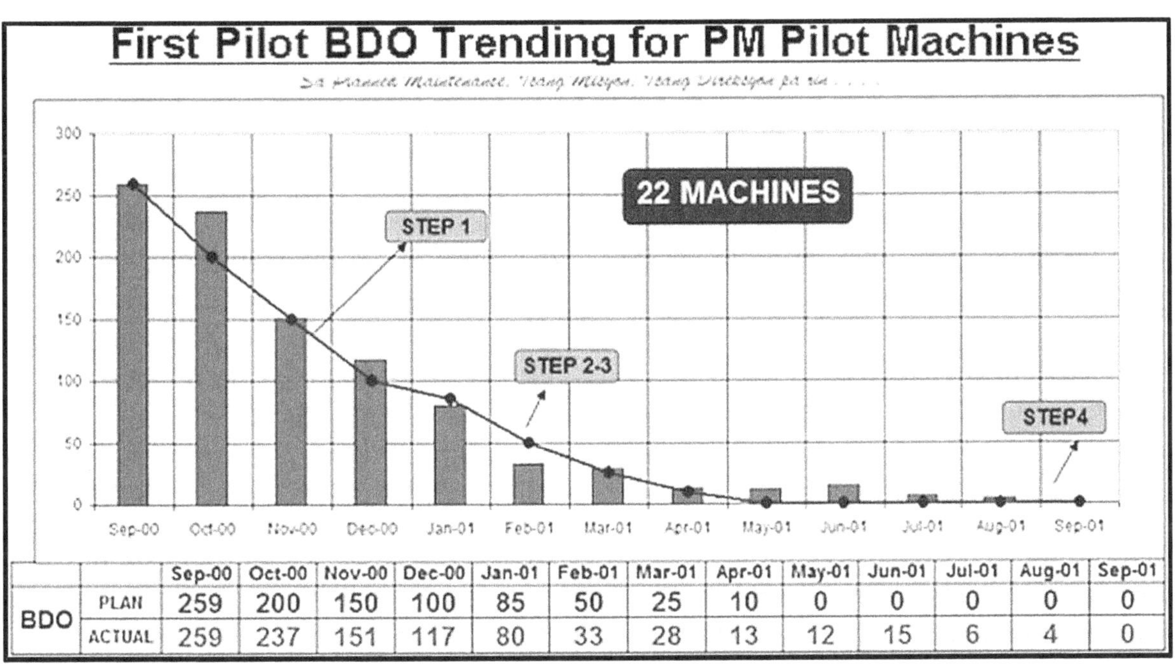

BDO		Sep-00	Oct-00	Nov-00	Dec-00	Jan-01	Feb-01	Mar-01	Apr-01	May-01	Jun-01	Jul-01	Aug-01	Sep-01
	PLAN	259	200	150	100	85	50	25	10	0	0	0	0	0
	ACTUAL	259	237	151	117	80	33	28	13	12	15	6	4	0

Figure 1.8: Reduction of Breakdown in Our Planned Maintenance Activities

When I was employed with a large semiconductor here in the Philippines, our President decided that TPM will be implemented. It was a top-down approach that started from him. We bought several books on TPM and read them. Our role in the TPM office is to interpret the book and adapt it to our plant. We failed miserably during the first few years of implementing TPM. We decided to hire a JIPM consultant who guided us meticulously. The Japanese JIPM (Japan Institute of Plant Maintenance) consultant guided us all way on what to do and what not to do. We followed him to the letter, and our maintenance indices started to improve slowly. The Planned Maintenance teams piloted 22 Rank A Worst machines composing of different types across divisions and successfully reduced the Breakdown Recurrence (BDO). Our goal was to totally reduce to zero all the unplanned breakdowns for the 22 PM Pilot Machines as we complete Step's 1 – 3 which we did, almost. Figure 1.8 was our actual breakdown record for the 22 Pilot Machines.

After reducing the breakdown, our industry became known. Several industries came to our plant to benchmark how we did it, in which they were really impressed. We provided an hour or two hours long presentations and a line tour to see how clean and organize our equipment and people were. However, we have one clear instruction from our manager that we need to follow strictly, and what is that? We just show them the good stuff. The instructions given to us was not to provide these benchmarkers what our problem, failures, depressions, or struggles were. The instructions to us were crystal clear and vivid, show them only what they want to see. If they asked how we achieve that, tell them that we also benchmark in the beginning. In benchmarking other industries, there is one unique thing that you need to ask before you can implement what they have implemented. That question is, is their culture and your culture exactly identical or different? The answer here is obvious, it's different. In fact, there are no two industries with the same culture. You cannot copy one's culture and implement it in your plant.

Culture is like our fingerprints, there are no two persons with identical fingerprints, and each of our fingers has different fingerprints. This means that in any organization, the culture from one function or department might not exactly be identical, just like our fingers. Meaning the culture in the warehouse department might be different from the Finance, which is also different from the Maintenance, which is also different from the production, and so on. If what you have benchmarked did not work in your industry, it means that your culture did not accept it. It's more important for industries to focus more on their failures and learn from them rather than benchmark or copy other industries' success. If you know the song Let It Be, by The Beatles, Paul McCartney sang that song. Try to check the version of Ray Charles on the same song on youtube, I find it to be better than the original. Ray Charles sang it with pure heart and soul. TPM is not culture-bound, which means that it was originally adopted for Japanese industries, but rather TPM recommends industries to adopt it the way their culture would accept it. It means flex it a little bit and do not apply it the way the Japanese adopt it.

Success may mean differently from one person to another depending on your goal in life, whether you want to be the fastest man like Usain Bolt, be the best football player, or be the first man to live on Mars and so on. The same goes for industries, either we want to achieve a World Class Maintenance Management stage, be the best in this category, be the most profitable, and so on. We can only achieve this feat if we challenge failures and learn from them.

1.9: Why MTTR and RCFA are The Opposite of Both Worlds

For industries, failure is just the tip of the iceberg. Typically, when a failure occurs on our equipment, it is critical for maintenance to restore the equipment as soon as possible. It can be said that 95% of maintenance people in industries will just repair the failure, while 5% or less will try to analyze the failure and attempt to learn from what went wrong with their equipment. If we talked about MTTR or the Mean Time to Repair, this is the average time required to repair a given component or equipment. Other terms used for this would be Mean Time to Restore or Mean Time to Recover. The trend that we want for MTTR will be the lower, or the shorter the time to repair, the better. MTTR may also be defined as the time it will take to bring a failed equipment back to its available or operating condition once more. It analyzes how long repairs and maintenance tasks will take in the event of an equipment failure. The unit mostly used to measure MTTR will be in hours.

The formula for MTTR will be the total repair time divided by the number of breakdowns requiring repair during a given period. On the other hand, Root Cause Failure Analysis will try to understand why something went wrong so that industries can finally learn from the things that go wrong. It will identify the true cause or origin of the problem based on the evidence unfolded by separating the facts from the other probable causes.

I have several friends who worked in industries, and one of them is Charlie. If I recall him right, he had been working in the plant for many years. 25 years to be exact and is still working in that same plant up to this point of writing this book. There are several rotating equipment in their plant. Charlie is one of the maintenance guys that maintain their equipment. He told me that when he was new at that plant, it took him several hours to fix the equipment by experimenting on what to do because you are usually on your own. When a bearing fails, it usually took him a couple of hours to replace the failed bearing. 25 years had passed, and today when the same bearing failed, it will just take him around 20 minutes or even less to replace it.

My question is, is this a good thing, or a bad thing? The answer here depends on whether we're discussing RCFA or MTTR. If we speak about MTTR or the Mean Time to Repair, Charlie's MTTR improved a lot throughout the years he had worked in the plant by gaining experience on repairing the bearing since the trend we want for MTTR is the lower, the better. On the other hand, if we speak about Root Cause Failure Analysis, Charlie cannot find the cause as to why the same bearing keeps on failing. Why? Because he never bothered to identify the root cause of the problem and believe me when I tell you that this is not a good thing. Why did Charlie become good at fixing bearings? It is because the bearing fails all the time. This is the reality of what is happening to industries right now. If the consequences of failure will be limited to the cost of failure alone, which means that if the motor that failed has a standby motor in place, then MTTR can either be prolonged to allow us the time to gather evidence on the part or component that failed.

However, if this is a stand-alone motor where the failure can affect some critical parts in the plant, then repair time should be kept at the minimum and shortest possible time. In most cases, performing an investigation through Root Cause Failure Analysis will be ignored. Although I do not recommend performing a Root Cause Failure Analysis for every single breakdown, what I am saying is that if the failure keeps on repeating itself, then it is time for maintenance to slow down and take the time to analyze the problem to understand the real cause of the problem. This means that we leave our hands alone for the meantime and use our brains to analyze what went wrong with the bearing. This is what I think separates maintenance from a mechanic. A mechanic will use the hands 100% of the time to repair the failure, while maintenance must balance the use of the brain and the hands. This means that maintenance must understand that if the problem keeps on repeating itself, it is time to slow down and investigate the problem. This means that in order to be fast, we need to take things slowly.

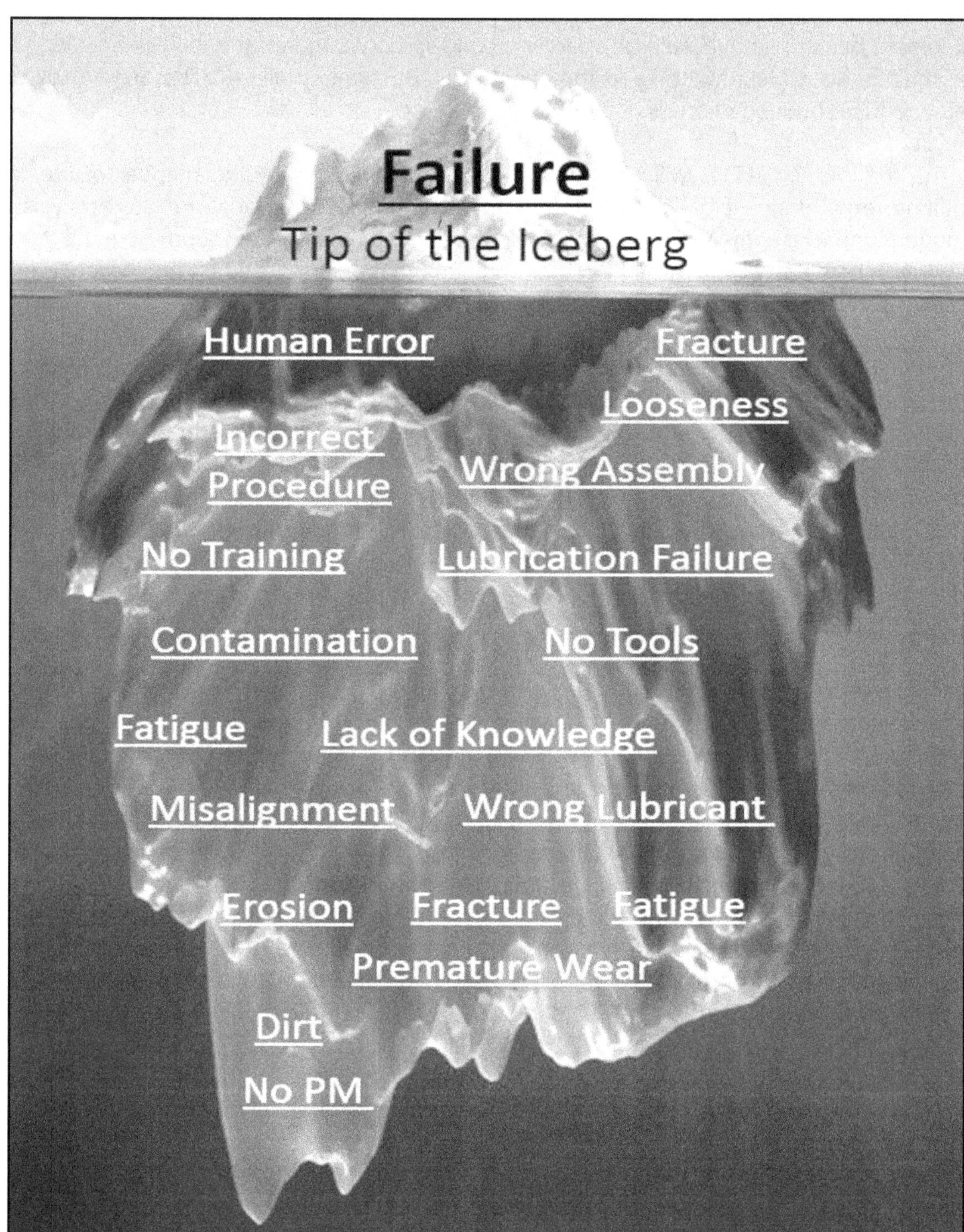

Figure 1.9: Failure is Just the Tip of the Iceberg

What is happening today in the majority of industries is that in most cases, maintenance will become so good at repairing failures. This is the skill they learned the hard way from experience and the knowledge they got from the University of Hard Knocks. But what industries failed to realize is that while these maintenance people become so good at repairing, time is also on their side. As maintenance mature in their repair skills, their knowledge has not been passed on to the new generation. It was just kept to themselves and one day these good old maintenance folks will go away and never return back as they reached their retirement. With this, industries will hire new fresh graduates with absolutely no experience and will work their way out, ending up again in the same old situation as before and the cycle goes on. As we have mentioned previously in this chapter, these successful people learned from their failures, and challenge them to become better. For industries, we will never avoid failures by doing the same thing again and again by merely repairing the failure. The truth of the matter is that failures simply repeat themselves because we did not take the time to investigate the root cause of the problem.

The majority of industries especially the top management side are stubborn. They do not want anything new as they think that this will be another added expense to the industry. The only time they will change their mind is when something hard hits them. Literally speaking, sometimes you need to get a rock, hit the rock in your management's forehead until he bleeds, and realize that what you are saying is important. But of course, we would not do that. What I meant was that the only time management will agree is if a disaster took place. Their thinking is that we have done this before without that so why do we need it now. And when the disaster came, again the management will blame us for not following it up with them.

1.10: The Slow and the Fast Train

When a failure occurs on the equipment, there are actually two teams that will be coming. First, we have a fast team. They are fast and will take a ride on the fast train. The fast team will always be ahead on the scene since they are fast. They will be the restoration or repair team that will be responsible for bringing the equipment up and running in the shortest possible time. They need to do things fast and quickly since operations people will monitor and watch them while repairing the equipment most of the time. The restoration team knows that time is always of the essence, and the clock is ticking. Tic-Tac. Once the equipment is repaired, the operator will again operate the equipment, and now the slow team arrived. Who are these people? These people will be responsible for analyzing the root cause of the problem, and they will be taking a ride on the slow train. They will take things on a slow basis by picking up debris, taking photographs of the failed equipment, interviewing operators and people nearest to the asset during the time of failure. However, when these people arrived at the failure scene, the fast team had already restored the equipment. So what is wrong with this situation? My God, everything is so wrong! Why? *You simply cannot perform a Root Cause Failure Analysis if the equipment was already restored or repaired.* Why? Because every single shred of evidence had already been washed out by the fast team and when the boss finally asks for the root cause, what we usually do is a guess, witch hunt, based it on experience or instinct or declare a probable cause. The most important thing that I have learned in performing a thorough Root Cause Failure Analysis investigation is to freeze the evidence, but in this case, the evidence had already been completely washed out and destroyed by the fast team. This is the case in the majority of the industries I have served. Then how on the planet can maintenance perform a Root Cause Failure Analysis investigation? My response is, you just simply can't. And most of the time, managers complain that their maintenance people lack root cause analysis skills. Oh, come on, give me a break!

<u>**Investigating Equipment Failures Through Root Cause Failure Analysis**</u>

Many years ago, I conducted this survey question with my lists in which the question asked was if you are the maintenance manager and you will be hiring a couple of maintenance in your department, which one will you hire?

• A: I will hire a person who is good at analyzing equipment-related problems and failures.
• B: I will hire a person who is good and fast in repairing equipment failures and breakdowns.

Figure 1.10: The Fast Train and the Slow Train

I received several responses, and the result was a landslide. 82 % have chosen letter B, in which they will hire a person who is good at repairing or fixing failures, while 18 % will choose maintenance who knows how to analyze problems. No wonder why the majority of industries are reactive. It's not just about the culture, but it's about their attitude and mentality.

If we perform a thorough Root Cause Failure Analysis on equipment-related failures or assets that failed, MTTR will be prolonged since we need to pick up whatever evidence that can be found on the equipment before it can be repaired by the restoration team. Since the countdown will begin when the equipment stop working until the equipment runs again. I don't think operations people will like that very much. I know of one Key Top Management person from a manufacturing plant who almost lost his job because one of their biggest customers almost pulled out all their work in their plant when they cannot explain the root cause of an equipment failure that delayed their shipment. The corrective action simply indicates that they repair the failure, but when the customer asked what caused the part to fail, they have no idea, and their customer was not very happy about that either.

• Is failure good or bad? Failure is not that bad after all.
• Don't we learn from failure? Failure can be our greatest teacher.
• Can we succeed without failing? No, we need to overcome failure first.
• Is failure telling us something? Yes, there are lessons to be learned from failure.

• Is it okay to fail? It is failure that makes people better. The question is, do we learn from the failure or simply do the same things always.

We can learn from failures if we take the time to analyze them. It is an opportunity for us to begin more intelligently. Therefore, failures tell us something, whether from the lessons of life or problems that an industry encounters. It tells us to slow down and take time on how we do things. We need to learn and find its causes and not the symptoms of the problems. Addressing the symptoms will just create a repeat of the failure. Simply fixing failure will not stop them from recurring. What we need is to understand the causes so we can learn from the things that go wrong

The lamest excuse most maintenance will provide is that operations keep on pressuring us to fix the failure. Hence, we really have no time to investigate and analyze the problem. My question that often pops up in my mind is whether it is really the lack of time or is it the lack of knowledge and skills to analyze a problem.

Chapter **2**

Understanding Root Cause Failure Analysis

> *Troubleshooting is no longer an effective strategy. Remember that when your people become really good at repairing failures, then something is definitely wrong with your maintenance organization. Why? Because they are doing it much too often, but when we expect a different result from the same things that we are doing, it's just not possible; the Chinese called this "insanity."*

2.1: Root Cause Failure Analysis Explained

For the reader's sake, both the term Root Cause Analysis and Root Cause Failure Analysis will be used throughout this book. Our main distinction here is that Root Cause Failure Analysis will be used for equipment-related failures, and Root Cause Analysis will be used for non-equipment-related failures. Now, let us ask a question. Does your company perform Root Cause? Others will say, we perform 5-why's, others will say we do the fishbone analysis, while some will say we do the 80/20 Pareto Analysis or FMEA, and so on.

First, let us explain, Root Cause Failure Analysis is about trying to understand why something went wrong with our equipment and assets. It will try to identify the basic source or the origin of the problem so that industries can finally learn from the failure and the things that go wrong themselves. Both RCA and RCFA will provide a methodology for understanding, investigating, and categorizing the origin of the problem, and incidents concerning safety, quality, reliability, equipment failures, and manufacturing the process losses. This means that identifying the Root Cause will allow our people to finally understand what happened, how the failure happened, and why the failure happened. The objective of conducting an RCA or RCFA investigation is to;

• First, to learn the truth behind the reason for the failure.
• Second, to understand the faults, mistakes, and human errors that had been committed.
• Finally, to decide the changes to be done on the cause of the failure so as not to repeat itself.

Proper Root Cause Failure Analysis will identify the basic source or the origin of the problem. Every system, item, spare or component failure happens for a specific reason. There is always a succession of events that will eventually lead to failure. Failure will be the last event to happen in a sequence of events. Performing a root cause investigation is just rewinding the events that eventually lead to the final event, which is the root cause. If we try to walk 10 steps, performing root cause is just like walking backward from Step 10 to 9, then from Step 9 to 8, then from step 8 to 7, and so on until we reach the first step, which is why the failure, or incident occurred. RCFA will follow the cause-and-effect path from the final failure back to its origin. Root Cause Failure Analysis methodology will provide a specific and solid foundation for preventing the recurrence of the same cause of the problem or failure. It is a tool to better explain what happened, determine how the failure happened, and better understand why the failure happened. Once we understand the failure, the next step is to determine what are we going to do about it.

The good thing about conducting Root Cause Failure Analysis is that it will separate the facts from hearsay or fiction. It is not about trial and error or seeing what works and not. While there are many techniques in analyzing a problem that provides a quick answer, it does not mean that the answer is correct every time. This means that all probable or most likely causes will be verified based on the evidence uncovered so we can move on to the next level. A true and meaningful Root Cause Failure Analysis will take the time to prove that what we say is fact and supports our hypothesis and conclusion with evidence before we spend our money to improve the design of the equipment. Therefore, when the facts are finally backed up by evidence and science, and they are separated from the fiction, we now have a better understanding of the real Root cause of the problem. In root cause, we are not basing our judgment on instinct, brainstorming, experience, history, or any other means. Our basis of judgment will be based on one thing and one thing alone, and that is evidence.

2.2: Case Study: Evidence of the Missing Money

A car wash manufacturer sold one of his complete turn-key car wash units to a client in Maryland, USA. This includes an automatic machine for people who wished to have their car wash automatically. After operating for quite some time, the owner realized that he was losing a significant amount of money from this change machine every time he opened it and tried to collect the money from the machine. The owner insinuated that the manufacturer's employees have a spare key and were stealing the money when he was away. The problem started when the new franchise owner complained to Bill, the car wash franchise owner, that he was losing a significant amount of money from this coin machine every time he opened it and insisted that his people were stealing the money when he was not around. Bill could not believe that his employees were stealing the money because Bill had worked with them for a very long time. These people were honest and loyal to him. So, upon Bill's advice, the owner decided to install a hidden CCTV camera to determine who the real culprit was and determine who was actually stealing the money.

In conducting a root cause investigation, the first thing we need to do is to write the problem as precisely as possible. Once the problem had been written, we need to write the probabilities on why the incident happened. In Root Cause Analysis, we called them a hypothesis. Hypotheses are the possible or probable reasons why the failure happened. It is an assumption made that may or may not be true. A hypothesis is an assumption, an idea that is proposed for the sake of argument so that it can be tested to see if it might be true. In root cause, a hypothesis is constructed to guess the probability of failure. It is usually tentative and is derived strictly for the objective of being verified. In RCM, these are called failure modes. In a crime scene, they are called suspects. The first level in the RCA logic tree diagram will be to determine the probabilities. These are possible causes that may eventually lead to the problem itself, but these probabilities need to be verified as to which caused the problem to occur. In this case study, the video surveillance indicates that the customers entering the car wash were actually paying; hence their hypothesis or theory that customers were not paying was not the cause of the problem. In the RCA logic tree diagram, this will be disregarded. Next, the owner tried to simulate the machine by placing some coins in them, and the machine was working properly; hence, the change machine malfunction was not the cause that led to the problem. Again, in the RCA logic tree diagram, it will be disregarded. Now, we only have one hypothesis left. It was clear to the owner that someone was stealing the money, but the question was who?

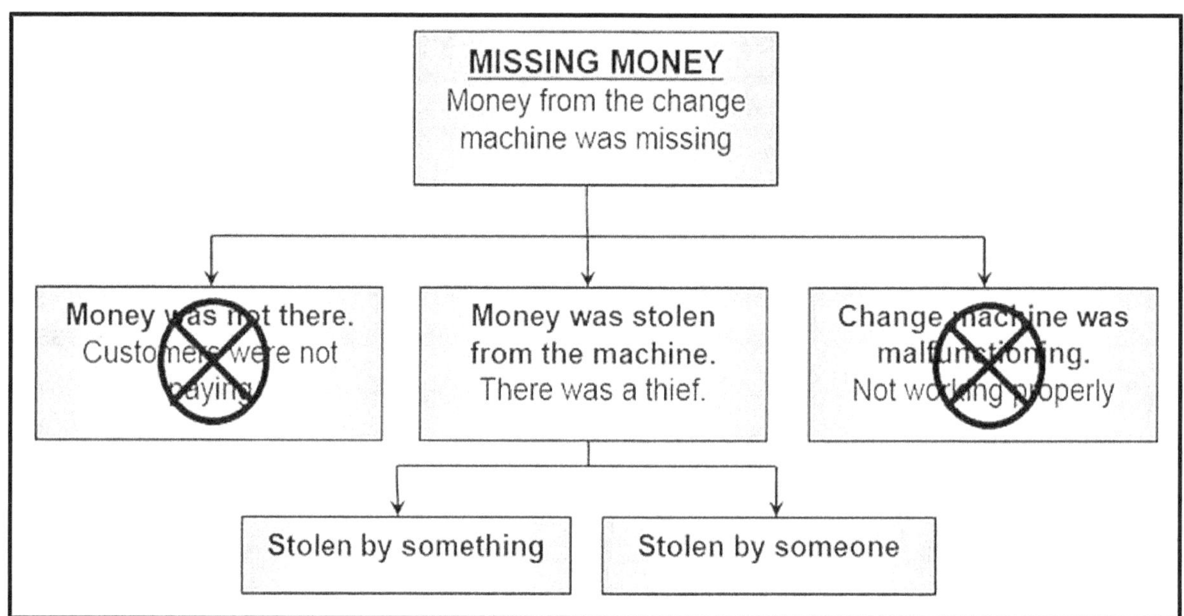

Figure 2.1: RCFA Logic Tree Diagram on the Missing Money

The owner monitored the surveillance camera non-stop. The surveillance camera clearly indicated that it was something and not someone that was stealing the money. Whoa, it was a bird. The owner followed where the birds went and found around USD 4,000.00 in the roof of his car wash and more under a nearby tree. The new owner realized that it was not the employees who were stealing the money. He apologized to the people he blamed, and the people accepted the owner's apology. Everybody was happy; therefore, the case of the stolen money was solved. Thanks to Root Cause Analysis.

This is a simple case of root cause investigation that answers why the money on the car wash system was missing. This is not about brainstorming, doing a Pareto Analysis, or using any conventional problem-solving tools. Rather, everything was based on the evidence that was unfolded. This is the most important part of any root cause or root cause failure analysis investigation process. We shall discuss the subject of evidence gathering in the succeeding chapter of this book.

That's a bird sitting on the change slot of the machine and it had to go down into the machine but why ?

That's 3 quarters he has in his beak, another amazing thing is that it was not just one bird but several of them

There goes another bird this time taking only 1 quarter

Figure 2.2: The Evidence on the Missing Money Now Known

Every failure has a pattern, and we must realize that failure is actually being looked back in reverse. The root cause is actually the point at which the failure began. The event is merely the result of the root cause or how the failure manifests itself. I usually asked my students on RCFA, which comes first, does misalignment caused the high vibration or the high vibration caused the misalignment. Some will say that it is the first, while others insist that it will be the latter. Actually, this is a case of a chicken and egg situation. What is important is that there will always be a cause-and-effect relationship associated with the pattern of failure. The key in conducting a root cause investigation is for the investigator to follow the cause and effect based on the evidence uncovered. To analyze a problem, we have to ask ourselves the following:

• What is the problem? In its precise terms, we first need to define the problem precisely.
• Why did the problem occur?
• What are the possible causes of the problem?
• How did the problem occur?

The RCFA principal investigator must explore why people make decisions that result in the undesirable outcome of the failure. The Principal Investigator (PI) will simulate the problem, like In a crIme scene where the role of the forensIcs and detectives is to determine the science on how the event occurred. In Root Cause Failure Analysis, we are simply not interested in who caused the problem unless this is a clear case of sabotage.

The root cause investigator or analyst acts like detectives when analyzing failures since failures will leave some sort of clues when they happen. These clues are just like the many pieces of the puzzle. Every equipment failure happens for a specific reason or reasons. There is a definite progression of actions and events that will finally lead to the failure. Failure will be the last event to happen. Root Cause Failure Analysis investigation will trace the cause and effect trail starting from the failure back to the root cause, much like a detective trying to solve the forensics of a crime. Detectives will tend to freeze the scene and collect as much evidence as possible for later analysis. They take photographs, fingerprints, search for clues, interview witnesses, and the like.

Collecting the evidence that led to the failure will be a critical process in the root cause investigation since these are like pieces of the puzzle that need to be solved to see the bigger picture. Once the evidence is gathered, only can we perform a sequence of events on how the failure started to see the bigger picture that eventually led to the problem. If we take this sample in figure 2.3 and arrange the sequence, we can clearly identify the problem.

Figure 2.3: Arranging the Sequence 1

In figure 2.3, the problem, in this case was, the kingdom was lost. By asking the question of why the kingdom was lost? Our answer is that the king was killed. The next question to raise is, why was the king killed? Our answer is that the king was killed because he fell on the horse. Why did the king fell off the horse? It is because the horseshoe comes off. Why did the horseshoe come off? It is because one nail is missing from the horseshoe. Why is one nail missing on the horseshoe? Because there was a shortage of nails since it was used on all the knight's horses. This means that the king's horseshoe was the last to be replaced. Once the sequence had been arranged, a story will unfold in our mind as the what really happened on why the kingdom was lost.

And so the story goes that before an important battle, the story is told that a king sent his horse with a groomsman to the blacksmith for shoeing. But the blacksmith had used all the nails shoeing the knight's horses for the battle and was one short. The groomsman tells the blacksmith to do as good a job possible, but the blacksmith reasons out that the missing nail may allow the shoe to come off. The king rides into battle, not knowing of the missing horseshoe nail. Amid the battle, he rides towards the enemy. As he approaches them, the horseshoe comes off the horse hoof, causing it to stumble, and the king falls to the ground. The enemy is quick unto him and kills him. The king's troops see the death, gave up the fight, and retreat. The enemy surges into the city and captures the kingdom. The kingdom was lost all because of a missing horseshoe nail.

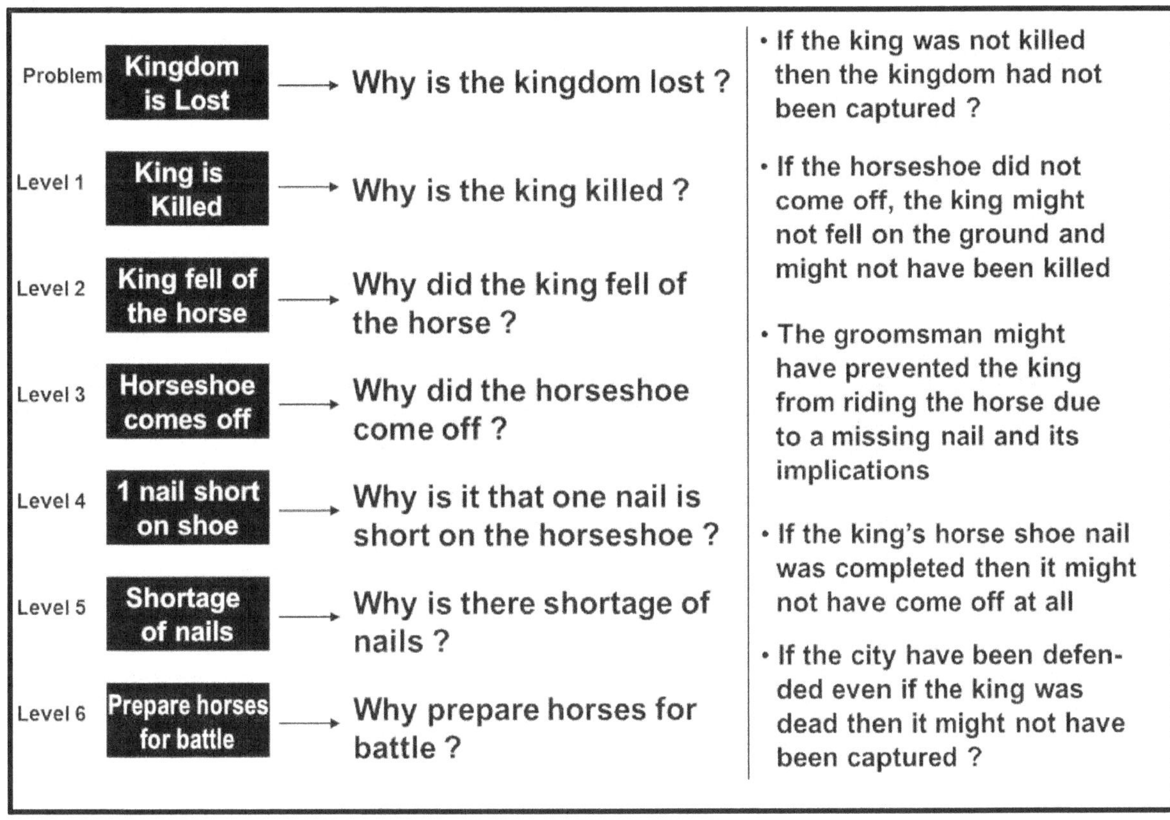

Figure 2.4: Arranging the Sequence 2

In performing RCA or RCFA, I would consider Level 4 as the reason why the kingdom was lost. During my RCFA class, I usually ask my students how they can relate this simple exercise to the causes of their day-to-day problems. The message here is quite loud and clear: Small problems are just the beginning of bigger problems in our plant. The problem in most cases is that we only take care of the problem when it is big, but when the problem seems to be small, no one reacts to it, and when the problem gets big, it will get our attention. These catastrophic failures that we experience on our equipment are just merely an accumulation of small problems that had been left unattended and neglected over time. As my good friend Bob Nelms from Failsafe-Network said, small things that go wrong are the most valuable phenomena of life. Big things go wrong because we do not act on small things. Unresolved small problems are usually the causes of big and catastrophic failures. This means that waiting to do root cause on big problems will just assure us of a continuance of bigger problems. TPM also has the same concept and that before anything else, address the basic equipment condition as this is what usually leads us to failures and breakdowns in our assets, machines, and equipment.

2.3: Difference Between RCFA, RCA, and Failure Analysis

Most industries are confused about what Root Cause Failure Analysis is all about and how to thoroughly investigate equipment failures. Both RCA and RCFA are simply based on evidence. Without evidence, it is impossible to conduct a root cause investigation to determine the equipment failure. One could only guess, speculate, witch hunt, or simply based it on their own experience and judgment. Root Cause Failure Analysis can only be done if the evidence had been well-preserved. While most problem-solving techniques and analytical problem tools will base their analysis on data and history, Root Cause is simply based on one thing and one thing alone, and that is evidence. Hence, here is a distinction between Root Cause Failure Analysis, Root Cause Analysis, and Failure Analysis.

Failure Analysis: If we speak about failure analysis, we stop our probe or investigation on the parts or component level or the failure's physical cause itself. We refer to this as the metallurgical aspect as to why the failure occurs. When we send a failed bearing to the metallurgical lab, they will check the raceway, perhaps conclude and write a report that the bearing failed due to fatigue. The physical cause is the reason how the parts failed technically. This is the technical explanation of why things broke or failed. This mostly explains the metallurgical factor of why and how the failure occurred. Failure Analysis is an engineering approach in determining how and why a particular part, equipment, or component failed. The goal of failure analysis is to understand the physical cause of the failure. Failure Analysis will be considered the first part of any Root Cause Failure Analysis investigation, which is to determine the physical cause of the failure. This means that failure analysis is just a part of Root Cause Failure Analysis.

Root Cause Analysis: This is much broader in scope because this can be applied to medicine, crime investigations, accident investigation, aircraft accident investigation, etc. For industry-related issues, this is applied to safety, environmental issues, process-related incidents, and quality-related issues and defects.

An example of a root cause analysis happened in one of the hospitals here in the Philippines. A patient died in one of the leading hospitals in the Philippines because he was given the wrong type of blood. As the story goes, according to witnesses. One cancer patient walked to another patient whose bed was in front and befriended him. He asked his permission if they can switch beds because he wanted to be near a window so that he can see a glimpse of what is happening on the outside. He said to the other patient that he has cancer and has a few months to live. The other patient showed his compassion by agreeing to the demands of the cancer patient so they exchanged beds. But the problem was that they switch beds, but they did not switch their records. The nurse went in read the records and was about to give a blood transfusion to the cancer patient, but the patient was sleeping on the opposite bed. He injected the blood into the person sleeping in the bed thinking that it was the patient with cancer. The nurse thought that the person sleeping on this bed was the patient to receive the blood transfusion. Eventually, the person died after a few hours. This case will warrant an RCA investigation. Both RCA and RCFA will have the same procedure, which will be done by determining the physical cause, human cause, system, and the latent cause of the problem.

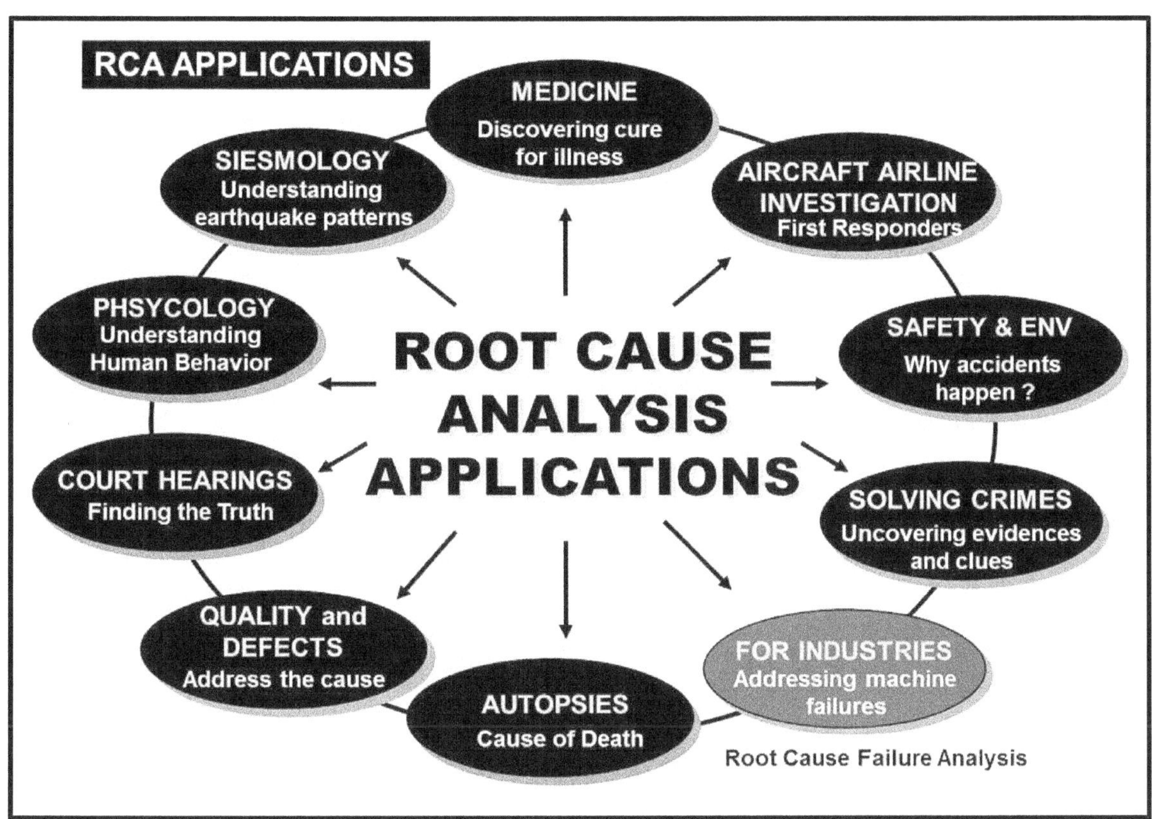

Figure 2.5: Difference Between RCA, RCFA, and Failure Analysis

Root Cause Failure Analysis: Both RCFA and RCA are performed by determining the problem's physical, human, system, and latent causes. The distinction is on the word failure; when we include the word failure in Root Cause Analysis, we refer to equipment-related failures. But we need to be specific in this case; RCFA will not only be limited to mechanical failures but will also include electrical and electronic failures for as long as this is equipment-related. For quality defects caused by equipment-related problems such as raw materials, this will be under RCA. Hence, if we speak about Root Cause Failure Analysis, we are trying to understand why something went wrong with our equipment and assets. It will identify and address the basic source or origin of the problem so that industries can learn from the failure itself. The reason for conducting a Root Cause Failure Analysis is for industries to learn from the things that go wrong. RCFA will provide a structured methodology for investigating, categorizing, and eliminating the root cause of equipment-related incidents. Root Cause Failure Analysis is not interested in who caused the problem since there are much deeper causes on why people commit mistakes and errors. The goal of any Root Cause Failure Analysis investigation will be to identify and expose these causes, which are known as the latent cause of the problem.

2.4: Chronic and Sporadic Failures

Sporadic failures often indicate sudden large and abrupt deviations from the norm. In contrast, chronic failures indicate smaller but frequent deviations that, most of the time, had been perceived as usual. Sporadic failures occur suddenly that mostly result from a single cause. They are easy to solve. Once the cause had been identified, sporadic failures are eliminated. **Chronic failures** are usually made up of a wide variety and combination of causes. Frequently, maintenance can provide temporary relief or a containment plan, but as the situation gets worst, chronic failures resurface once again from time to time. Addressing chronic failures requires innovative and breakthrough approaches.

Why Do Chronic Losses Persist? There are many reasons why chronic problems and defects recur. Phenomena are insufficiently stratified and analyzed. Defects and failures are not carefully observed and stratified. This means that people do not notice the pattern. Some factors related to phenomena are overlooked and uncontrolled. Another reason is that hidden abnormalities in individual factors are not addressed since people are more alert and respond to large problems since they appear more significant and the smaller the problem the more likely they are to be ignored. These slight abnormalities include dirt, dust, vibration, looseness, slight wear, leaks, and dirt. Sporadic failures are expensive compared to chronic failures but chronic failures may tend to be more expensive in the long run if the failure recurs frequently.

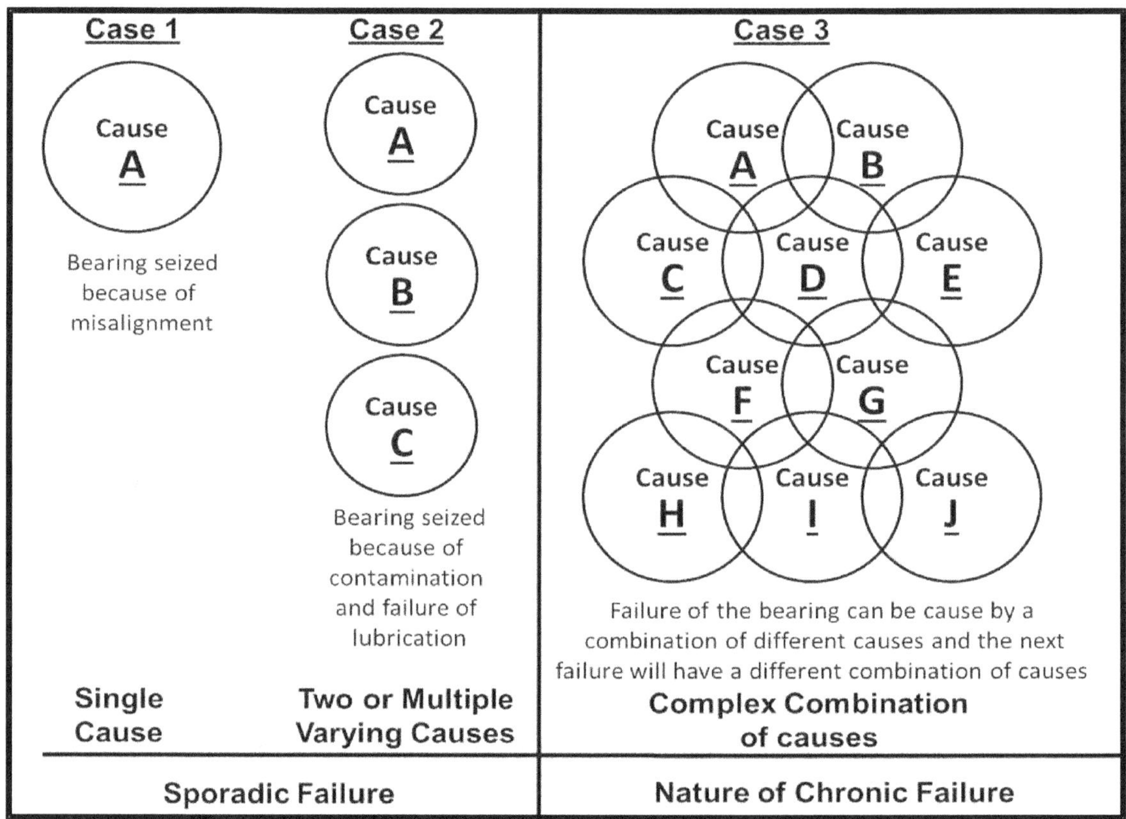

Figure 2.6: Sporadic and the Nature of Chronic Losses

To address chronic failures is to understand the nature of chronic losses fully. Figure 2.6, Case 3, is a classic case of chronic failure. In this case, we need to understand and expose every single possible cause that can affect the problem. Root Cause Analysis may not be recommended for chronic losses since the root cause will only address one problem at a time, depending on the evidence that caused the problem, especially for quality defects since most of them are chronic by nature. Quality defects are better off using an analytical problem-solving tool such as P-M Analysis. Addressing only a single cause will not eliminate chronic failures and will just repeat itself. A tool highly recommended for addressing chronic losses for defects will be P-M Analysis by Shirose, Kimura, and Kaneda. P stands for Physical and Phenomenon, while M stands for Mechanism, and the 4M (man, machine, method, and materials). However, we can use Root Cause Failure Analysis for chronic failures and breakdowns. For sporadic failures, the effect will be dramatic, costly, and catastrophic. While in the majority of industries, attention is given mostly to sporadic failures since most of the causes of this failure are just a single cause, which is easy to address. Sporadic failures can be said as dramatic deviations from the norm. Expect immediate damage when they occur. Chronic failures, on the other side, are characterized by their low cost, but high-frequency occurrence. These failures are difficult to eliminate due to their complex combination of causes when they occur on their own. They can be more costly than sporadic failures in the long run if they continue to recur.

2.5: The Golden Rule on Root Cause Failure Analysis

First, before applying or initiating any Root Cause Failure Analysis investigation in your plant, you need to adopt the "Golden Rule" of Root Cause Failure Analysis, and that is "**amnesty**." Nobody should be punished or disciplined due to any Root Cause Failure Analysis probe or investigation effort unless this is a clear case of sabotage. If the reason for conducting a Root Cause Failure Analysis is to find out the guilty person and to punish them outright or pin them down, then all your collective efforts on having the chance to learn from the problem or from the things that go wrong had been deemed unsuccessful and useless. Perhaps if your intention in performing a Root Cause Failure Analysis is to put the blame or punish someone, then you need to call your analysis "who-who analysis" instead of the root cause. Remember that when a person commits an error or mistake, it is not always the fault of the person who committed the mistake or error. There is always a deeper reason or latency that needs to be exposed before we totally condemn anyone in our industry.

This is a quote from my good old friend, C. Robert Nelms. The issues of blame, discipline, fear, punishment, and accountability are near the root cause of why industries don't learn from the things that go wrong. There is most certainly a need for rules, regulations, policies, and procedures in our businesses. Likewise, there is most certainly a need to enforce them by applying discipline to those who do not abide. However, it is vitally important to acknowledge the reason for our rules. They exist to help prevent incidents. The time for discipline should be before an incident, and not after it happened. Unfortunately, most organizations do it backward. They do not consistently apply their disciplinary policies ahead of time. Instead, they will wait for something awful to happen and then go after the person. This is outrageous, illogical, and immoral. And this is perhaps the reason why most industries do not learn from their own failures.

[1]From the book of Keith Mobley on Root Cause Failure Analysis (page 4), he quotes that the purpose of RCFA is to resolve problems that affect a plant's performance. It should not be an attempt to fix the blame for the incident. This must be clearly understood by the investigating

---

[1] R. Keith Mobley, **Root Cause Failure Analysis**, (Butterworth Heinemann, 1999), Page 4

team and those involve in the process. Understanding that the investigation is not an attempt to fix blame is important for two reasons. First, the investigating team must understand that the real benefit of the analytical methodology is plant improvement. Second, those involved in the incident generally adopt a self-preservation attitude and assume that the investigation is intended to find and punish the person responsible for the incident. Therefore the investigators need to alleviate this fear and replace it with the positive team effort required to resolve it.

Suppose people know that someone will be held responsible for the incident; in that case, people will definitely be defiant and hesitant to help in the investigation, especially during the interview process. To these people, it is better to practice their right to remain silent or speak as little as possible to avoid being involved in the problem. These people will not be very enthusiastic about sharing what they know on why the incident took place. But in most industries, when the impact of the failure is high, the insurance companies, lawyers, the media frenzy, and those fat-crooked old politicians are involved. These crooked old politicians will always give some reactive remarks after an incident of major interest happens and not before. They will be interviewed by the media, and they will state a remark that industries should have done this and that. Why are people so dumb in voting these crooked and corrupt politicians? No more comments, and back to the root cause.

Therefore, before applying the RCFA process, it must be explicit that nobody should be blamed or punished for the incident unless a crime is committed, or this is simply a clear case of sabotage. In short, for those who want to perform Root Cause Failure Analysis, there must be an amnesty policy. Before we start putting fingers or placing blame on someone else, kindly answer one simple and honest question, are we convinced beyond a reasonable doubt that if we were in the shoes of the person who committed the mistake have we done the same thing, or have we done it differently given the same circumstances, and conditions? If your answer is doing the same thing, you need to probe further beyond the human cause because there is a much deeper cause which is the latent cause of the problem. And this latent cause of the problem is where people involved usually wash their hands, just like when Pontius Pilot condemns Jesus to death. These people involved with the latencies are usually invincible, and the funny thing is sometimes they do not even know that they are also part of the problem.

Last July 14 to 16, 2010, I conducted three days of Master Class on Root Cause Failure Analysis. The first day was really tough because of a typhoon that hit the country. Day 1 of Root Cause Failure Analysis was also signal number one in the Philippines, and three delegates did not make it on the first day; however, two of them were able to finally come on the last two days of the RCFA training class. I had a participant in that class that shared his experiences. He said that previously when an equipment failure or problem hit their plant, all maintenance people would come down running as fast as they could towards the failure or failed equipment to extend their help to the maintenance on duty which is a very noble gesture indeed on the part of the maintenance crew. That was their culture. Nobody was left behind; they will always help because they wanted to involve themselves and resolve the problem together. These people understand the word team. They work to resolve the problem completely, even if they belong to different functions of the plant. One day, a new management team was placed in the plant and started changing SOPs and policies. After management starts changing their policies and rules, they started suspending people involved in the failure incident. Things changed dramatically in this plant because, unlike before, these very same people started walking away from the failure instead of involving themselves and extending a helping hand as they graciously did previously. They do not want to involve themselves anymore. It's not my problem anyway, so I do not want to get involved since I might end up getting caught in the problem too, or management might accuse me of the problem. Perhaps if the person who did this policy himself is perfect in every way and has not yet committed an error in his entire lifespan, then I will succumb and submit to that policy.

When we accuse people or blame them, people's lips automatically begin to seal. You will unlikely get to the root cause of the problem because people will practice their right to remain silent. They would rather not talk about it, and they will become more defensive and defiant when you interviewed them. Ask me why; it is simply a case of human nature. Try to do something good on your plant, and nobody will talk about it, but when you commit just one mistake, you will be the talk of the town until the next incident comes. Am I right? If you asked me why just listen to the song of Michael Jackson on Human Nature, and you will find the answer.

While it is true that people err and commit errors, many cases had been stated so far that people commit errors and mistakes way beyond their control. If this is the case, then we must dig deeper into the real reason why the person committed that error so we can finally learn from the things that go wrong. There underlies a deeper reason why people commit mistakes and errors. A true Root Cause Failure Analysis will uncover that fact. What both Root Cause and Root Cause Failure Analysis believe is that all physical failures are triggered by humans, but humans are negatively influenced by latent forces. Therefore, in any Root Cause Failure Analysis investigation, the goal must be to expose and identify these latent causes.

The difference between humans and machines is that people can sense, change, hear, smell, see, feel, or taste something different and take the necessary actions to correct the anomaly. Humans are flexible, while machines are frequently not. Research by Dr. James Reasons of the University of Manchester in England found that humans commit an average of 6 errors per week. The latest official current world population estimates that as of August 11, 2021, the world's total population was estimated to reach about 7,885,428,998. This means that, on average, around 6,758,939,141 errors are being committed daily. And with all the errors occurring and all the ways we could destroy ourselves, life is still preserved. Why? Because humans can sense change and break the error chain. In short, we cannot eliminate human error; but we can only reduce them or learn from them. I would prefer the latter and learn from the failure itself because whether from the lessons of life or from industries problems, it makes us better people if we can learn from the problem.

If any industry wants to perform a thorough Root Cause Failure Analysis, do it for the right reasons. If the reason why you want to perform a Root Cause Failure Analysis is to put the blame on someone and find a culprit, then we are not learning from anything. There must be complete "amnesty" at the very beginning when performing a Root Cause Failure Analysis. The people involved in the investigation should understand that there is a much deeper cause why people commit mistakes, which is the latent cause of the problem. The only way that we can learn from our own failures, whether from our equipment failures or from the lessons of life, is when we set aside our position, pride, ego, sarcasm, pessimism, negativity, and once in for all, try to look ourselves in the mirror and ask ourselves that we too are also part of the problem. Whether we like it or not, humans will continue to make errors and mistakes; that's why the need to analyze them is a must simply because failures will allow us to begin again more intelligently.

When John F. Kennedy was assassinated on November 22, 1963, Lee Harvey Oswald was convicted of the crime; Oswald said he was just a patsy or a fall guy. The truth was never known who really pulled the trigger and who the mastermind behind it all was. Oswald was killed by Jack Ruby, who later on died in his prison cell by hanging himself, or perhaps Oswald and Ruby were just pieces of the puzzle that still remain unresolved on why Kennedy was killed, although there were many conspiracy theories behind the assassination. All human beings commit mistakes and this is part of human nature and we cannot eliminate it. Human error will always be inevitable but the good news is that we can manage human errors but not by punishing people. The only exemption from this rule is if the incident is a case of sabotage, or if corruption is involved.

To all ▮ Vendors,

During our Vendor Assembly this year, we discussed with you our move towards Business Practice Excellence. We shared with you our Code of Conduct and Business Ethics Policy, Conflict of Interest Policy, as well as our Guidelines on Giving and Receiving of Corporate Gifts. Under ▮ established Code of Conduct and Business Ethics, employees shall consistently apply high standards of business and personal ethics in discharging their duties and responsibilities with other employees, business partners, and the general public.

In this regard, ***we would like to emphasize that the giving of gifts during this Holiday Season is strictly discouraged***. We appreciate your support to ▮ this year and expect your understanding and compliance in this direction. Please refer to the attached letter signed by our Vice President of Supply Chain, ▮ ▮ ▮.

My Reply,

Thank you for informing me as I am in the process of wrapping my books as a gift to your Maintenance people. I guess I will just give this book to my other clients who do not posses the same policy as yours.

Rolly Angeles
Maintenance Preacher and Author

2.7: Sometimes Company's Policy Can Be Funny

One more thing I would add to this is that the bigger the industry, then expect more problems to happen. People write so many policies and procedures on what to do and what not to do. There are rules, policies, procedures, restrictions, and so on, and sometimes industry's rules or policies are a bit funny as I do not want to use the S-word. If errors are made because of a wrong or weird policy then we need to update them and sometimes we also need to flex it a little, which means that at times, there must be some exemptions. When the months of "**<u>ber</u>**" come, I usually received an email just like in figure 2.7, reminding me of their policies.

2.6: Why Many Root Cause Initiatives Fail

Many industries have some form of Root Cause Analysis or problem-solving tools they use to analyze problems, yet many seem to be misguided about what Root Cause Failure Analysis really is all about. I admit that I had only known its true meaning when my good old friend, C. Robert Nelms, invited me to attend his Latent Cause Analysis training in March of 2007. This book will clarify what Root Cause Failure Analysis is all about and why most industries fail in their Root Cause Failure Analysis initiatives. There are some reasons why many Root Cause Failure initiatives fail.

1) Simply, I don't have the time to do a Root Cause Failure Analysis. People complain a lot about having no time to analyze problems. People do not have time to perform RCFA because they are busy reacting to the needs of putting out the fire, which means fixing the failure first. We simply have no time to analyze failures, but we always have time to fix the problem repeatedly. In fact, we have been doing it for a very long time. People know that it is important to do Root Cause Failure Analysis, but we cannot do maintenance and RCFA together; there are only 24 hours in a day. How often do we see a problem resurface once more, and we decide to modify or improve the system without really analyzing the real root cause of the problem? Sometimes we think that we have solved the problem only to realize that a new problem emerged because of the modification and improvements. Have we really eliminated the problem, or what we have performed is just a simple case of an improvement failure? People tend to go to the solution without first analyzing the root cause of the failure.

2) Root Cause is not just about data: Many people believe that we must have data to perform Root Cause Failure Analysis. I heard this message many times, and I just cannot emphasize how these people are so wrong in every way. The lifeblood of any Root Cause Failure Analysis is evidence and not data. When something goes wrong, there is a cause behind every bit of the problem. To determine the cause, these bits of puzzles are gathered so that the analysts can determine what the big picture really is. Data are just part of the evidence, but when you tie up the things together, we can understand what caused the problem and what the Root Cause is all about. To understand the cause of the failure is to envision an iceberg. Failures are just the tip of the iceberg which is clearly evident, but underneath the iceberg lies the hidden causes. Our only link on what is underneath the failure will be the pieces of evidence unfolded.

3) Most of the pieces of evidence are washed up by the fast team. When an equipment failure occurs, there are actually two teams addressing the failure. As explained in Chapter 1, these are the fast team and the slow team. The fast team refers to the restoration team, which will restore or repair the equipment so that operations can be up and running once again in the shortest possible time. They are fast. Spares must always be at hand. These people will be timed about how fast they can repair the equipment and respond to failure; otherwise, the longer they take to repair, the more shadows will be watching them at their back asking them how much more time? There is pressure on the fast team. Improving the MTTR or Mean Time to Repair would be their goal, and they need to do things quickly. These people are considered heroes because, without them, operations will be delayed. Does this sound familiar to you? On the other hand, we have a slow team, the team that will analyze the failures. The problem is that when the slow team arrives, the equipment is already running, and most of the evidence has already been washed up and evaporated. In this case, my question is, how on the planet can we perform a Root Cause Failure Analysis in this situation when the fast team always destroys the evidence at hand? Many will disagree with me, but I live by my principles and what I believe is right. MTTR or Mean Time to Repair can never be a measure of reliability. In fact, when people become too good at repairing, it only means one thing that the failure keeps on repeating itself. It would be much better to take the time to understand and learn from this failure rather than keep on repairing the same failure all the time. My recommendation for this situation is simple. Before fixing the failure, have the restoration team take photographs of the equipment inside and out and a close-up photograph of the equipment's affected part or component. Secure the part or components that failed and place them in a plastic bag together with any foreign debris that can be found near the part that failed and endorse them to the slow team in charge of the physical evidence. Make this a habit before repairing the equipment, especially if the failure warrants a Root Cause Failure Investigation.

4) How deep should we probe on Root Cause Analysis efforts is unknown? A true and meaningful Root Cause Analysis always believes that all failures have a physical cause. Still, all physical causes are triggered by humans, which means that there is always a human error committed that eventually leads to the problem's physical cause. Still, humans are negatively influenced by latent causes. Somewhere between the human and latent cause is the system cause. Therefore the goal of any Root Cause Failure Analysis is to expose and identify these latent causes of failures. The depth of probe on any Root Cause Analysis efforts should end once the Latent Causes are reached. Surely, when we probe our Root Cause analysis, it will have something to do with the oil that causes the compressor to fail, but as said, all failures are triggered by humans, and when we visit the oil storage, the oil drum lids are either missing or open not only on this drum but on the other drums as well. Perhaps the maintenance has forgotten to close the lid after getting some oil in the drum. Again my second question is, do we blame the maintenance for not closing the container's lid that led to the compressor failure? I guess that will be the case. But when we probe deeper into the Latent Cause of the problem, the drums and barrels are stacked out in the open, exposed to sun, dust, rain, and moisture. I bet that even if you cover all the container lids, there will still be moisture present due to condensation effects, most especially during the early morning, since this is not the correct way to store lubricants. The drums are piled up on one side of the building, and every single employee that enters and leaves the plant may have noticed this small problem that some of the lids of the oil container are either missing and that no one seems to care or no one seems to understand the effect that causes the failure of the air compressor which paralyzed their plant a couple of weeks ago, so if you asked why no one reacts, it is because no one really understands about lubrication. After all, nobody was ever trained on this subject despite several proposals by maintenance requesting for this training only to be denied by management due to their cost-cutting schemes. Latency lies in each of us, yet sad to say that when the person involved in the root cause is known, most organizations end their probe, and the guilty one is condemned and the case is dismissed. I cannot overemphasize it more.

5) Root Cause takes much more than 24 hours to complete. Root Cause is not about some fancy and cool software that can provide you a list of failure modes to select, so you can click here and there. To undergo a thorough Root Cause Failure Analysis is a slow and painful process. People undergoing the investigation will require time to complete and absorb the learning. In fact, we can only learn from failure if we acknowledge that each of us is part of the problem. Depending upon the severity of the problem, Root Cause Failure Analysis cannot be completed in 24 hours as other industries claim; some cases will take more than a week or even months to complete. This will take time, most especially in verifying the problem's causes based on the evidence unfolded. If the evidence doesn't show up, we need to dig more evidence until we verify each possible cause of the problem. I worked previously in the manufacturing industry and spent quite several years there, and most manufacturing industries are required to complete their Customer Complaint Report in 24 hours, most especially if their customer complains to them about a defective product. The engineers will use an 8-Discipline technique and are forced to complete their analysis within 24 hours. The industry wants to impress their customers on how fast they can resolve and analyze problems, but even if we ask the experts on Root Cause, the analysis simply cannot be completed in just 24 hours. Again, Root Cause is a slow and painful process, and we must bear with it to the end to get to the truth and the true cause of the problem.

6) Root Cause is not about the probability of failure and failure modes. There is some confusion regarding what methods and tools would determine a Root Cause Failure Analysis, and some believe that these tools will eventually lead them to the root cause of the problem. FMEA, Pareto Analysis, Fishbone Diagram, 8-D Analysis, 7 Q.C. Tools, P-M Analysis, and others are called problems or analytical problem-solving tools. They are not meant to derive the exact root cause but only the problems most likely or probable cause. The root cause and probable cause are not synonymous or identical words. A probable cause may or may not be the root cause that led to the problem, and the only way of knowing the cause is to determine the pieces of evidence that led to the problem. This is fully explained in detail in Chapter 3 of this book.

7) The focus on the root cause should be on small problems and not on big problems. Big problems are caused by small problems; in fact, these small problems accumulate over time and end up causing chaos and catastrophic failures in the organization. Remember my example above about the air compressor causing a certain portion of the plant to be shut down and where the plant loses revenue because the lid on the oil container is missing allowing moisture to penetrate the oil, which eventually causes the air compressor to fail. A power plant was shut down for days. After all, its triple redundancy protective devices all failed over time because the people in the plant never inspected them and were confident that this would not cause them any problem because the protective device had two more additional back-ups. In a manufacturing plant I previously worked with, one of the main customers of this plant pulled out all its work because the products they shipped were off-specs, and it was traced later on to a tiny proximity sensor that was clogged due to dirt. This sensor determines whether the product is off-specs or not. I can go on and on. These things have one thing in common. Big problems are just an accumulation of small problems. Root Cause is reactive because we can only perform a root cause analysis if the failure has occurred. On the other hand, valuable lessons that we learn from root cause allow us to reflect that we can only learn from the things that go wrong when we truly understand and expose the latencies of the problems that require people to change. When people start to change, then this is what makes Root Cause Analysis Proactive.

8) Root Cause is not about who caused the problem. Root Cause is not designed to determine who causes the problem but rather to understand why and how it manifests itself. While it is true that there is always a human cause for every failure, there is also a hidden cause that causes this person to commit the mistake. In fact, according to C. Robert Nelms, the Golden Rule of any Root Cause is to understand to such an extent that we are convinced that we would have done the same thing if we were that person. It must be clear to all people in the organization that no one should be punished for any root cause analysis investigation especially if the human error is unintentional. Our point is to understand what leads that person to commit that mistake so that we can all learn from it before punishing someone else. Are we certainly sure that we have done differently if we were in their shoes? Blame and finger-pointing have no room for any root cause investigation. Suppose our intent for any Root Cause Analysis is to punish people; in that case, our efforts will be entirely useless, and the industry will never learn from their problems since it will just repeat itself again.

9) Lack of Management Support and Commitment on RCFA. If management does not support the teams performing RCFA activities, the team efforts will fail. Management must provide technical resources such as sending failed parts to an independent metallurgical laboratory to determine the failure's physical cause for the team to verify its hypothesis. Management must make changes in the work order system or procedures to ensure proactive recommendations. If modification requires some budget, management needs to approve them. Management must allow the team to meet and perform RCFA and must gain cooperation from other departments. Frequently it is not only support that is likely needed from management but also commitment. The teams can only complete their analysis if management allows them to do it. Otherwise, the team goes back again to its wicked old job of repairing failures. Suppose the corrective action requires a budget and falls on management deaf ears; in that case, everything will just be a complete waste of time, and nothing will ever change.

10) We Praise Our People for being Reactive. Why do we spend so much time working on the problem (reactive work) instead of being proactive? The reason is that production supersedes all other concerns. Maintenance is praised not because they can analyze the problem but how fast he can repair and replace the failed parts. We value people who can give a quick fix or band-aid therapy and can have the equipment up and running once again. In many cases, a good backlog of maintenance work will ensure their job security. They always feel that if we analyze failures and prevent the problem's recurrence, we may have nothing more to fix, and there will be no more overtime pay for me. This may be a valid fear by the maintenance personnel, but they also need to realize that if their industry is reactive, their industry is simply losing revenue which is bad for business.

11) The Biggest Barrier on Change will Always be Cultural in Nature. This is the way we do things around here. Resistance to change is quite normal, especially for the older workforce. The more the program is pushed, the more resistance is expected from the users themselves. People do not want to improve because they think that improvement programs are fads and aim to reduce the workforce in the long run. They usually observe the program for six months, and if it's not heard after that date, the program has died a natural death.

2.7: Top Reasons Why We Need to Perform RCFA

Every industry is familiar with the word root cause, but not all understand what it really is and how it is being done. We perform root cause failure analysis to understand why things go wrong so that we can finally learn from the root cause of the failure. We need to understand the root cause so that the cause of the failure does not repeat itself. Although I do not recommend performing root cause failure analysis on every failure that occurs, I recommend performing root cause if the failure keeps recurring.

1) Failures simply won't go away by fixing them: Failures struck us right in our face every time we deal with them to keep the equipment up and running for operations to continue with their production. We encounter the same failures repeatedly, and we always perform the same thing of fixing them all the time. RCFA helps us understand why these failures occur so that maintenance can focus more on improving their equipment and assets. We can only learn from the failures themselves if we follow the evidence and perform RCFA. Without evidence, one can only guess, speculate, based it on experience, or simply address the most likely cause of the problem.

2) We perform RCFA to arrive at the correct solution to our equipment problems: RCFA is not about addressing all the probable causes of the problem, but rather failures are being looked back in reverse to determine what really caused the problem to occur. In performing RCFA, each hypothesis is verified until we have gathered enough evidence to support the actual facts that eventually lead to failure. In performing Root Cause Failure Analysis, it is important to address the physical cause, the human, system, and latent cause of the problem.

3) Being proactive will give me a sense of job security: Many maintenance people believe that a good backlog of maintenance work and being reactive will ensure their job security by doing a lot of overtime work. This is not the right mindset for maintenance since traditional maintenance people are confined to repairs and fixing failures, but the scope of maintenance is beyond boundaries. The true job of maintenance is to improve equipment reliability. The best position an industry can have will always belong to the maintenance function; however, maintenance people are often treated as repair people or mechanics in most industries and are often seen as a cost center. The true value of maintenance can be seen not in how fast they can repair the failure, but in how they were able to analyze equipment-related failures.

4) People learn from the failure itself: For every failure that occurred on the equipment and that had been thoroughly analyzed through RCFA, there is a learning that we can all gain from these experiences to prevent the recurrence of the cause of the failure itself. Sometimes, failures speak to us in a different language. It is like when we feel pain, and when we don't remedy it, it will worsen until the equipment has finally given up. Remember that there is always a lesson that we can learn from the failure itself.

5) There are failures where the consequences far exceed the cost of failure itself: Not all failures are equal. There are failures, that when it happens, will have no effect, but there are also failures that can have very severe consequences once it happens in operations. When the impact or consequences of the failure can have safety or environmental consequences, we really need to put all our collective effort into eliminating these types of failures. Industries cannot afford to encounter a failure that can paralyze their operations, even for a single minute.

6) RCFA will save us money: Unexpected failures cost us money not only on the amount of damage or cost of spare as there are other costs beyond the spare parts itself, such as downtime cost, cost of shipment delay, labor cost, cost of loss production, secondary damages, etc. Most of the time, the majority of these causes are based on deficient organizational systems, which includes maintenance and operating procedures, policies, guidelines, and training systems since most people will make decisions based on these systems, and most of the time, we end up wasting money because we tend to spend the money on equipment and not on investing on the knowledge of our people. Investments must be spent on making our people make better decisions. Remember that it will be the people that will improve the equipment and not the other way around. This will provide us far greater benefits and returns than many equipment investments. The saying that people are the company's greatest asset is not always true, what is true and correct is that the right people are the company's greatest investment, and we can only have the right people if these people are equipped with the right knowledge to do their jobs respectively.

7) To change our people's mindset: The new paradigm is that failures must be analyzed. People can learn from failure if they take the time to take things slowly and analyze them. The true job of maintenance is not to repair and fix them all the time. While I do not recommend performing Root Cause Failure Analysis on every single failure that can occur on the equipment, but if the failure keeps recurring, it is time to slow down and perform a Root Cause. Remember that troubleshooting is no longer an effective strategy. In today's competitive world, the "Analysts" find real solutions to our problems; that is why performing Root Cause Failure Analysis is necessary. This simply means that if your maintenance people become so good and fast at repairing failures, then something is definitely wrong with your maintenance organization because the failure simply keeps on repeating itself, and maintenance has not analyzed why it keeps on failing. Hence, let us leave the fixing to the mechanics and the thinking to the maintenance people.

8) There are cases where the consequences of failure are more important than the failure itself: Perhaps a classic case of this will be the Challenger Disaster; what simply failed was the O-ring on the Right Solid Rocket Booster, attached to the main tank causing a leak which ignites the main tank causing it to explode, destroying the Challenger Shuttle and killing all astronauts including Christa Mc Aullife, the winner of the Teacher in Space Program which was watched by the whole wide world when it took off last January 28, 1986. The problem with industries is that most maintenance needs are neglected or subject to cost-cutting schemes. The industry only reacts after a painful incident occurred in which their industry will be hurt and devastated. The main reason for conducting a Root Cause Failure Analysis is for industries to learn from the things that go wrong, but this can only happen if people working in industries are brave enough to face themselves in the mirror and admit to the fact that we too are also part of the problem. Fixing failures will just assure a continuance of the same problems, and the only way to get rid of the problem is to analyze and learn from it.

9) To Manage Our MRO Spare Parts More Intelligently: Making our equipment more reliable simply means lesser failure and downtime on the equipment. As I have written in my book on Problems and Solutions on MRO Spare Parts and storeroom that improving the MRO spare parts and reducing the inventory should be done both inside and outside the storeroom. Performing root cause means lesser downtime and lesser downtime means the lesser acquisition of MRO spare parts. The logic goes that the more reactive the industry, the more difficult to manage the storeroom since in reality, there is a lot of emergency buying going on, and we really do not know what to stock.

2.8: Is Root Cause Failure Analysis Reactive or Proactive?

Many are asking whether Root Cause Failure Analysis is reactive or proactive. This is like answering the question of which comes first, the chicken or the egg. Although there is no scientific proof that the chicken emerges from a single cell and just came out of the farm a couple of thousand years ago. All we know is that the chicken comes from the egg and the egg comes from the chicken, but whichever comes first, I think, I leave it up to the reader.

Although Root Cause is both reactive and proactive, I believe that in Root Cause, we need to be reactive first to be proactive, just like we need to fail first before we can succeed. Root Cause Failure Analysis can only be performed if there is an actual incident or a failure. Nobody can perform a Root Cause Failure Analysis if no failure actually happens. A failure must first occur before a Root Cause Failure Analysis investigation can be performed. Without a failure, then how in the heck can we perform a Root Cause Failure Analysis? Likewise, the failure must be fresh, meaning it just happened. It is highly unlikely to perform a Root Cause Failure Analysis if the failure happened a year ago or 6 months ago. Why? Because in performing a Root Cause Failure Analysis, the evidence must be frozen.

When Is Root Cause Reactive? Root Cause will always be reactive at the start since a failure needs to be in place before anyone can perform an investigation. For example, a ball bearing had failed that caused the equipment to stop functioning. There was no stock in the storeroom for the part; hence downtime was enormous. Management warranted a root cause investigation of why the bearing had failed, so an investigation team composed of a Principal Investigator and Evidence Gathering Team had been formed to investigate the failed phenomenon. We can only perform an analysis when something had failed, or a problem erupts that causes the operation to halt. The team is interested in understanding the underlying causes of the problem so that the people involved can learn from the things that go wrong in their industry.

<u>Investigating Equipment Failures Through Root Cause Failure Analysis</u>

I recall one time, I was called to a plant to provide a brief presentation with their management team about Root Cause Failure Analysis since one of their leading customers got very irritated, which almost pulled out their business in that plant. One piece of equipment failed that eventually caused a delay in their shipment. When the customer asked the reason for the delay, the director said that a bearing failed in the equipment, which eventually stopped the equipment. When the customer asked what did they do to address the problem? The director said that they replaced the bearing with a new one. The customer was so appalled and almost pulled out their business with this plant because they do not answer why the bearing failed. This plant, which I worked on previously, called me and asked me to present what courses I can teach. As I was presenting, someone interrupted me since what they want is a total one-time approach to capture all the failures on this equipment. I lost that opportunity when I told them that Root Cause was not designed to handle that and provided my recommendation to do RCM. In my mind, these people do not know what Root Cause Failure Analysis is all about. It is not something that will address all the equipment's problems, but rather it is used to understand the cause of a particular failure that actually occurred in the equipment. You see, the people above wanted to address every bit of probable cause that can occur on the equipment; I think what they need was some sort of FMEA or RCM and not RCFA. The difference between the two is that FMEA is designed to capture the probable cause and not the root cause of the failure and prioritize the failure according to its severity, occurrence, and detection, while RCFA is designed to address the true cause of failure based on the evidence unfolded so that we can learn from the things that go wrong.

When is Root Cause Proactive? First, Root Cause Failure Analysis can only be Proactive if we can finally learn and understand that big problems are just an accumulation of small problems neglected in our equipment and assets. Most Root Cause initiatives and efforts seem to address catastrophic and big failures. The important thing that matters is why not address the problem when it is still small. Second, most people say that performing Root Cause Failure Analysis will prevent a recurrence of the problem. Let me rephrase this. It is much more likely that performing a root cause failure analysis will help us prevent the same cause of the failure from occurring. When we have taken care of a single or couple of causes, there is still a probability that the failure can occur again in the future, but due to a different cause. Root Cause is not designed to address every single possible cause in a single investigation unless the evidence concludes that the failure has a variety of causes, but rather we are only interested in the cause of the problem based on the evidence unfolded.

Let us assume that the RCFA analysis shows traces of silicon contaminants and metal debris on the bearing's raceway, causing the bearing to suffer from fatigue and spalling, and the investigating team indicates that the oil was too contaminated with metal and dust contaminants. A particle count indicates that the ISO range was way out of the standard. Good filtration practices were provided, and contamination control practices were adopted on this equipment and on the rest of the assets with lubrication. How lubricants were stored was addressed, and contamination control was initiated. But all these efforts were done as a result of finding out the cause of the failed bearing. After initiating these corrective measures, they experienced smooth operations due to their proactive efforts. They addressed that single equipment that failed and the other could probably fail due to a similar cause. This means that after performing the corrective measures, it is unlikely that the bearing will fail because of contamination, but it is possible for the bearing to fail due to a different cause, such as misalignment, electrical pitting, false brinelling, and other possible causes.

Third, Root Cause Failure Analysis can only be proactive if we can learn from the failure itself. This is easy to say, but it is much more difficult to do in the real world. There is always a deeper underlying cause behind the physical and human cause, which is the latent cause. Many industries think that when they find out the physical cause of the problem, they stop, only realizing that the failure will revert back to them in the least unexpected time. Leo Tolstoy quotes, everyone thinks of changing the world, but no one thinks of changing himself. The late "King of Pop," Michael Jackson, in his song "Man in the Mirror," sang, "If you want to make the world a better place, take a look at yourself and make that change." Latencies are hidden causes that need to be exposed. It is not just about system causes, rules, and procedures that we ought to follow, but it is about how each of us has contributed to the problem. To address the latent cause of the problem, we need to look ourselves in the mirror and answer the following questions:

• What is it about the way I am that contributed to the problem? (About you)
• What is it about the way we are that contributed to the problem? (Organization)

• Space Shuttle Challenger exploded because the O-ring on the right solid rocket (SRB) booster leaked, which came in contact with the flammable external tank. The Roger Commission concludes that the explosion was caused by the solid rocket booster joints' faulty design. Morton Thiokol engineers already knew this problem and had informed their management about this. However, Morton Thiokol's management team also knows that their billion-dollar contract with NASA was nearing to an end and needed to be renewed soon so they can once again be in business with NASA. Richard Feynman, a novel prize winner, physicist, and part of the Roger Commission that probes the Challenger Disaster investigation said that engineers and managers were not communicating effectively. NASA was not communicating well with its suppliers. This is a classic example of a latent cause.

• An electrician rewind the motor backward. The motor was set up in a critical location on the plant. Management decided to discipline and suspend the electrician without pay for a month, which eventually wrecked the turbine. Later on, a deeper probe into the investigation revealed that the electrician had already been working for 24 hours straight because the person who should relieve him was on sick leave. The operations manager requested the electrician to extend his shift as they badly needed the motor, to which the electrician agreed. Was it really the fault of the electrician? Was it the fault of the operations manager since he was focusing on shipping their products on time? If you are in the shoes of the electrician, would you sacrifice to work 24 hours since the motor is badly needed? Can you stay focused while working for 24 hours straight? I leave that good judgment to the reader.

• The team probing the failure of the pump found out that misalignment caused the pump to fail. Likewise, they disciplined the person who performs the misalignment. When the investigative team probed deeper into the cause of the problem, they found out that this person only used his eyesight to perform an alignment. No instruments were being used; no training was ever provided to this guy. They learned that the instrument had already been requisitioned 5x but was disapproved by higher management 5x due to cost-cutting schemes. Was it really the person's fault who performed the alignment, or was there a much deeper cause?

I can go on and on and give more cases, but let me stop at this point. Whether we accept it or not, latent causes exist, and they must be exposed. Change can only occur if we have the guts and courage to look ourselves in the mirror and ask ourselves, what is it about the way I am that contributed to this problem, and what is it about the way we are as an organization that contributed to the problem. Therefore to answer the question is Root Cause Failure Analysis Reactive or Proactive. Root Cause Failure Analysis is both Reactive and Proactive. Still, Root Cause will start on a reactive scale. This means that something has to fail first before we can perform a Root Cause investigation. Once we learn from the problem and provide the necessary corrective measures in which the problem no longer recur, then this is when Root Cause becomes proactive. Root Cause is proactive if we can address the small things and expose the latencies of the problem, and I think that's all I have to say about that.

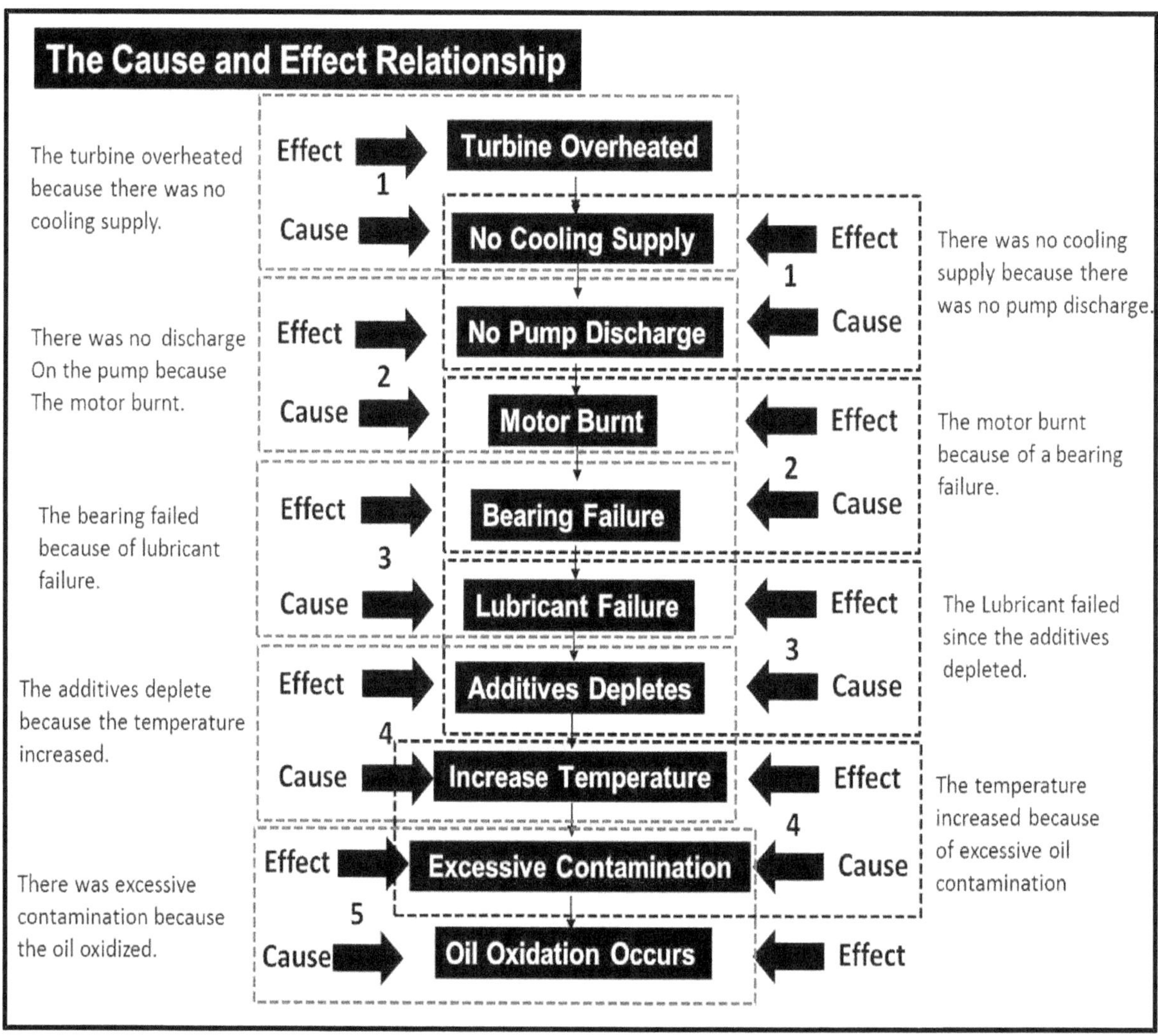

Figure 2.8: The Problem will Always be the Effect

2.9: Defining the Problem (The Cause and Effect Relationship)

The first step in conducting the RCFA investigation is to precisely define the problem explicitly as possible. To establish a cause-effect relationship, the cause has to occur before the effect. Whenever the cause happens, the effect will take place. Respectively, if the cause does not happen, then the effect will not take place. The problem that has been defined will be considered the effect. Once the problem has been identified, there will be several possible causes that lead to the problem. Each of these causes should be verified based on the evidence gathered. In figure 2.8, the problem is stated as the turbine overheated. This will be considered as the effect.

Let us assumed that the evidence verifies that there was no cooling supply, then no cooling supply will be the cause that led to the problem in which the turbine overheated. However, this can change depending on how we declare the problem. For example, the problem can be declared as; turbine failure or breakdown. If we define the problem as turbine failure, then this will be the effect, and the turbine overheat will be the cause that led to the turbine failure. What is important is that the problem stated will always be the effect, and level 1 of the RCFA logic tree diagram will be considered as the possible causes until it is verified with evidence.

Chapter **3**

Different Analytical Problem Solving Tools and RCFA

> *We can only learn from failures if we take the time to analyze them. It is an opportunity for us to begin more intelligently. Whether from the lessons of life or from the industry's problems, failures will tell us something. It tells us that to be fast, we need to take things slowly. Fixing the symptoms will just cause a recurrence of the problem.*

3.1: A Different Root Cause Failure Analysis Experience

We often hear the word Root Cause Failure Analysis, yet I wonder if we really know what it truly means. Almost every industry has its own unique way of doing Root Cause. I have been teaching industries about reliability and maintenance courses, and one of the courses I truly love to teach is about Root Cause Failure Analysis. Last July 2007, I attended my good old friend course on The Latent Cause Experience, and after the training ended, I just can't help thinking that for the first time, it seems clear to me what Root Cause Analysis truly means, which I thought I thoroughly understand. It allows me to reflect on my perspective and the basic lessons of what life is all about. During my employment days, we solve problems, thinking that we are pinpointing the root cause of the problem. After hearing from Bob Nelms, I came to realized that I knew nothing about root cause analysis.

Some of the basic questions being raised that need an answer are: How far should we probe on our analysis? Although there is a wide range of analytical problem-solving tools that exist, such as Ishikawa or Fishbone Diagram, FMEA-FMECA, Pareto Analysis, Eight Disciplines, TOPS, P-M Analysis, 5-Whys, Fault Tree, and so on, the real question raised is, are these tools really meant to address the root cause of the problem or only the most likely or probable cause? Each of these tools will claim yes, but I really doubt if they do.

Root Cause Analysis (RCA) is an investigative method to identify the origin of problems or events. RCA's practice is predicated because problems are best solved by correcting or eliminating the root causes instead of merely addressing the immediate and the most obvious symptoms. By directing corrective measures at root causes, it is hoped that the likelihood of the problem recurrence can be eliminated if not mitigated. *However, it is also recognized that a single intervention is not always possible to prevent a recurrence.*

Figure 3.1: The Latent Cause Analysis Experience with Bob Nelms

Complete elimination of failure in our equipment is impossible; what maintenance can do is to delay, prolong, prevent, predict, manage, or merely anticipate its recurrence. In its technical sense, failures are inevitable and are meant to happen. If a part wears out, then technically, it had actually failed. The question is, did we catch the failure first, or it just happened, and maintenance performs repair on their equipment.

Root Cause is not a silver bullet or rocket science strategy designed to eliminate all known problems, defects, and failures. This tool can only be useful if we truly understand what its intent and purpose is all about. Yet, most industries abuse the word root cause, thinking that it is meant to end all problems. All I can say is that in every way, they are wrong. We need to understand that the root cause is being performed so that industries can finally learn from the things that go wrong.

• In Preventive Maintenance, we perform these tasks to prevent failures.
• In Predictive Maintenance, we perform these tasks to predict failures.
• In Failure Finding Tasks, we perform these tasks to avoid the chances of multiple failures where both the protective and protected functions are in a failed state.
• In Run to Fail, we allow some failures to occur since the consequences are low, or the operations will not be affected since we have some form of redundancy or backup.
• In Root Cause Failure Analysis, we also allow the failure first, and then challenge them.

The question is, have we actually learned from these failures by allowing them to repeatedly happen in our equipment? The problem lies in understanding the problem and what problem-solving tools we used to analyze it. Pareto will say that 80% of the effects come from 20% of the causes. Are 20% of the causes really the root cause or just some probable causes? The answer is it may or it may not. Five-why states that by performing why-why five times and finds it hard to answer, the bottom line is that the root cause had been defined. But worse than that, when we find the guilty person, who performed the error, the root cause was finally identified. Given all of these, let me shed some light on what a true and meaningful Root Cause Analysis is all about.

<u>**Investigating Equipment Failures Through Root Cause Failure Analysis**</u>

First: Root Cause is not about failure modes or probable causes. It is always based on facts, and the facts are always based on the evidence gathered. Once the evidence is in place, we can perform a sequence of events that ultimately lead to the failure. Root Cause will always have to be based on pure evidence. Every failure will leave some sort of clue as to why it occurred. Did we talked and interviewed people who were involved in the problem? Did we examine the part that failed? Did we find anything unusual or odd about the event that took place before the failure? Are our RCA efforts based purely on evidence or not?

Second: Root Cause is about learning from the things that go wrong. This statement had to change the way I think about failures. Let me put it this way. Have you ever heard the word benchmarking about other industries' success? I am not a fan of benchmarking other industries and just copying the good stuff in your plant. I have seen and experience many industries that benchmark other industries ending up once again in their old reactive ways of doing maintenance. I have been involved in benchmarking. When I was still employed in a large semiconductor industry in the Philippines, many industries benchmark us. The problem is we only show the good stuff. We never show them our struggles and failures on how we achieve that stage. Do we really learn from other industries' success, or do we learn from our own failure? Let me get a little soft and philosophical about this matter so we can understand it better. Have you ever thought about what it takes for these industries to reach that level of excellence? Is your industry's culture the same as theirs? Rethinking about failure is not bad after all. Whether from the lessons of life or equipment failures, it can be our greatest teacher if we can learn from them. The same principle applies in Root Cause Failure Analysis.

Third: People commit mistakes and errors, but all failures lead to an error or mistake done by a human being. Even with the best system in place, people will still err, and the analyst must realize that not all mistakes people make are within their control. In this regard, I would strongly emphasize that root cause is not a tool to blame and punish someone. This will only make people more defensive. Some industries that truly understand root cause have some amnesty program and emphasized it clearly at the beginning of any RCA investigation process that we only would perform a thorough Root Cause Analysis because we want to learn from the things that go wrong and not to blame or punish someone. We all contribute to problems, yet management and decision-makers in industries are like Pontius Pilate, washing their hands most of the time, thinking that we ain't part of the problem. Their thinking is that if we just discipline the person who committed the problem, then the problem is finally solved.

Fourth: A true and meaningful Root Cause Analysis is done on three levels. First, determine the Physical Cause, then analyze the Human Cause and finally determine the latent cause of the problem. There will also be cases that between the Human and Latent causes are the system causes.

Fifth: The greatest lesson so far, one can benefit from truly understanding the root cause, is that it is always better to analyze failures than fixing them every time. If we become good at fixing failures, then something is definitely wrong with our organization. Why? Simple, because we are doing it much too often. Troubleshooting is no longer an effective strategy. In the real world, mostly in manufacturing, or whatever type of industry, we need people who can analyze problems and not just fix them. This eventually makes us understand the difference between a maintenance person from a mere mechanic. A mechanic mostly uses his hands to fix problems, while maintenance mostly uses a balance of his brain and mind to understand and analyze the problem instead of using his hands too often. Sometimes I find it funny how most industries think. Most industry often yells that they have no time to perform an RCA or RCFA Analysis, yet they have all the time in the world to fix the problem repeatedly, which makes their people really good at fixing failures. Industries with this mindset will never learn from their failures.

Sixth: Root Cause can only guarantee that the same cause of the problem will not be repeated. It is not a guarantee that the problem will entirely be eliminated. A bearing can fail in many ways. A root cause investigation shows that the physical cause of the bearing was due to a lubrication failure which was evident on the raceways of the bearing. Corrective action had been performed in which they upgraded their lubricant so it can withstand a much higher temperature. This does not mean that the bearing will no longer fail since it can also fail due to other causes such as misalignment, and other causes. In root cause, we are only interested in addressing the cause of the failure based on the evidence gathered to prevent a repeat of the same cause of the problem.

Here are questions that can allow us to determine your basic RCA requirements.

- Does everyone in the organization understand the objective of performing a Root Cause Analysis Investigation? Are they united in their purpose, or do they have their own agenda?
- Is management willing to be trained in Root Cause Analysis? Does management truly understand what RCA can and cannot do?
- Will the people investigating the problem complete the RCA analysis, and will their recommendation be implemented or fall on deaf ears of management?
- Is 3rd party consultation being required in the initial process of RCA implementation?
- Are we willing to learn from the things that go wrong and be part of the learning process?

3.2: Difference Between Analytical Problem Solving Tools and RCFA

Although both Root Cause Failure Analysis and these Analytical Problem Solving tools will both address the problem or incident. There are differences between the two. Listed here are some of the difference

1) The main difference between an analytical problem-solving tool and RCFA is that RCFA relies on 100% evidence. RCFA can only be performed when the evidence of the failure had been preserved. A Root Cause Failure Analysis always depends on the evidence found to determine the precise cause of the problem. It is not about a selection of the trivial few. It is not about brainstorming. It is not about the probability of failures. It is not about experience. It is not about mere gut feeling, but rather root cause is all about evidence that we must unfold so that eventually it will lead us to the truth behind the cause of the failure.

Without evidence, one can only guess, speculate, hypothesize, brainstorm, base it on experience, theorize, or simply address the most probable or likely cause of the failure mode.

Although the majority of these problem-solving and analytical tools will claim to address the root cause of the problem. My question is how can we derive the cause of the failure without any evidence? Root Cause Failure Analysis will always be based on the evidence and takes the time to verify each hypothesis mode to separate from those that actually happened. The end goal of RCFA is to expose the latencies. In performing Root Cause Failure Analysis, we are interested in knowing the real cause of a particular failure by verifying each hypothesis until we reach the final cause. RCFA only concludes once the latent cause had been identified.

2. These analytical problem-solving tools such as Ishikawa Diagram, FMEA/FMECA, Pareto Analysis, Five-Why, P-M Analysis, 8-Disciplines, and others are only meant to address the most probable or likely cause of the problem. These tools are not designed to address the root cause since they are only a product of brainstorming and probabilities. For example, once a fishbone diagram has been completed by the team, it is unlikely that the sequence of events can be unfolded that led to the root cause.

3. To conduct a Root Cause Failure Analysis Investigation means that the failure should need to be fresh. This means that the failure just recently happened. We cannot perform a Root Cause Failure Analysis Investigation on a piece of equipment that happened 3 or 6 months ago, but it is possible to perform analytical problem-solving tools if data is available.

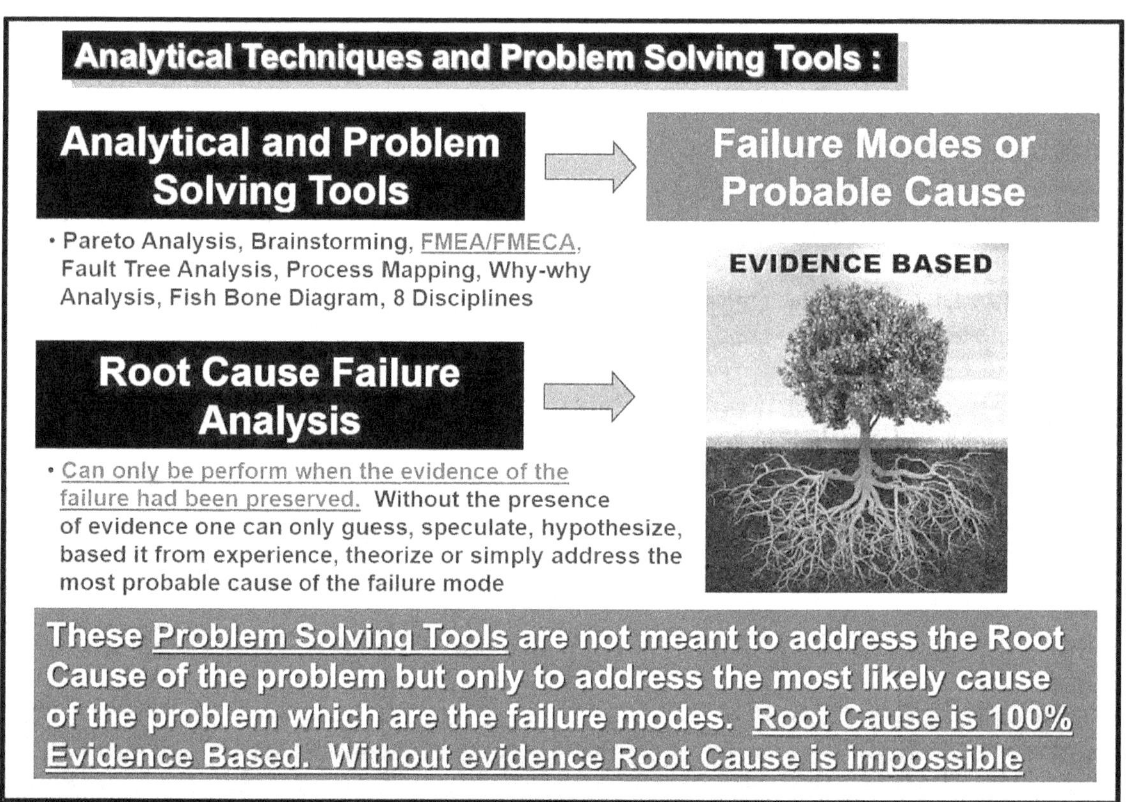

Figure 3.2: Difference Between Problem Solving Tools and RCFA

4. Majority of these analytical problem-solving tools will address the physical, human, and in some cases the system cause of the problem. The most important distinction between a

problem-solving tool and a root cause is the depth of probe being conducted. These problem-solving tools will either end on the physical, human, and sometimes the system cause. To conduct a Root Cause Failure Analysis, the principal investigator and the evidence gathering team together with the stakeholders should address the physical, human, system, and latent cause of the problem. The distinction, in this case, is that these analytical problem-solving tools are not designed to address the latent cause of the problem.

5. Determining the root cause and the probable cause are entirely different. A bearing can fail in many ways such as fatigue, misalignment, poor lubrication, loose or tight fits, overheating, contamination, false brinelling, electrical pitting, and so on. Performing an analytical problem-solving tool to determine the root cause may base the judgment of the team on personal experience, brainstorming if they have data. In conducting the root cause, once the bearing fails, both the inner and outer raceways, ball/roller, and cages need to be seen to determine the precise reason why it failed, yet this is only the physical cause of the bearing failure, the principal investigator together with the evidence gathering team should also determine the human, system and the latent cause of the problem,

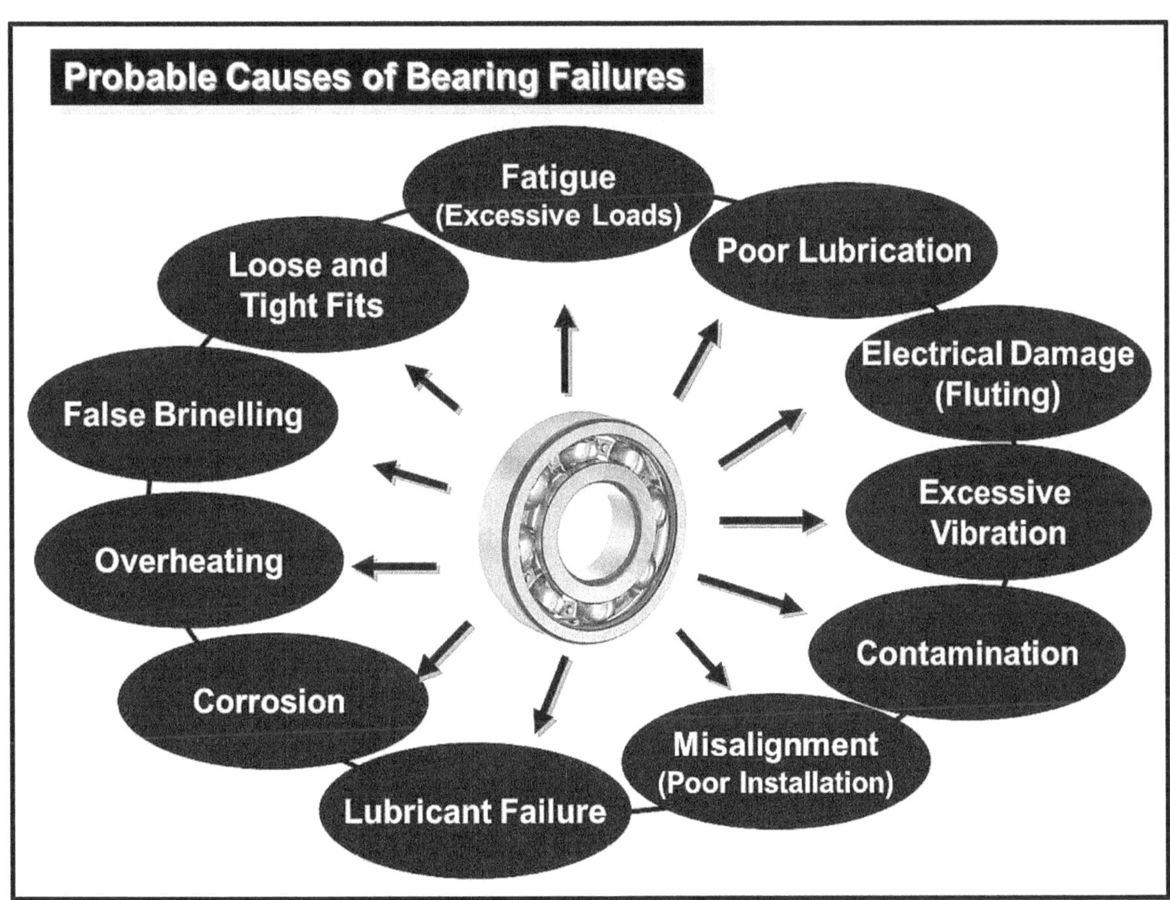

Figure 3.3: Probable Causes of Bearing Failures

6. Typically, the team conducting an analytical problem-solving tool will require a cross-selection of members to form a team, these people can come from maintenance, process, quality, safety, operations, engineering, and other functions of the organization. Conducting a root cause investigation will require a Principal Investigator (PI) and evidence-gathering, team.

It is highly recommended that both the Physical Investigator and Evidence-Gathering team should not come from the same function where the incident took place to avoid bias and any cover-ups. The only way to perform a root cause investigation is to freeze the evidence to determine the real cause of the problem.

Difference	ANALYTICAL PROBLEM SOLVING TOOLS	ROOT CAUSE FAILURE ANALYSIS
Who will Compose	Cross Functional Team	Principal Investigator and the Evidence Gathering Team
Analysis Based On	Data, Brainstorming, History Records	100% Evidence-Based
To be Done On	Failures that happened in the past	Fresh Failures
Will Determine	Probable or Most Likely Causes	Root Cause based on Evidence
Pattern	Most Problem Solving Tools will Not Provide a Sequence of Events	Will Provide a Sequence of Events
When to Apply	Evidence have not been preserved	If the evidence have been preserved
Investigation Includes	Physical, Human and sometimes the system cause	Physical, Human, System, and The Latent Cause of the Problem

Figure 3.4: Difference Between Problem Solving Tools and RCFA

7. Another distinction I can provide between a problem-solving or analytical tool and the root cause is that in RCA or RCFA, is that RCA will handle one cause at a time, while a problem-solving tool can take several probable causes at a glance. A root cause is like a rifle where the sniper will take his target one at a time, aim and fire with precision accuracy, while a problem-solving tool is like a machine gun where it will fire at random, which may or may kill their enemy. One clear distinction between an analytical problem-solving tool will be the depth of probe or when do we end with the analysis, and analytical problem-solving tool may end with the physical cause but with root cause investigation, this will end up in the latent cause of the failure.

3.3: Fishbone or Ishikawa Diagram

The cause-and-effect diagrams were developed by Kauro Ishikawa of Tokyo University in 1943 and are often called Ishikawa Diagrams. They are also known as fishbone diagrams because of their appearance. The cause-and-effect diagrams are used to list all the different probable causes attributed to a problem or an effect. The cause-and-effect diagram can aid in identifying the probable causes why a process goes out of control or a machine fails.

An Ishikawa diagram is typically the result of a brainstorming session in which the group members offer ideas on how to improve a product, process, service, failure, or problem. The main goal is represented by the trunk of the diagram, and primary factors are represented as branches. Secondary factors are then added as stems, and so on. Creating the diagram stimulates discussion and often leads to increased understanding of a complex problem. A fishbone diagram is an analytical problem tool that provides a systematic way of looking at effects and the causes that create or contribute to those effects. A fishbone diagram will include the probable causes generated by the following:

- Methods: Includes the details of the process, system, and specific requirements for doing it, such as procedures, SOP, rules, policies, regulations, and laws.
- Man: They are the ones who are involved in the process. This will include the human error committed by the person.
- Machines: Includes the probable causes of the equipment problem, machine, or assets involved in the analysis and brainstorming.
- Measurements: This includes any deviation or anomaly from the data, parameters, or settings generated from the process used to evaluate its quality.
- Materials: These are the parts, raw materials, and items used to produce the final product.
- Environment: These refer to the actual conditions, time, location, humidity, or the temperature in which the equipment operates and is exposed.

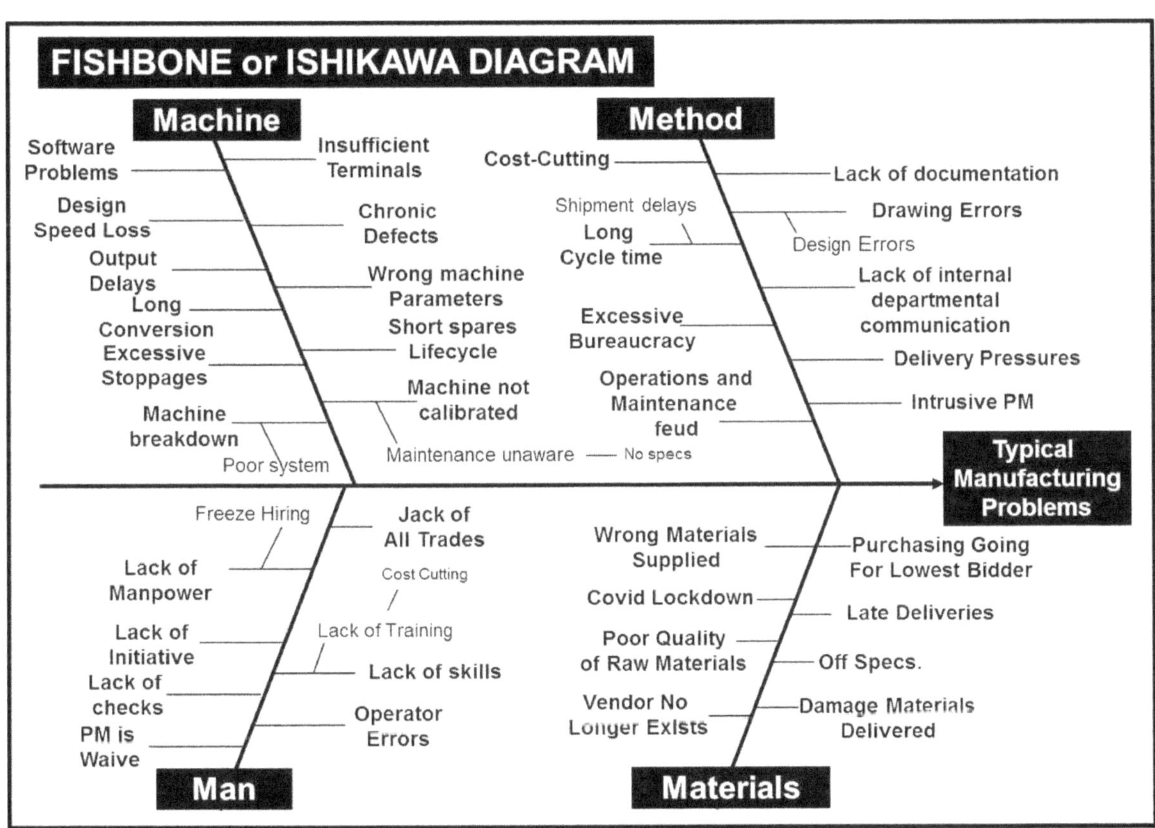

Figure 3.5: How Ishikawa or Fishbone Diagram is Constructed

The first step involves a clear definition of the effect or symptom for which the causes must be identified. The second step involves brainstorming or a rational step-by-step approach to identify the possible causes of the problem being analyzed. This involves the use of 4m and 1E, 4p's, etc. For manufacturing, the 4m's includes man, method, machine, materials, and environment. It can also include an additional M for measurements. The 4p's include policies, procedures, people, and place. After identifying the probable causes, the group will select one of the causes and work on it systematically, identifying as many secondary causes as possible. The following are the steps in conducting a fishbone diagram.

1. Draw the fishbone diagram.
2. List the issue, or problem to be analyzed in the head of the fish.
3. Label each bone of the fish.
4. Use an idea-generating technique such as brainstorming to identify the factors within each category that may affect the problem being analyzed.
5. Repeat this procedure with each factor under the category to produce sub-factors and ask why it is happening. The analysis may include secondary or higher levels under each probable cause.
6. Analyze the results of the fishbone after team members agree that an adequate amount of detail has been provided under each major category. Do this by looking for those items that appear in more than one category. These become the more likely or the most probable causes.
7. For items identified as the most likely causes team should reach a consensus on listing those items in priority order.
8. Generated corrective actions for the probable causes selected by the team.

Authors comment if it will address the root cause: Ishikawa or Fishbone diagram will not address the root cause but only the probable or most likely cause of the problem. The limitation of the fishbone or cause and effect diagram is that it provides no clear sequence of events that leads to the failure. Instead, it displays all the possible causes that may have contributed to the problem. While this is useful, it does not isolate the specific factors that caused the event to occur. Other approaches provide the means to isolate specific changes, omissions, or actions that caused the failure, accident, or other investigated event. This problem-solving tool is not based on evidence but only on a product of brainstorming. This means that the final cause the team selected may or may not be the root cause of the problem, but the good thing is that it will lessen the problems on the equipment if the corrective actions are implemented successfully.

3.4: FMEA/FMECA Explained

Failure modes and effects analysis (FMEA) is a step-by-step approach for identifying all possible failure modes in either the design, equipment, process, system, service, or product. Failure modes mean the ways, or modes, in which something might fail. However, the coverage of this guideline will only be for equipment FMEA. Depending on the type of FMEA, the tables on the severity, occurrence, and detection might differ, respectively, but the process of conducting FMEA will be the same whether this is a machine, service, product FMEA and so on. Failures are any breakdowns or defects that will affect the equipment and assets of the plant. Failure Mode and Effect Analysis (FMEA) is a proactive tool for evaluating failure modes and probable causes. Note that FMEA/FMECA will be limited only in determining the most likely cause or probable cause and not the Root Cause of the Failure itself. It helps prioritize critical failure modes and recommends countermeasures for avoiding catastrophic failures to improve maintainability, safety, quality, and reliability. It is also commonly defined as a systematic process for identifying potential equipment failures before they are likely to occur, with the intent to eliminate or reduce the risk associated with the failure.

Figure 3.6: Prioritizing Corrective Actions on FMEA

Equipment or Machinery FMEA/FMECA is a risk analysis tool used to identify all possible failure modes that have happened in the past and the possibility of happening in the future. The FMEA result highlights failure modes with relatively high probability and severity of consequences. Although FMEA will be using RPN (Risk Priority Number), FMECA will be using a criticality analysis. Also, it should be noted that the most critical failure mode usually has a low occurrence and may not have a high RPN number as in figure 3.6. Priority in determining corrective actions should be based on the severity of the failure mode itself, and not merely on the highest RPN. Doing FMEA/FMECA can be conducted on a component level, equipment level, sub-assembly level, system or sub-system level, or even through a parts level, depending on the FMEA/FMECA Analysis team. History of FMEA and FMECA includes:

- **1949** - Standards for conducting FMECA were described in the US Armed Forces Military Procedures Document MIL-P-1629 and revised in 1980 as MIL-STD-1629A.
- **1960's** - Contractors for NASA were using variations of FMECA/FMEA under various names. NASA programs using FMEA variants included Apollo, Viking, Voyager, Magellan, Galileo, and Skylab.
- **1967** - The civil aviation industry was an early user of FMEA. The Society for Automotive Engineers (SAE) published ARP926. This standard had been replaced by ARP4761, which is now broadly used today in civil aviation.
- **1970's** - The automotive industry began to use FMEA. Ford Motor Company introduced FMEA to the automotive industry for safety considerations and applied Process FMEA to consider potential process-induced failures before production.
- **1971** - NASA prepared a report for the U.S. Geological Survey recommending the use of FMEA in the assessment of offshore petroleum exploration.
- **1973** - Automotive Industry Action Group (AIAG) first published an FMEA standard for the automotive industry, which is now in its fourth edition.
- **1994** - SAE first published related standard J1739. This standard is also in its fourth edition.

Failure Mode and Effects Analysis (FMEA) provides only qualitative information, while Failure Modes and Effect Criticality Analysis (FMECA) will provide both qualitative and limited quantitative information, which means that FMECA can be measured. It provides a level of criticality to the failure modes. FMECA is just an extension of FMEA. To perform FMECA, the team must perform FMEA, followed by a criticality analysis. FMEA identifies failure modes of a product, design, process, or equipment and the effects of the failure, while criticality analysis ranks those failure modes in order of importance, according to the failure rate and severity.

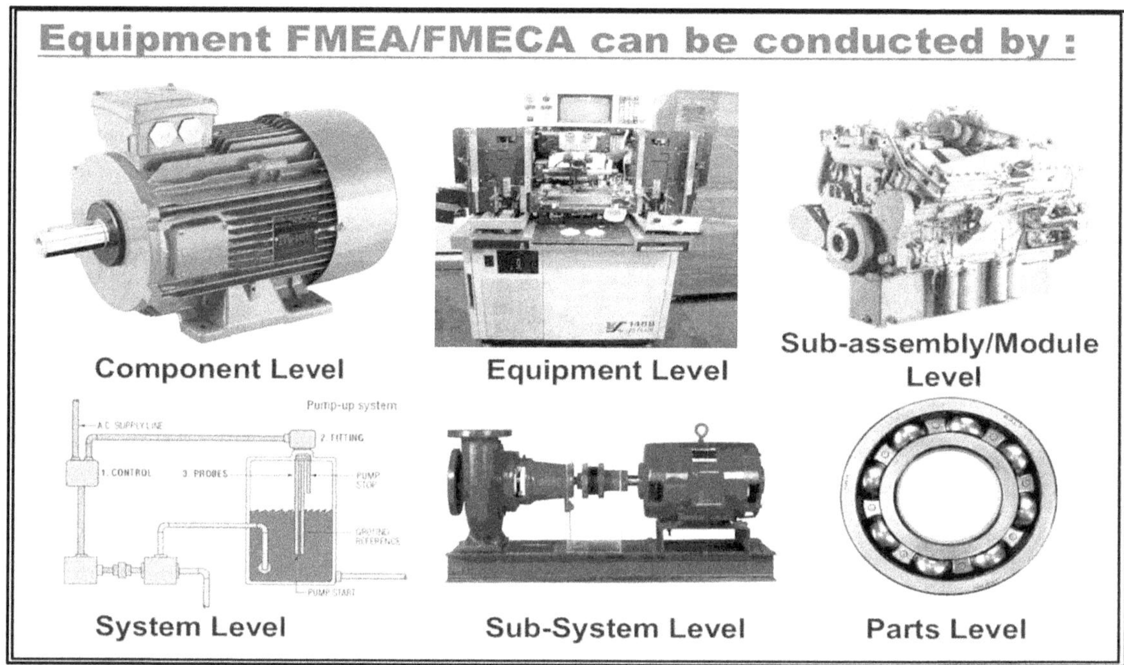

Figure 3.7: How FMEA/FMECA will be conducted

The criticality part of FMEA consists of assigning a number from 1-10, where 1 has the lowest criticality, and 10 will be the highest criticality of failure. The criticality of the failure mode will be done by multiplying the severity (consequences) by the occurrence (probability) and detection (confidence), which will obtain values from 1 to 1000. The detection or confidence is largely based on the ability to catch the failure mode in time. If we are doing FMECA, we estimate the probability that gives us the risk of failure. Although there is a problem of definition, in FMECA, the RPN will be termed as criticality analysis or CA, while in other industries, we understand and use the word Risk Priority Number or RPN. Listed are the steps to conduct FMEA

1. Write the FMEA/FMECA Header:
2. Define the function of the asset being analyzed through FMEA/FMECA
3. List all possible Potential Failure Modes of the Equipment
4. Determine the Failure Effect
5. FMEA Team to rate the severity of the potential failure mode and effects
6. Place CC for Classification if the severity is 9 to 10:
7. Determine the potential causes of the failure
8. Determine the occurrence of the failure
9. Determine current design machine control
10. Assign a detection ranking
11. Determine the Risk Priority Number (RPN)
12. Lists all recommended corrective actions

13. Assign responsibility and target completion
14. List down corrective actions taken
15. Recalculate the final RPN

Recommended Actions	Responsibility and Target Completion Dates	Actions Taken	Action Results					RPN
			SEVERITY	OCCURRENCE	DETECTION	RPN BEFORE		
- Contamination - Introduce contamination control awareness campaign to all maintenance involved - Regular cleaning and inspection	Mick Jagger 7/14/2014	- Have conducted training on contamination awareness control for all maintenance involved	3	1	10	80		30
- Lubricant Failure - improve the choice of lubricant and study feasibility of using synthetic oil.	Mick Jagger 7/14/2014	- Conduct a feasibility study on the best type of synthetic oil to be used for all motors - Standardize all lubricants for motors	2	1	8	64		16
- Misalignment - Have contractor perform lazer alignment during scheduled outage on the components	Keith Richards 12/15/2014	- Discuss with vendor to perform alignment practices whenever the components will be disassembled - Still Ongoing	8	1	10	80		80
- Contamination - redesign the system by placing pressure guages before and after the strainer and conduct a regular monitoring on the drop in pressure. To be done during the scheduled outage for the motor and pump components	Larry Watts 7/15/2014	- Pressure gauges in place	2	1	2	80		4
- Strainer Mesh too large - Modification of strainer mesh from 400 microns to 150 microns and study flow rate if it will be affected	Ronnie Woods 7/25/2014	- Strainer already modified and flow of water is normal	2	1	2	80		4

Figure 3.8: FMEA Comparing Before and After RPN

Authors comment if it will address the root cause: FMEA/FMECA is not and will not address the root cause but only the probable failure modes that led to the failure. As the FMEA or FMECA, methodology will involve listing all the possible failure modes and probable causes that are likely to happen in the equipment or asset. This is not designed to derive the root cause itself. One thing the team conducting FMEA should note is that do not always base your priority on the highest RPN but on the severity of the failure. The RPN gives us a relative risk ranking. Theoretically the higher the RPN, the higher the potential risk. We should likewise be very careful when prioritizing corrective actions as there are failures with very high impact and high consequences that are of very low occurrences. Following the concept of FMEA most critical failure will have a very low occurrence yet there are easy to correct failures that will occur frequently. Priority on corrective actions should provide a balance between those critical and high risks failures and those failures that occur frequently. Remember that failures with high consequences are usually of very low occurrence, do not ignore them in the FMEA/FMECA analysis.

3.5: Five Why Analysis

Sakichi Toyoda, the founder of Toyota Industries, developed the 5-why technique in the 1930s which became popular in the 70s and is still being used today by Toyota Motor Corporation to find solutions to their day-to-day problems. The approach uses a systematic questionnaire technique to search for the probable causes of a problem. The team uses a tool by asking why at least five times as you work through the various levels of detail. Once the team finds it difficult to respond to why the probable cause of the problem may have been identified. As you trace the "why" back to their root causes, you will find yourself confronting issues that affect the original symptom and the entire organization. To be effective, the team's answer to the 5-whys must steer away from blaming individuals. Blaming people leaves you with no option but to punish them, leaving no room for substantial change. The focus is on the process of the problem and not the person involved, since frequently answering why will lead you to the person responsible for the problem. Here are the steps in conducting a why-why analysis.

Step 1: Assemble the team and identify the problem

• Select a problem to work with.

• Define the problem associated based on the information known to the team.

Step 2: Ask the 1st why to the team

• Determine and ask the first why. The team will end up with several probable or possible causes.

• Use a whiteboard or index card and place them on the wall where each team member can see them.

Step 3: Ask four more successive whys

• Repeat step 2, and post each answer near its parent.

• Continue to ask why beyond the arbitrary 5 why's if necessary to get to the probable cause of the problem.

Step 4: Determine the possible causes for the problem under consideration

• Circle those causes that are more likely to be the most probable cause of the problem and discuss them with the team.

• See if you can limit them to just a couple of causes and finalize the analysis with the team

• Use the causes listed, especially those on the last level of the diagram.

• Generate corrective actions based on the causes selected.

Once you have mastered the five why techniques and an experienced facilitator is in place, the method becomes quick and easy to use. The five-whys can help initiate a thinking process that will ultimately solve the problem at hand. This leads the team to a deeper understanding of the issue they are seeking to confront. Using the 5-why's can lead the team to the most important cause of the problem and help the team members to challenge the group's perception of the problem. Sometimes a combination of fishbone and five whys is required to help solve more complex problems. The key to performing the five why analysis requires strong facilitation assistance that can keep the team on track and avoid diversions. If the analysis is diverted, it can become what is known as analysis paralysis, what is key is knowing which will be the last why and if this is the ultimate cause of the problem at hand.

Authors comment if it will address the root cause: My personal thoughts on why-why or 5 why analysis are that this tool is not designed to address the root cause. Again, in this case, the causes selected are not based on evidence but only on the team's brainstorming. This means that the final cause may or may not be the actual cause that eventually leads to the problem. There are limitations in conducting a 5-why or other analytical problem-solving tools mentioned in this section as the approach may be subjective especially if the analysis was not based on evidence. Even if the why was based on evidence, a Root Cause Failure Analysis Investigation should not only answer the question of why but also the what and how the failure or incident occurred.

3.6: P-M Analysis

Physical Analysis analyzes chronic losses according to the inherent principles and naturals laws that govern them. It was developed by Mitsugu Kaneda, Shirose Kunio, and Yoshifumi Kimura. P stands for Phenomena and Physical. A phenomenon is a deviation from a normal to an abnormal state. M stands for Mechanism and the 4Ms (man, machine, method, and materials). The mechanism is about understanding the mechanics of the abnormal phenomena or how they are being produced. It also needs to understand the relationship between the abnormal event and the four production inputs or 4Ms. Its principle asks in precise physical terms what happens when a machine breaks down or produces bad parts or defects and how it happens. P-M Analysis clarifies the mechanics of their occurrence and the conditions that must be controlled to prevent them.

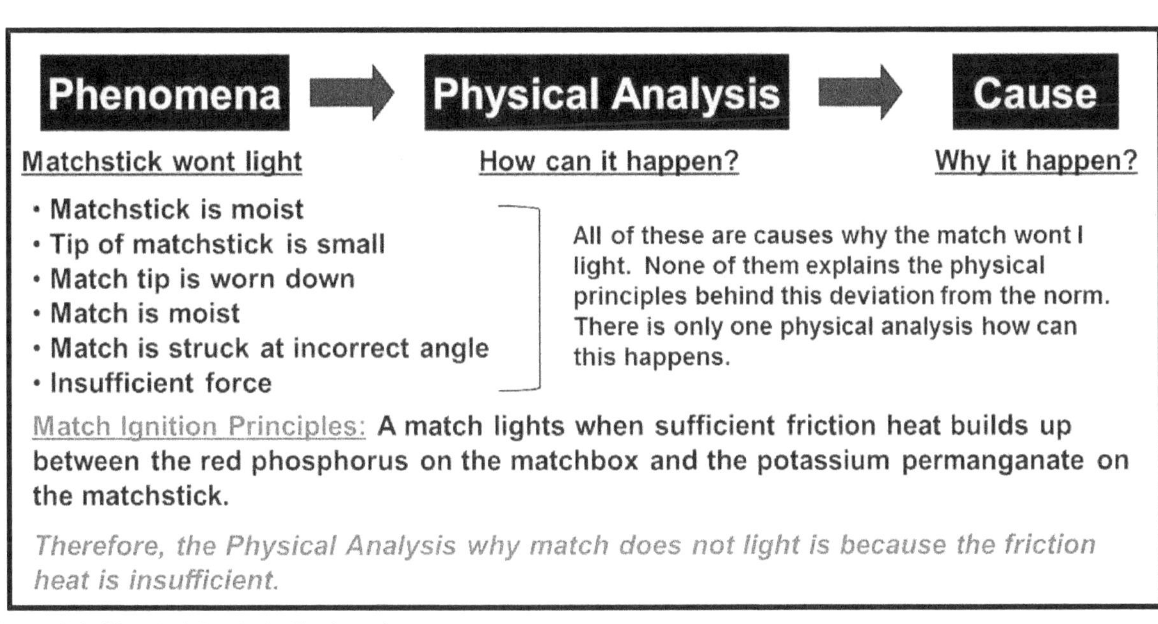

Figure 3.9: Physical Analysis Explanation

According to the machine's inherent operating principles, P-M analysis will analyze chronic problems such as defects and failures. It is a logical investigation of phenomena such as defects or breakdowns that explain how the phenomena occur regarding their physical principles. To analyze means to break down a whole into several parts and learn their nature and relationships. Physical Analysis uses a machine's operating principles to clarify how various parts interact to generate abnormal phenomena. It means logically explaining how phenomena occurred. It does not necessarily explain why. It is a bridge that helps us draw a logical relationship between phenomena and their potential causes.

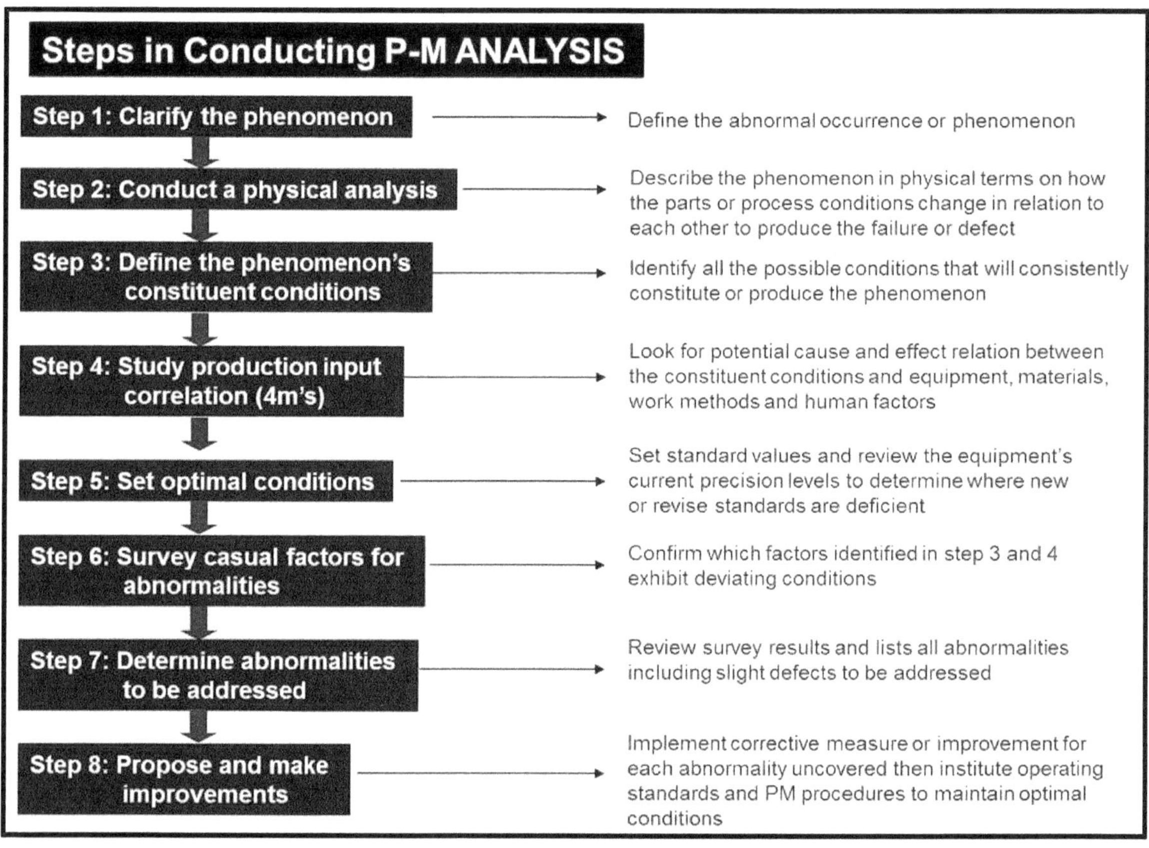

Figure 3.10: Comparison Between P-M Analysis and Common Problem Solving Tools

Figure 3.11: Steps in Conducting P-M Analysis

According to P-M Analysis, chronic problems persist because of failure to understand the nature of chronic losses or using ineffective approaches in dealing with them. This means that defects and failures are not carefully observed and stratified. People do not notice the defect pattern. Another reason is that some factors related to the phenomena are overlooked, which means that hidden abnormalities in individual factors are not addressed or failing to identify and respond to abnormal conditions. People are more alert to large and catastrophic failures since they appear more significant, and the smaller the problem, the more likely they are to be

82

ignored. These slight abnormalities include dirt, dust, vibration, looseness, slight wear, dirty contact, and leaks.

These big and catastrophic problems will be more likely to be expensive initially, but chronic losses may tend to be more expensive in the long run if they repeatedly occur in our assets. In P-M Analysis, we do not attempt to isolate the most influential factors. Instead of prioritizing the causes, we need to consider all possible and logical causes and emphasize each factor that might impact the chronic problem. In each possible factor, we identify all the possible abnormalities, no matter how small, and arrive at an appropriate countermeasures or corrective actions for each of them. Before attempting to use P-M Analysis, it is highly recommended to use conventional methods such as Pareto Analysis or cause and effect to lower the defects and breakdowns. When the frequency of these defects is around 0.1 to 1%, then P-M Analysis will be used. This means that when the rate of failure or defects is high, do not use P-M Analysis, but rather try to use the conventional problem-solving tools first.

Authors comment if it will address the root cause: Although this is a unique tool to address chronic defects and failures. Again, this tool is not meant to address the root cause since this method will address all the probable causes as to why and how the failure occurred. If a particular defect has 20 causes, P-M Analysis is designed to address all the 20 causes. However, the good point of using P-M Analysis is that it can zero out all defects on the equipment. For failures, definitely, this tool can improve the MTBF of the equipment.

3.7: Pareto's 80/20 Rule

Dr. Joseph Juran, one of the pioneers of quality, worked in the US from 1930 to 1940. Juran recognized a universal principle which he called the vital few and trivial many. In an early work, a lack of precision on Juran's part made it appear that he applied Pareto's observations about economics to broader work scope. As a result, Dr. Juran's observation of the vital few and trivial many, where 20 percent of something is always responsible for 80 percent of the results, now became known as Pareto's Principle, the 80/20 Rule. This is also known as the vital few and trivial many.

The origin of the Pareto Principle started in the relationship between wealth and population. According to Vilfredo Pareto's observation, 80% of the land in Italy was owned by 20% of its population. After researching the situation in other countries, Vilfredo Pareto found the same situation. The Pareto Principle is an observation that things in life are not always distributed evenly. This means that 80% of consequences come from 20% of the causes, asserting an unequal relationship between inputs and outputs. The Pareto Principle can be applied in a wide range of manufacturing, management, and human resources. For instance, the efforts of 20% of an industry's staff could drive 80% of the company's profits. Pareto Analysis will be a good problem-solving tool if data is given

Authors comment if it will address the root cause: No, this will not address the root cause of the problem as this is not based on evidence but rather on the possible 20% based on the data. Determining the root cause is not about the 20% trivial few that contribute to 80% of the problem. This will only address the most likely or probable cause of the failure but not the root cause.

3.8: Fault Tree Analysis

[2]Fault Tree Analysis was developed in the 1950s by BOEING Aerospace Engineer for use in the design process's development stages. It is a mathematical tool and will yield probabilities. The purpose of a Fault Tree Analysis is to predict the probability of a specific failure. Fault Tree Analysis is not actually a Root Cause Analysis tool but rather a design tool.

FTA was first used in 1962 for the US Air Force by Bell Telephone Laboratories on the Minuteman Weapon System. It has been adopted by many industries in the field of reliability engineering. It received extensive coverage at a 1965 System Safety Symposium in Seattle sponsored by Boeing and the University of Washington. Boeing first began using FTA for civil aviation aircraft design in 1966. It provides a basis for analyzing design by understanding the most common failure modes that can affect the design of a system. FTA is analyzed using Boolean logic, which combines a series of lower-level events. This method is mainly used in safety and reliability engineering to understand how a system can fail and identify the best ways to reduce the risk associated with the failure. Boolean symbols and gates are used in FTA to identify to most likely cause of the problem.

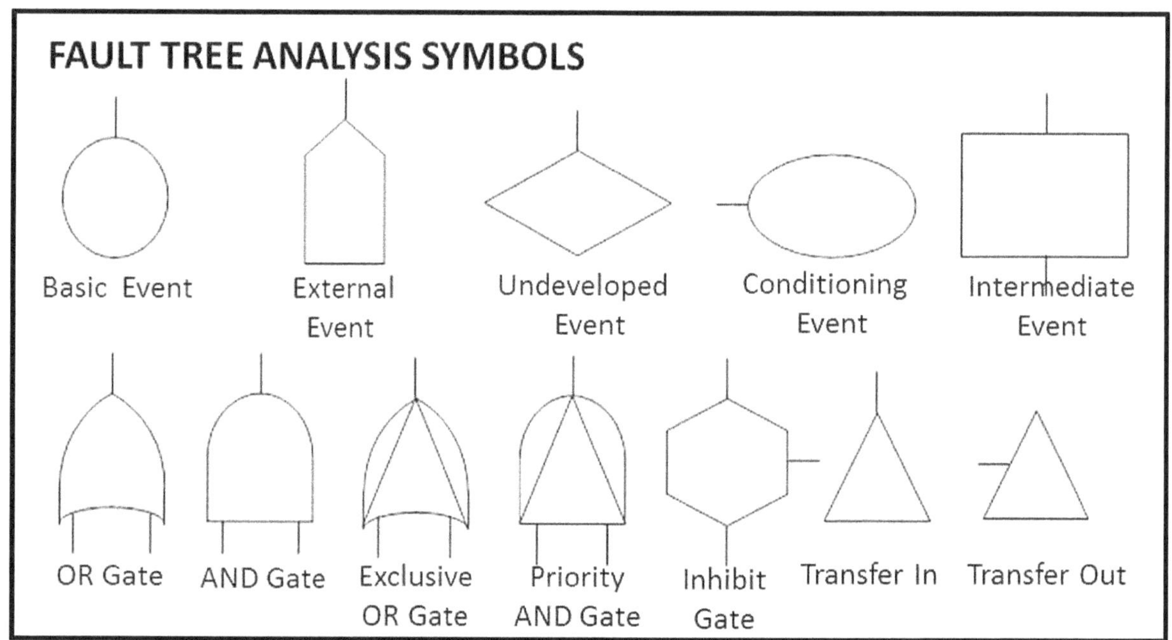

Figure 3.12: Fault Tree Analysis Symbols

• Basic event failure or error in a system component or element.
• External events that are normally expected to occur.

---

[2] C. Robert Nelms, __What You Can Learn from the Things That Go Wrong, A Guidebook to the Root Cases of Failure__, (Virginia, Failsafe Network Inc., USA: 2007), Pages 101

• Undeveloped event is an event about which insufficient information is available.

• Conditioning events are conditions that restrict or affect logic gates.

• OR gate the output occurs if any input occurs.

• AND gate the output occurs only if all inputs occur (inputs are independent).

• Exclusive OR gate the output occurs if exactly one input occurs.

• Priority AND gate the output occurs if the inputs occur in a specific sequence specified by a conditioning event.

• Inhibit gate the output occurs if the input occurs under an enabling condition specified by a conditioning event.

Authors comment if it will address the root cause: No, Fault Tree Analysis is a design tool and again is not design to address the root cause of the failure. This is a tool that aims to forecast the probability of failure and is used in the design of a product or equipment. This problem-solving tool is used during the design stage, and not during operations. The failures on the current operating equipment are listed and are used to design better products and equipment.

3.9: 8-Disciplines

The 8-Disciplines is an analytical problem-solving tool designed to determine the problem's probable or most likely cause. Again the process of conducting 8-Disciplines is not based on evidence. Hence we cannot conclude that it will derive the root cause of the problem. This method can be applied to defects and equipment-related failures. This method establishes a permanent corrective action based on data and provides the probable cause of the problem. The corrective and preventive measures generated should address both the defects and system causes. These actions should be proliferated to a family of products or similar processes to prevent a recurrence of the same defect or problem in the future. The 8-Discipline steps will include the following;

D0: Preparation for the 8-Discipline Process
• Will 8-D be the correct methodology to use?

• Is the problem a common cause or a special cause?

• Will the 8-D method induce new problems when initiated?

• Do we have the technical capability to address the problem?

D1: Organize the 8-D Team
• Select the most qualified people to assemble the team.

• Are the people affected by the problem included in the team?

• Does the team have sufficient knowledge to address the problem?

• Is the customer or client viewpoint being considered?

• Does the team members have the authority to make the necessary changes?

D2: Describe the Problem
• What is the problem in its precise term?

• Is the problem crystal clear to the team?

• When did the problem start?

• How was it detected?

- When did the problem occur?
- How much impact does the problem provides in terms of production?

D3: Develop Interim Containment Plan and Actions

- Have all alternative actions been evaluated?
- Are the responsibilities of those involved for the containment action clear?
- Is the required support available?
- Are the products, raw materials, WIP, and inventory stores been checked for the affected product?
- Has the material been put on hold to prevent processing the product and reaching the customers?
- Have the containment plan alerted the equipment engineers, process, production, maintenance, quality, and other functions?

D4: Define and Verify the Root Cause

- Has the process flow diagram been drawn, and have all of the sources of variation been identified?
- Is the potential cause challenged with the question why and followed with because to construct alternatives?
- Has a comparative analysis been completed to determine if similar problems existed in the past related defects, processes, or failure? If yes, are the documents available?
- Has the most likely or probable cause been verified?
- Have all possible corrective actions been identified?
- Have contingency actions been identified in case the original corrective action does not work as intended?

D5: Verify Permanent Corrections Actions

- Were the solutions, and corrective actions verified that the expectations of the customers have been met?
- Have the cost and time been considered in selecting the corrective action?
- Have monitors been set and identified to determine the long-term effectiveness of the corrective action?
- Have data been monitored to prove that actions are effective and no undesirable side effects are generated?
- Is the corrective action fool-proof?

D6: Implement and Validate Corrective Actions

- Do the actions taken represent the best possible long-term solution from the customer's viewpoint?
- Do the actions make sense concerning the cycle time for the products?
- Have the corrective action plans been coordinated with the clients and customers?
- Has the effectiveness of the corrective action been validated successfully?
- Has the containment plan been removed?

D7: Take Measures to Prevent Recurrence of the Problem

- Are affected personnel been notified regarding the permanent corrective actions?
- Has the system which will prevent the problem in the future been documented and standardized?
- Is the information available to all departments and augmented to continuous improvement efforts?
- Have all other possible processes or products been reviewed to prevent a recurrence of the same problem?
- Have the improvement been replicated and fan-out to similar equipment and machines such as BKM or Best Known Method?
- Are the control plans, FMEA, standards, SOP, procedures been revised?

D8: Congratulate the team

- Has the team been recognized for its efforts?
- Has the team shared the lessons from the 8D developed to other departments that may pose the same problem in the future.

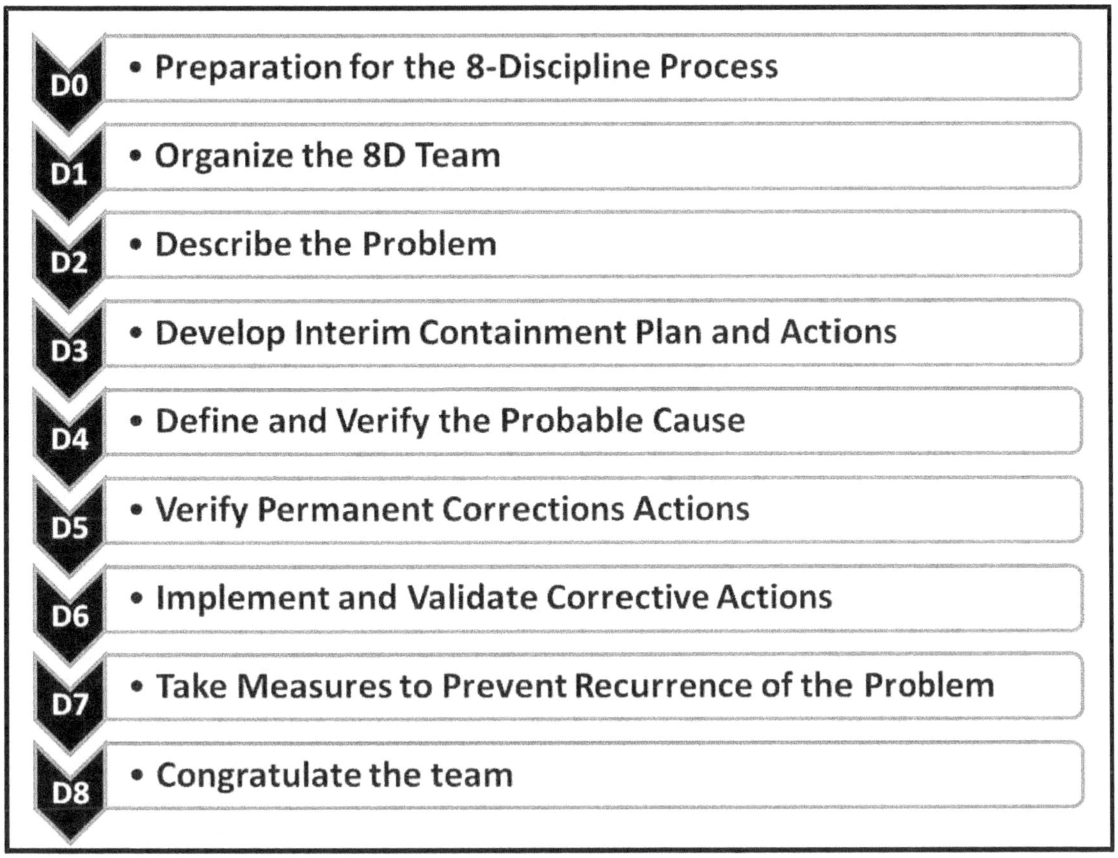

- **D0** • Preparation for the 8-Discipline Process
- **D1** • Organize the 8D Team
- **D2** • Describe the Problem
- **D3** • Develop Interim Containment Plan and Actions
- **D4** • Define and Verify the Probable Cause
- **D5** • Verify Permanent Corrections Actions
- **D6** • Implement and Validate Corrective Actions
- **D7** • Take Measures to Prevent Recurrence of the Problem
- **D8** • Congratulate the team

Figure 3.13: The 8 Disciplines Process

Authors comment if it will address the root cause: This is a tool we used previously when I was still employed at a large semiconductor industry in the Philippines to deal with customer issues, shipment delays, and equipment-related problems. Although the 4th Discipline states about verifying the root cause, doing 8-Disciplines are not based on evidence but on data and brainstorming about the most probable cause of the problem, hence, this cannot be considered as a tool to address the root cause but only the most likely or probable cause of the failure.

3.10: Kepner-Tregoe

Kepner-Tregoe is a problem-solving tool derived from its founders, Charles Kepner and Benjamin Tregoe. They authored their first book in 1965, which was titled, The Rational Manager. In 1981, they republished a revision of the book titled The New Rational Manager. Through the years, the Kepner-Tregoe method researched and identified the troubleshooting skills of people. Through the years, people have learned to deal with the complexity of incidents and problems which occur happen mostly in industries. Often, there is a pressure of time to solve the problems and arrive at a solution to the problem, whether the solution is correct or not. The Kepner Tregoe method or KT method is a problem-solving and analytical tool in which the problem is first disconnected from the decision. The Kepner Tregoe method will define the following:

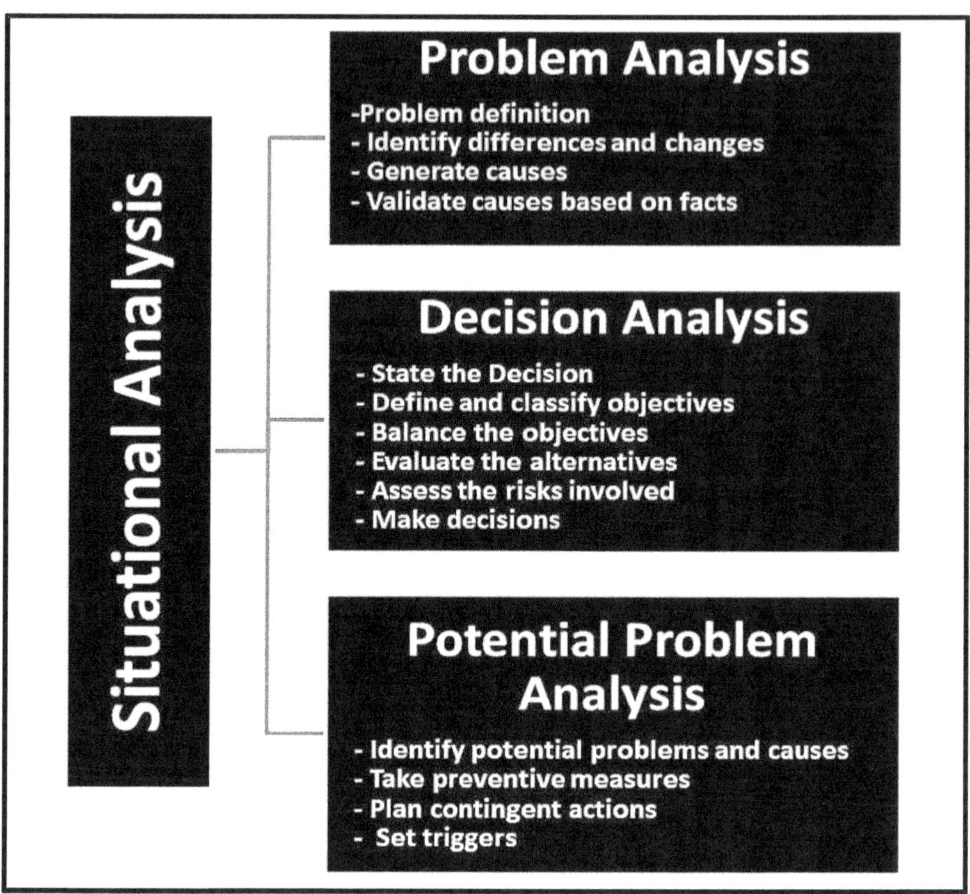

Figure 3.14: The Kepner-Tregoe Method

1) **Situation Analysis:** This clarifies the problem situation. The team clarifies the problem that matters in complex situations and determines how to prioritize and handle it. A detailed plan is developed for dealing with each issue and includes how to deal with each situation and who needs to be involved. This answers the question of what happened or what is going on?

2) **Problem Analysis:** Refers to uncovering the actual cause of the problem together with the relationship between the cause and effect. The team will determine the cause by organizing and analyzing information about the problem. Possible causes are identified and then tested against these facts. The team verifies the true cause of the problem and considers where both the cause and any fixes may have additional impact. Problem Analysis ensures that the cause is known before fixes are implemented. This will answer why the incident happen.

3) **Decision Analysis:** This will be based on the decision-making criteria. Several choices are made to arrive at the potential problem resolutions. The team clarifies the purpose of their decisions and form conclusions. The team evaluates different alternatives and assesses the risks involved before arriving at a decision. Decisions made ensures that maximum benefits and lesser risks. This will answer the question of how we should act or what we should do about the problem.

4) Potential Problem Analysis: Discuss the potential future problems that are anticipated, and develop corrective actions. The team anticipates threats to the success of the corrective actions. They consider the possible causes for each potential problem and how to prevent them. Contingency plans and actions are done if the problems recur. Potential Problem Analysis helps prepare in advance and avoid unplanned reactive actions. This will answer the question of what will the results be or what lies ahead?

Authors comment if it will address the root cause: The K-T method will provide a systematic way to derive what they think is the root cause of the problem. However, performing the K-T method is not based on the evidence that caused the failure to occur but a product of brainstorming, hence any analysis not based on evidence may or may not derive the root cause but only the most probable cause of the problem. I consider this method to derive the most probable cause and not the root cause of the problem.

3.11: Difference between a Probable Cause and The Root Cause

When equipment fails, there is a wide variety of possible causes as to why it fails, and when we speak of this vast amount of causes that prompt the equipment to fail, we speak about its failure modes. As the late John Moubray in his book on RCMII said, a failure mode refers to an event that could cause a functional failure. These refer to the probable causes of failures that might have occurred before or could possibly occur in the future. On the other hand, to understand the root cause is to reconstruct the pieces of evidence to find out what caused the problem. The depth of the analysis in Root Cause will always tell us that for every failure, there is always a corresponding Human Cause as well as a hidden cause or Latent Cause behind the problem. Failure modes and root cause are two different terms. Its difference lies in the depth and breadth of the analysis.

• In a Reactive Industry, we accept failures by fixing the failure modes every time they occur.
• In industries implementing RCM, we prepare for the failure by providing the most feasible tasks to deal with each failure mode.
• In industries implementing RCFA, we challenge the failure by investigating what, why, and how it happens so we can learn from it.

Addressing the failure Modes are like having a machine gun, where Al Capone is shooting in all directions in the dark, while a root cause can be compared to a rifle equipped with a laser and scope hitting the target with precision accuracy one at a time. This can be compared to a sniper in the military force.

RCFA will use the word hypothesis instead of failure modes. Each hypothesis is being verified so we can proceed to the next level. In a Crime Scene investigation, these failure modes can be compared to the different suspects involved in the crime. Just like in any crime scene, the root cause will always be based on the evidence found during the investigation. Any analysis which tends to end up at the part or component level is considered a physical cause of failure analysis. Let me provide an example to clarify this, a pump failed to fulfill its function since it is not discharging any fluid. This failure to discharge is known as a functional failure. Let's think of the different reasons or probable causes of why this pump is not discharging any fluid. In this case, we are speaking about failure modes and not the root cause. Therefore the failure modes of the pump on why it is not discharging fluid may be as follows;

• Valve is totally closed
• Motor totally burnt out
• Bearing seizure (stuck-up)
• Strainer totally clogged
• Broken pump impeller
• Supply tank is empty
• Clogged strainer
• Driver imbalance
• Foreign particle trapped at the tip of the impeller
• Power failure
• Broken Coupling
• Insufficient suction pressure

I could continue on with the list of failure modes or probable causes that could warrant a pump not to discharge fluid at all, but let's assume all the failure modes had been listed so far. Failure modes are the probable causes that will affect the pump from not discharging any fluid at all. You may even include lighting to strike the pump or an earthquake that severely damage the pump. On the other hand, to determine the root cause, we need shreds of evidence to conclude the reason for its failure. We can consider the physical cause related to the clogged strainer because there is a pressure drop or we can consider a broken coupling because visibly the coupling is broken. This means that the pump must actually fail so we can determine the pieces of evidence to define the causes of failure. Second, the failure should be fresh, which means that it just happened recently. We cannot get the root cause of this pump whose failure happened two years ago, but we can guess the probable reasons. This is what makes Root Cause Failure Analysis reactive since a failure must occur before conducting an investigation.

On the other hand, Root Cause Failure Analysis is, at the same time, Proactive because by knowing the cause of the failure, we can learn from it and address similar situations in the future when it arises. Therefore, by analyzing the pump's failure, we need to verify what failure mode actually happened during the time the pump had failed, and when the analyst had verified each failure mode and found out, for example, that the actual cause of the pump's failure to fulfill its function is a bearing seizure then we need to analyze what caused the bearing to seized, the other failure modes will be disregarded. Corrective action will be done for the physical cause, human cause, and system cause of the problem. After we analyze the cause of the bearing seizure, a failure analysis will be written.

Failure Analysis: Bearing Seizure due to lack of lubricant in the raceway. Failure analysis will stop at the component level. However, a Root Cause Failure Analysis will still proceed with the investigation. It will only conclude when the Latent Cause of the problem had been identified. Hence, a complete Root Cause Failure Analysis investigation will be written as follows:

Root Cause Failure Analysis Includes:
• Level 1: Physical Cause will be a bearing seizure due to lack of lubricant in the raceway
• Level 2: Human Cause was due to the wrong lubricant used by the technician.
• Level 3: System Causes indicate that there was no procedure on how to lubricate and the type of lubricant to use since the people were never trained.
• Level 4: Latent Cause will ask the organization their contribution to the problem. Why did they put a restriction on the lubrication training if this is what the people want so that they can do their jobs?

We have differentiated failure modes from failure analysis and Root Cause Failure Analysis; the question raised is: does RCFA fit in the RCM Strategy? Some say that RCFA can fit into some failure modes that are already the subject of some Root Cause Failure Analysis investigation. I disagree with this since if the failure mode is subject to investigation, then the correct term is Failure Analysis and not Root Cause Failure Analysis. RCFA will go much deeper and would expose the hidden causes of failure. Second, if a Root Cause Analysis had already been performed in any of the failure modes, then it should not be written in the RCM Analysis lists of failure modes since a recurrence is unlikely to happen if a successful RCFA had been concluded so far. The last chapter will explain how to integrate RCFA in RCM.

John Moubray, the RCMII author, asks that which comes first, redesign or maintenance? He simply stated that maintenance should come first before the redesign. I agree with his statement since there is always a temptation to modify or redesign the system or asset without reviewing the current maintenance tasks at hand. On the other hand, most people are tempted to redesign or modify the equipment without even performing a thorough RCFA investigation, leading to new problems with the equipment. RCM is being performed on an asset so that the people involved, which are the operators and maintenance, can better understand the different failure modes that can affect their asset, its consequences when the failure occurs and the most feasible maintenance tasks to address the particular failure mode.

There is a limitation to what maintenance can do as a task. This is seen in the last question on the RCM's seven basic questions. What should be done if a suitable proactive task cannot be found? One of the options here is to redesign. A redesign is warranted and feasible to use if the consequences of failure are simply not acceptable to the user. RCFA is used after performing an RCM Analysis and not before starting an RCM Analysis. I believe this is where RCFA will fit in the RCM process. However, as previously stated, we need the failure to occur first before we can redesign so we can determine the correct cause of the problem. Hence, after performing an RCM Analysis, and there are still failure modes that keep repeating themselves in which the consequences are not acceptable, an RCFA should be conducted. The second point is that equipment failures attributed to human errors should be included in the Root Cause Failure Analysis Investigation.

I have friends who provide consultation to both RCM and RCFA respectively. Sometimes, I see conflicting views in both of them. I believe that both strategies are not perfect, but if we can combine them both will create a much more powerful strategy for the maintenance function, but this can only be possible when used from the right perspective. Again, they should not contradict each other, but rather they should complement and supplement one another.

3.12: Steps in Conducting Focused Improvement

- Focused Improvement are various improvement activities used to accomplish maximum efficiency of individual facilities, equipment, and manufacturing processes and the entire plant by thoroughly eliminating losses and improving its performance. *From JIPM Materials*

- Focused Improvement includes all activities that maximize the Overall Equipment Effectiveness of equipment processes and plants through uncompromising elimination of equipment losses and performance improvement. *From Tokutaro Suzuki*

The Focused Improvement team comprises of individuals who are both knowledgeable on the losses to be improved, consisting of anyone in the organization. The respective manager spearheads the Focused Improvement team. The goal of the Focused Improvement team is to maximize the Overall Equipment Effectiveness by eliminating or reducing the 8 Major losses encountered in their equipment. Focused Improvement teams are like an elite force or special ops in the military with a specific mission: eliminating equipment losses in their equipment and assets. These teams must be heavily equipped with several problem-solving and analytical tools to analyze the problem and on how to address these equipment losses. The composition of the Focused Improvement team will include a cross-selection of people from different departments with knowledge of the losses they are about to address. Below is a step-by-step guideline for conducting Focused Improvement activities to address equipment failures, losses, quality defects, and other administrative problems in the plant.

1. Form the Focused Improvement Team: Membership of the Focused Improvement team will compose of a cross-selection of people from different areas knowledgeable on the problem to be addressed. Minimum of 4 but not to exceed 8 members will compose the Focused Improvement team. A focused improvement team should be knowledgeable on at least a minimum of three analytical and problem-solving techniques. Recommended analytical and problem-solving tools for Focused Improvement include FMEA/FMECA, Ishikawa or Fishbone Diagram, P-M Analysis for chronic problems, Pareto Analysis, process mapping, seven QC tools. 8-Disciplines and so on. The type of problem-solving or analytical tool used by the Focused Improvement team will depend on the problem being addressed.

2. Decide on the Improvement Topic: The improvement topic to be addressed by the Focused Improvement team should focus on reducing or eliminating any of the 16 big losses currently existing in the industry. The topic for improvement should be related to their current problems. Register the improvement topic with the TPM Office.

3. Understand the Situation: The Focused Improvement team should analyze and understand the current situation and identify the major losses and bottlenecks in the overall process. Losses may either be for equipment, manpower, or other losses. Provide initial measures and KPI as a baseline to set up targets. The Focused Improvement team should backtrack equipment's initial KPI and measurements to determine the success or failure of their effort. One of the measurements they will be tracking will be the OEE or Overall Equipment Effectiveness for addressing equipment losses. However, this will only be for equipment-related problems and these can even be breakdown into the type of losses the equipment is suffering. For example, a good KPI or indicator for breakdowns will be MTBF, while a good KPI for equipment with a lot of minor stoppages will be MTBA or Mean Time Between Assists.

Optimal Conditions

• A thorough elimination of slight abnormalities is a prerequisite for achieving zero defects and breakdowns

OPTIMAL = Necessary + Desirable

Mechanism	Necessary	Desirable
Pulley arrangement accommodating all three belts	• At least one V-Belt must be installed for correct operation	• All 3 V-Belts should be installed for operation • All 3 V-Belts should have equal tension • The belts should be free of cracks & grease • Pulley should be free of abrasion • The motor and speed reducer should be aligned properly at all times
Grease supply	• Grease must be supplied at specific locations	• Grease nipple should be kept clean • Area around the grease fitting should be wiped clean after each application • The condition and volume level of used lubricant should be checked constantly • Grease container should be kept clean • Used lubricant should be disposed of properly • The number of days for the lubricant to reach the end of piping should be estimated

Figure 3.15, Equipment Optimal Conditions

4. Expose and Eliminate Abnormalities: For equipment losses, construct a picture of the optimal conditions of the equipment or process. Before applying any basic analytical techniques, the team should address minor flaws, deviations, abnormalities, irregularities, and defects to ensure that the basic equipment condition is well in place.

5. Analyze the Probable Causes of the Problem: Focused Improvement team uses basic analytical problem-solving tools to analyze the possible or the more likely causes of the problem depending on the type of losses to be addressed. Specify the problem-solving or analytical tool that will be used by the team. For RCFA, the RCFA Core Team or Council will assign a Principal Investigator to address the problem. There will also be cases where each loss will require two or more analytical and problem-solving tools.

6. Plan for the Improvement: During the development of proposals for improvement, formulate several alternatives. Never dismiss any ideas at this stage. For best results, do not limit participation to one or two members of the Focused Improvement team, but rather all members should participate in this initiative. If a budget is required, the Focused Improvement team needs to prepare a feasibility or economic study and calculate the return on investment (ROI) to have the budget approved by their top management. The Focused Improvement team needs to guard carefully those improvements that may result in fresh and new problems on the equipment. For equipment-related losses, the team will mostly result from redesigning or modification. Refer to the questions below before proceeding with modification or redesign. If the team answers yes to all these questions, then proceed with the redesign or modification. Also, note that if a spare or part will be involved in the redesign or modification, the Focused Improvement team needs to check the actual inventory of the original parts in the storeroom and make an economic decision on whether or not to consume these parts first or proceed with the modification or redesign. If the team decides to proceed with the redesign or modification, then these original parts will become obsolete, and the storeroom people should be informed so that these parts will be removed from the storeroom.

• Does the failure involved cost and have major operational consequences?
• Is the cost or breakdown maintenance high?
• Are there specific costs that can be eliminated by the design change?
• Does the design have no harmful effects which can be generated afterward?
• Is there an economic trade-off study on expected cost savings?
• Is the asset to stay or be used for a long time and not be decommissioned?

7. Implement the Improvement and Evaluate the Results: Once the plan had been completed, implement the improvement, and evaluate the results on the equipment. If the improvement is successful, it will be replicated to other equipment with the same operating context or problem. It is not recommended to bring the improvement to similar equipment which does not possess the same problem. Check whether targets have been achieved; if not, the team will again proceed to Step 5 and analyze the causes once again.

8. Monitor the Results and KPI: KPI should be based on any of this PQCDSM. (Production, Quality, Costs, Delivery, Safety, and Morale) If equipment losses are being addressed, the Focused Improvement team should monitor the OEE of the equipment (Overall Equipment Effectiveness). If the indices and KPI improves, the Focused Improvement team needs to develop the procedures necessary and disseminate this information to concerned areas and people in the plant.

9. Sustain the Improvement: Implement standardization and measures needed for preventing the recurrence of the problem. Draw up control standards to sustain results. Formulate work standards or routine maintenance to sustain these improvements.

10. Horizontal Replication: Fan-out or horizontally replicate successful improvements generated by the Focused Improvement team to other processes, equipment, system with the same operating context or problems. Similar equipment which does not possess any problems whatsoever should be excluded in the horizontal replication plan.

11. Create a Loopback to Early Equipment Management (EEM): Once the improvement had been concluded, create a loop back to IFCA (Initial Flow Control Activities) or EEM (Early Equipment Management) for the possibility of purchasing additional equipment in the future so that these improvements will already be included before purchasing the equipment with the vendor. The issue of the improvement patent should be discussed with the OEM or Vendor.

12. Team Celebration: If the team is successful in the reduction, mitigation, or elimination of the losses identified in their Focused Improvement activities, a simple celebration will commence. This would be a continuous process; once the team completes their analysis and corrective measures successfully, they disband and go on to their original work post, and other teams are formed to deal with the other problems.

Chapter **4**

The Lifeblood of Root Cause Failure Analysis

> *The message of Root Cause Failure Analysis is simple. We can only learn from the things that go wrong if we can have the courage and guts to look ourselves in the mirror and admit to the fact that we, too, are also part of the problem.*

4.1: Evidence: The Lifeblood of Root Cause Failure Analysis

There are two mistakes industries make in performing Root Cause Failure Analysis. First, people performing Root Cause Failure Analysis think that the most important part of conducting the investigation is having sufficient data. Without data, it is seemingly impossible to carry out such an analysis. Second, most industries, especially manufacturing, think they can finish their Root Cause Failure Analysis in 24 to 48 hours. If you seem trapped in these predicaments, then hear me out first and read on.

Those who rely on data too much missed out on the most important part of any Root Cause Failure Analysis Investigation. Hence, if data are not the most important factor in performing a Root Cause Failure Analysis, what is it? The answer is "**evidence,**" and uncovering the evidence is the most important factor in any Root Cause Failure Analysis investigation. Data is just a piece of evidence. The difference between data and evidence is data is just a part of the evidence. The data may contain numbers, statistics being measured and collected for analysis. We can have data to determine the average life of a human being per country, or how many people died because of covid 19 or a million other things. Data can be both quantitative and qualitative. If the data contain numbers, then we called it quantitative data. However if the data is made up of personal judgments, and other non-numerical information, we called this qualitative data. However, individual accounts, rumors, and gossips do not qualify as qualitative data. This can only be considered qualitative data if done on a large scale. The more data we have, the more likely it becomes conclusive. And when the data is conclusive and based on facts, then this can be part of the evidence. Evidence is a body of facts and information showing whether a hypothesis is true or not. Conclusive data forms can be part of the evidence. However, raw or incomplete data may not form a piece of evidence. Evidence is when data is used to try to prove or disprove a particular point. In fact, it can be said that evidence is the lifeblood of any Root Cause Failure Analysis. Without evidence, it is entirely impossible to carry on such a probe. A bearing has failed, and they can't seem to get going with their Root Cause Failure Analysis because data for bearing failures are insufficient. The bearing failure's physical cause can be known if the analyst can see the bearing's inner, outer raceway, balls, and cages. It will tell you precisely what went wrong with the bearing. So to analyze and conduct failure analysis on the bearing failure, you do not need an enormous amount of data. What you need is just a hacksaw.

First, let us define what evidence is all about. Evidence is a piece of information that supports a conclusion. Evidence in its broadest sense will include anything used to determine or demonstrate the truth of an assertion. Philosophically, it can include propositions presumed to be true and used to distinguish or isolate other propositions presumed to be just gossip or hearsay. The term has specialized meanings concerning specific fields, such as policy, scientific research, criminal investigations, and legal discourse. Like in a court of law, we cannot prove our point with just opinions, hypothesis, hearsays, fiction, and arguments. Without evidence present in our RCFA efforts, then all we got is just our opinion. All failures do not just happen without leaving a trace. It will always leave some sort of clue for the investigator or analyst to uncover. Root Cause Failure Analysis will separate the facts from fiction.

This is one of the main problems in the majority of industries I served. What most maintenance actually does is when a piece of equipment fails, they will restore it immediately since they will be pressured by operations. In this case, the evidence why the failure occurred had vanished since it was not preserved. When the boss asked for the root cause, we guess, speculate, or just charge it on our experience. Here is the thing, when we repair our equipment, we destroy the evidence, and it is unlikely to perform a Root Cause investigation in this situation, but we can perform an analytical problem-solving tool just like what we have discussed in Chapter 2 of this book. But again, it may only provide us the most probable cause and not the root cause of the failure.

4.2: Three Types of Evidence on RCFA

Nothing in our existence happens without it being known in one form or another. Evidence is the word we use to describe how we become aware of something. More specifically, evidence is the way an event or condition manifests itself. It is defined as that which tends to prove or disprove something, ground for belief, or proof. This can also include the things helpful in forming a conclusion or judgment. Evidence in its broadest sense includes everything used to determine or demonstrate the truth of an assertion. These pieces of evidence are simply small pieces of the puzzle that we need to analyze to see the bigger picture of why the failure happened. This means that when the facts are finally backed up by evidence, we can now understand what causes the failure to occur. The evidence contains everything we need to know about an undesired event. In fact, evidence includes everything we need to know about life itself.

Without evidence, Root Cause Failure Analysis is impossible because RCFA, by definition, is an evidence-driven endeavor. Without evidence, we can only suspect, hypothesize, theorize, speculate, guess, agree, disagree, or brainstorm on the causes of an event that caused the failure to happen. This is what separates root cause from any analytical problem-solving tools. Once all evidence is laid down, the sequence can be unfolded to see the bigger picture to determine the real cause of the problem.

In conducting a Root Cause Failure Analysis, three types of evidence need to be gathered. [3]According to C. Robert Nelms, the founder of Failsafe Network, one of the leading providers of Root Cause Analysis in the US, classifies pieces of evidence into three types. He named them the 3P's, which stand for People, Physical, and Paper evidences. These pieces of evidence are arranged according to their importance.

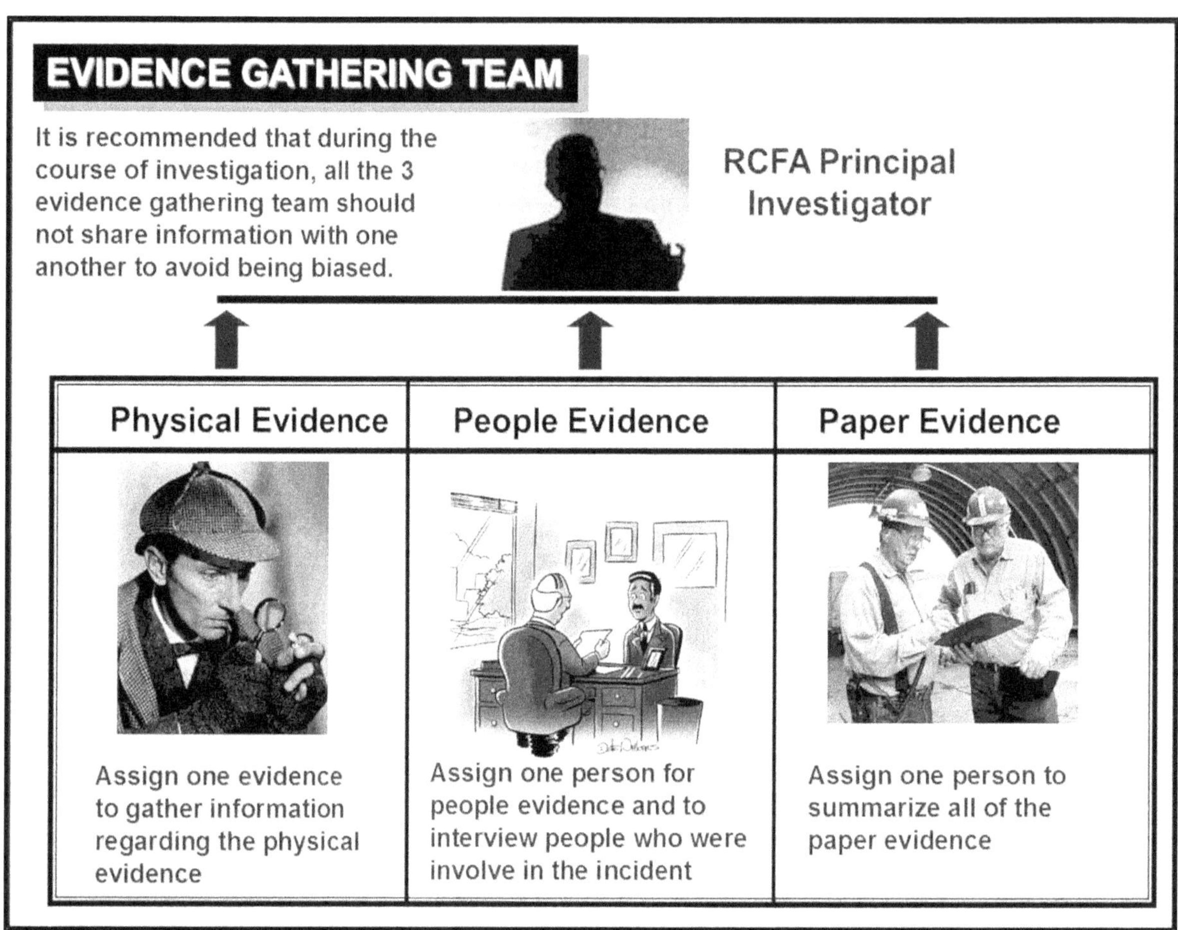

Figure 4.1: Types of Evidence for Root Cause

1) People Evidence: The most important evidence to consider is people. People who witness the failure should be interviewed immediately. An interview is a conversation intended to elicit information. Interviews are generally non-accusatory and should refrain from blaming, finger-pointing, and being biased. Try to interview people who have witnessed the failure themselves. Testimonies of people can strengthen and distinguish facts from fiction. We can also conduct an interview with people who have the knowledge or have experienced the incident in the past to gain a better perspective about the problem. Be objective and professional in conducting interviews as we are dealing with people in this case.

In any failure investigation, there is always a sequence of events that eventually lead to the

---

[3] C. Robert Nelms, *What You Can Learn from the Things That Go Wrong, A Guidebook to the Root Cases of Failure*, (Virginia, Failsafe Network Inc., USA: 2007), Pages 57 - 64

failure. It is the goal of the principal investigator to reconstruct this sequence of events. Once the conversation's tone has moved to an accusatory or interrogative mode, people will be defiant or hesitant to share what they know. It is, therefore, important that the investigator has a clear goal in mind when conducting an interview. How would we know if our interview is turning up into an interrogation? Use the 80/20 rule. This means that the interviewee should be the one talking most of the time. The role of the interviewer is to ask a specific question for the interviewee to answer. If the interviewer is talking most of the time, the interview now turns into an interrogation. Our main concern is to know what actually happened so that we can put the pieces of the puzzle together. You need to make the interviewees feel at ease to be more willing to share what they know regarding the incident. Be compassionate and friendly when interviewing people. The more relaxed the person is during the interview, the more information you can expect from them. Never interrogate people during the interview. They will just become defensive and practice their right to remain silent. The interviewer should have the trait of listening to the person they are interviewing. Never show facial reactions especially when you disagree with what they are saying. Be objective and let the interviewee speak without being interrupted.

Another thing is to remain focused and unbiased during the interview. No matter where the location will occur, some amount of privacy will be needed in conducting the interview. Make sure the room is comfortable. Conduct interviews with one person at a time. The information collected will tell us a great deal about the failure itself. The purpose of the interview is to find out why things happen. The interview preparation provides a foundation for understanding what happens and gaining the most important and relevant information from those directly involved with the incident. Amnesty must be in place when conducting a Root Cause Failure Analysis investigation. Our goal is to understand the cause of failure so we can all learn from it. Tell them that no one will be punished as a result of the interview. The main reason for conducting this interview is to know the truth as to why the failure occurred. People have senses; they see things, hear them, and feel them. They all have opinions as to what goes wrong. People also tend to forget things easily; that's why interviewing people should be prioritized after a failure has occurred. Remember that people's evidence is the easiest to evaporate. What is important is to interview and ask them the right questions, such as:

• Where were you exactly when the failure occurred?
• What did you hear, or how did it sound like?
• What did you do after the failure occurred?
• What happened first, and then what?
• Where their others with you that witnessed the failure?
• Were there any signs or symptoms before the failure occurred?
• What do you think went wrong with it?
• Did you hear or smell or sense something different before the failure occurred?
• What do you think people working here feel about their jobs?
• Have this failure happened before in the past?

If possible, interview as many people as you can, but one at a time. Have a set of questionnaires on hand and write down their responses. Tape recorders can only be used with the consent and approval of the person being interviewed. Ask permission from the person you are going to interview if you can record the conversation. If they refuse, then you need to write the important points manually. After interviewing several people, you will be quite surprised by the information you have gathered since a sequence of events will emerge from the interview. Remember that there are always two sides to a coin. When you have interviewed around four people or more, certain questions seem to share the same thoughts. These will help the investigator put up the pieces of the puzzle together so that the big picture can finally unfold. The following should be the people who will be subject for RCFA interview;

• Person who committed the error or mistake
• Operator involved in the incident
• Other operators who have previous experience with the failure
• Person who developed the procedure or SOP if they are still around
• Maintenance or technician during the shift the incident occurred
• Maintenance or technician who repaired the failure
• People indirectly involved such as those familiar with the incident that occurred.
• People directly involved in the failure
• Predictive or CBM group involved in monitoring the equipment that failed
• Supervisor of the person involved in the incident
• Other people we think might be involved or can provide information on the incident

The person in charge of the people's evidence (conducting the interview) will interview two sets of people. Those directly involved and those indirectly involved. Those directly involved are the people who had witnessed and are present when the failure occurred. Those indirectly involved are people who have experienced the failure or have some knowledge about the failure. You will be needing two sets of questionnaires, one for the people directly involved and the other for those indirectly involved. The questions should be consistent for those, directly and indirectly, involved in the failure. Upon completing the interview, if you noticed that two or more people provided the same answer to a specific question, then this is already considered as a piece of evidence.

2) Physical Evidence: Although the people's evidence is the easiest to evaporate, the physical evidence will be the most important of all since it will include the part or item that failed. That is why we cannot repair the equipment right after the failure. The physical evidence should be extracted from the equipment first before repairing. Once the equipment is repaired without the physical evidence, then it is unlikely to proceed with the root cause since the physical evidence has been washed out. The actual collection of evidence is a critical step in the investigation process. Each piece of evidence collected must be handled in a way that preserves its integrity. Once an object is identified as evidence, it must be tagged and preserved. Evidence tagging helps identify the collected item. The tag can consist of a sticker with the date, time, description, and control number. Failed parts, debris, should be collected for evidence, and positions should be noted. When we try to fix the equipment, we easily replace the part that failed and throws it away. The first rule on any evidence freezing process is collecting and preserving the part that failed, taking photos of the different parts of the equipment. Remember that we can only learn from failure if we take the time to takes things slowly. Physical evidence can conceivably include all or part of any object. One of the important things to consider in the physical evidence gathering will be to preserve the part that failed. Both operations and maintenance need to understand that the preservation of evidence is the key to Root Cause Failure Analysis which means that the equipment should not be repaired immediately. The evidence should be extracted before repairing the equipment. The physical evidence will include any of the following;

• Photos and pictures taken during the investigation on the asset
• Any Video or CCTV camera if it exists
• The mechanical part that failed
• Any electronic or electrical part that failed
• Fractured metal debris collected
• Metallurgical Laboratory results of the failed part or item
• Readings on charts, gauges, temperatures
• SPC (Statistical Process Control Charts) if existing
• Position of any unusual things on the equipment
• Position of the failed item
• Any item inspected by the person in charge of the Physical evidence
• Enlarge photos of the failed item at different angles and position
• Secondary or Tertiary damages if any
• Housekeeping practices

On the other hand, it is also important for the team to analyze and investigate the failure to freeze up the evidence as quickly as possible to uncover it while it is still there. The restoration team must understand that parts that failed should not be thrown away and should be preserved every time equipment fails; otherwise, parts evidence will totally be impossible for the investigative team. There will always be traces of evidence left on the part that failed. The actual collection of evidence is a critical step in the investigation process. Each piece of evidence collected must be handled in a way that preserves its integrity. Once an object is identified as evidence, it must be tagged and sealed. Evidence tagging helps identify the collected item. For system-based assets, the person responsible for the physical evidence must check all the components, assets, pipes, and protective devices for any signs of oddities, or abnormalities that may have caused the incident to occur.

Positional evidence will also be included in the physical evidence. This includes mapping out the position during the time of failure. What are the readings on the gages during the time of failure, temperature, what valves are open or close, where the leak seems to be coming out, what buttons were pushed, and so on? Map everything possible. As the restoration team tries to bring back the equipment up, most of the positional evidence will be lost. The investigative team needs to collect all physical evidence on the equipment that failed before restoring them back to normal. As much as possible, try to draw and record every piece of information you can when uncovering positional evidence.

Detective Varnike was playing golf in a hotel garden when he heard a shot in one of the hotel rooms. The detective appeared in one of the hotel rooms where the shot was heard. He slightly opened and squeezed himself into the room. "Help, someone wanted to kill me," cried addressing to Varnike, a well-known movie star who was in the hotel room. There was a man in a mask here. I was attacked by him. I protected myself as much as I could, then he fired. Probably, during the struggle, I snatched the gun out from his hands.

After that, he rushed to the door and disappeared into a corridor. "Please, call the police." After Detective Varnike observed the room, he said, "Maybe, I will invite the media people, photographers, and news reporters, since I think that is what you need. The police hardly should not be disturbed because of such worthless performance." Question: Why did Detective Varnike refuse to pursue the criminal? What piece of evidence in the picture convinced the detective that no one attacked the actress and that the mess was all her doing? The answer was the hairbrush. If the actress really claimed that an assailant was in her room and went out of the door, then the brush would not be in its original place since it would also be moved as the door was swayed. It was impossible for the **brush** not to move if someone went out of the door. This is known as positional evidence, which is why it is important for any root cause investigation not to repair the equipment right after the failure.

Figure 4.2: Can you find the evidence?

3) Paper Evidence: When we speak about paper evidence, this refers to all recorded documents that we can gather during the time of the incident. Paper evidence should be summarized by the person assigned doing the probe or investigation. The investigator should collect all paper evidence as much as possible for a thorough review of the process. This will include those of operations, maintenance, and personnel-related data. Determining which paper evidence is relevant during the incident is important for the principal investigator. Likewise, the principal investigator needs to have an RCFA tool kit, as shown in figure 4.3, on the process of collecting evidence on the failure. CBM or Predictive Maintenance such as Vibration Monitoring, Ultrasonic, Thermography, Oil Analysis, and other Non-Destructive instruments can also be a good source for paper evidence. An infrared thermography scan can indicate the exact source where the hot spot was exactly located. Paper evidence can include any of the following;

• Maintenance history logbooks

- Strip charts, operating temperature, gages
- Equipment Standard Operating Procedures (SOP)
- Operators and Maintenance Training records
- Operators Logsheet
- Electronic data recorded in computers
- CMMS or EAM Records
- Recent Preventive Maintenance Records
- Operating parameters such as temperatures, pressures, flow rates,
- Machine's settings and operating limits and set points
- Recent procurements on the equipment such as spares
- Equipment operating logs
- Equipment assembly drawings, parts lists
- Maintenance records and procedures
- MRO Spare Parts withdrawals and records
- Operators and maintenance inspection records
- Predictive or Condition monitoring reports and records
- Oil Analysis Reports
- SPC (Statistical Process Control) Charts if existing
- OEM or vendor's operating manuals
- Personnel competency requirements to operate and maintain
- Equipment drawings and schematic diagrams
- Any other documents you may find relevant

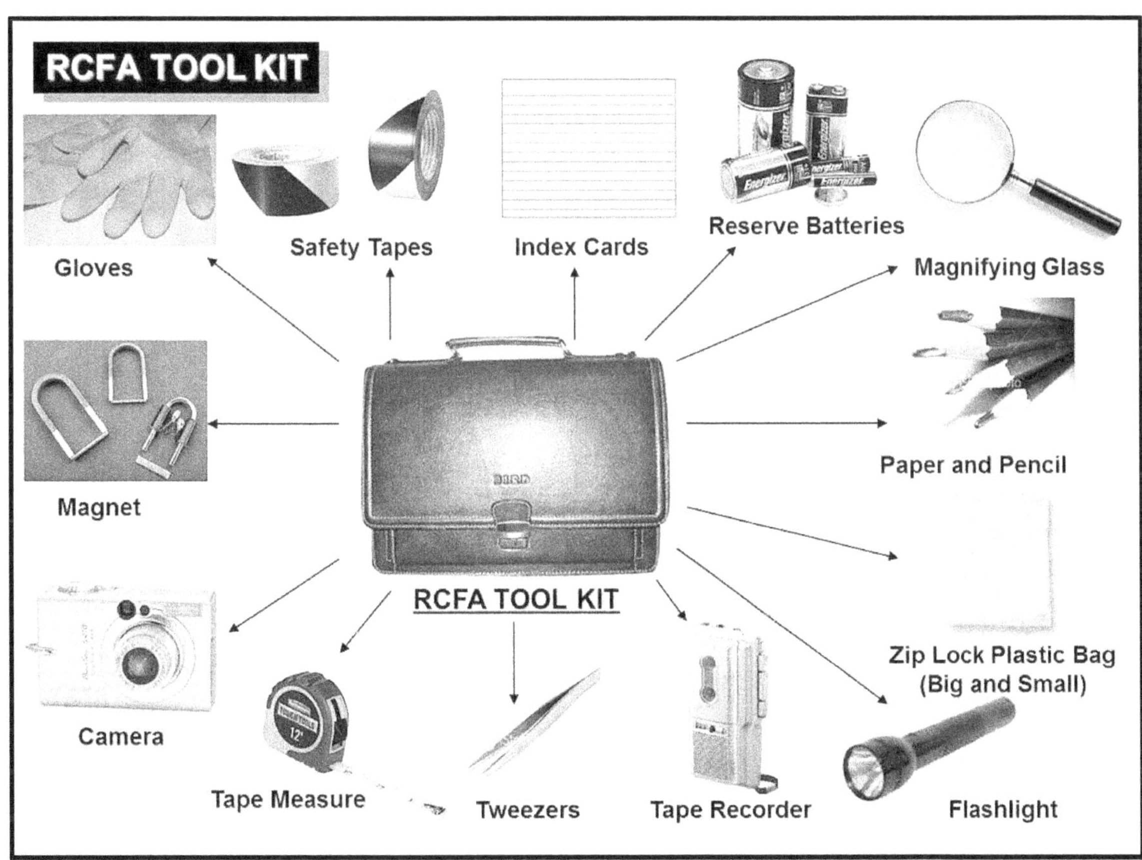

Figure 4.3: Initial RCFA Kit

Eighty percent of the time spent on Root Cause Failure Analysis must be based on the evidence gathering, including gathering the 3P's. Root Cause Failure Analysis is not about just about doing countermeasures, but rather understanding the problem so that we can all learn from the failure and understand how and why it occurred. I really find it difficult to conduct a Root Cause Failure Analysis in just a timeframe of 24 hours, or a couple of days and arrived with the Root Cause of the problem. Working in various industries such as shipping, assembly lines, manufacturing, and mining industries and with all the training I've been through in all these years on problem-solving allows me to think that Root Cause is all about solutions and countermeasures. I am very wrong in this regard. We all have a human intuition that when we encounter a failure, we want a solution fast to get rid of the problem, and most of us have the temptation to jump to modification without understanding the Root Cause of the problem. I know that we have many chronic failures in nature that continue to exist in the plant. We must understand that the most important factor in performing a Root Cause Failure Analysis is gathering evidence to truly understand why failures occur. We need to become one with failures and understand them. Most people think that failure is bad for business. Failure tells us something. It can be the best teacher we can ever have as long as we can all learn from it.

Although there are people assigned to collect the evidence, they should not restrict themselves to those that are assigned to them. For example, the person in charge of collecting the physical evidence should not restrict themselves to just collecting. They are free to ask questions to the operators, or maintenance and note them down since the person collecting the physical evidence might think that this is the job of the person collecting the people's evidence or the one that is conducting the interview. Collect as much evidence as possible as this is needed to construct the RCFA logic tree diagram. There might be some overlap when extracting the pieces of evidence and there is no problem with that as everything will be consolidated once the RCFA logic tree diagram will be constructed to determine the sequence of events. Meaning that the person in charge of collecting the people's evidence might also ask for paper documents, and the person getting the paper evidence may ask questions as well.

4.3: Which Failures will Warrant an RCFA Investigation

Although there are no general rules on which failures will warrant an RCFA investigation, conducting RCFA will not only be limited to big and catastrophic failures but also to small problems. It is also not practical and tedious to perform a Root Cause Failure Analysis on every single failure or breakdown that can occur on the equipment. The Golden Rule to adopt in this case is that maintenance and technicians must understand that if an RCFA investigation is warranted, they cannot repair the failure immediately. Rather, the physical evidence should be preserved. Typically, a principal investigator will be spearheading the RCFA investigation process. The Principal Investigator (PI) will assign a maximum of three people as evidence gathering team. One of the evidence gathering persons will be responsible for collecting all the physical evidence. An alternative is to teach our maintenance and technicians how to collect, preserve, and extract the physical evidence on the equipment that failed and surrender them to the RCFA Principal Investigator before repairing the equipment that failed. This will be a preliminary requirement that the principal investigator and the evidence-gathering team assigned to handle the case must undergo a full-blown RCFA training. The selection of the RCFA investigator will be done by a Centralized RCFA Core Team. If you are a big industry with several business units, the RCFA Core Team should be responsible for assigning the principal investigator to that specific business unit requesting for root cause investigation. Failures that will have the following impact, consequences, and aftermath will warrant a full Root Cause Failure Analysis Investigation

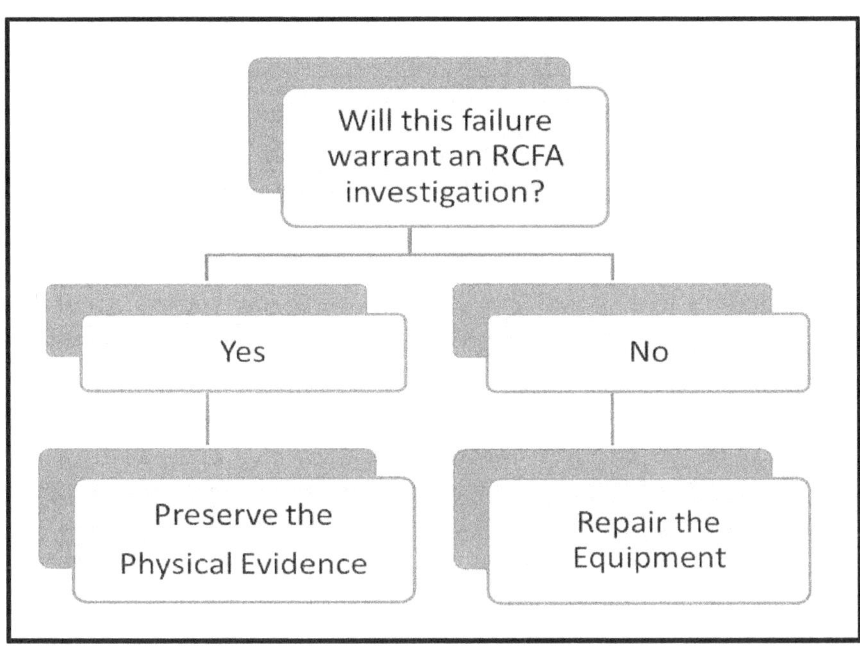

Figure 4.4: Will the Failure Warrant an RCFA Probe?

1) **Small-Scale Problems**: There is no restriction on when to perform a Root Cause Failure Analysis. In fact, RCFA can also be done on small-scale problems if the industry wants to. The only difference in approach is that for small-scale events, the principal investigator should also be the one to collect the three pieces of evidence and there will be no stakeholder meeting in this case. Once the RCFA Logic Tree diagram has been completed, and the corrective actions had been implemented, then the RCFA investigation is completed.

2) **Chronic Failures:** Failures that continue to repeat themselves in the past will be subject either to an analytical problem-solving tool or Root Cause Failure Analysis investigation. Most chronic failures are usually caused by a complex number of causes and not by a single cause. For complex causes or failures with several causes, an analytical problem-solving tool such as P-M Analysis will be much preferred than conducting a Root Cause Failure Analysis investigation. Initially, the cost of chronic failures is small compared to a sporadic or catastrophic failure, but over some time, if the failure continues to repeat itself, the cost can far exceed the cost of sporadic failures.

3) **Failures with Safety and Environmental Consequences:** According to the RCM process, a failure has environmental consequences if it causes a loss of function or other damage, leading to the breach of any known environmental standard or regulation. A failure mode has safety consequences if it causes a loss of function or other secondary or even

tertiary damage that could hurt, injure or even kill someone. When lawyers and insurance people are involved, the RCFA principal investigator is highly recommended to be a third party or someone not employed in the plant. However, the plant can still assign its own independent principal investigator and can compare its root cause investigation with the independent third-party investigator.

4) **Extremely High-Cost Failures:** When the cost of failure is extremely high, such as a complete shutdown of operations, a Root Cause Failure Analysis investigation will be warranted on the equipment or asset that failed.

5) **Parts with Inherent Design Weaknesses:** When a part has a short lifespan, either a root cause or a problem-solving analytical tool will be warranted to improve design flaws and weaknesses in our equipment to further lengthen the lifespan of the component or spare parts. The main difference on whether to use a root cause or problem-solving tool is that for the root cause, we need a failure to happen first and the evidence should be preserved, while for problem-solving tools, data must be sufficient that the part indeed has a short lifespan.

6) **Failures on Bottleneck or Constrained Equipment and Process:** A bottleneck is a constraint or a point of congestion in a production system that occurs when the products to be processed arrive too quickly for the production process to handle and build inventories. The inefficiencies brought about by the bottleneck often create production delays, longer cycle-time, and higher manufacturing costs. In figure 4.5, machine 3 will be considered the bottleneck equipment since inventory will be built on that equipment; since it has the lowest units per hour, to produce a product compared to the other machines.

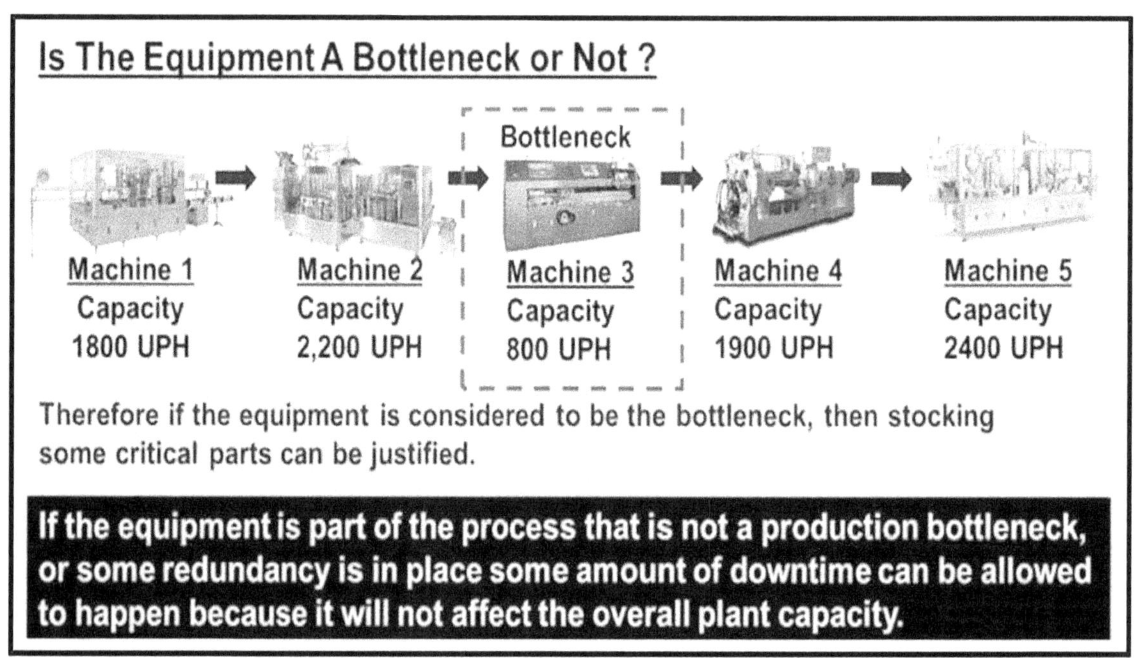

Figure 4.5: Bottleneck Equipment

7) **Product Defects:** For Quality-Related problems and defects a Root Cause Analysis can be used, however, if the defects are chronic and have a complex cause, it is highly recommended to use P-M Analysis or other problem-solving tools.

Again, conducting root cause failure analysis will not be limited to large-scale failures but also to small-scale failures or mini-event. We can classify failures in Root Cause as mini or small-scale failures, midi-event, or medium-scale failures, and maxi-event, or large-scale failures. This will be explained in Chapter 8 of this book.

4.4: How to Conduct the Interview Process in Root Cause

The person assigned to gather the people's evidence will be responsible for conducting the interview process. What is important for the person assigned to conduct the people interview is to have the same set of questionnaires for people directly involved and another set of questionnaires for the people indirectly involved in the incident. If you plan to record the conversation, ask permission first from the person you are interviewing; if they refuse, then everything will be done manually. After completing the interview process, the person in charge of people's evidence will summarize the answers of every person interviewed directly and indirectly. If two or more persons provide the same answer to a specific question, then this should be noted as this will be considered as a piece of evidence. Here is how to conduct the interview process.

Try to understand their perspective
• Introduce yourself and ask the name of the person you are interviewing
• Try to learn a little more about their background and how long they have been working?
• Ask them physically where they were during the time of the incident?
• Ask them how did they become aware of the incident?
• Ask them who else was present at the time of the incident?

Probe into their senses
• Let them lead you. What is important is not to lead them.
• Probe their senses and ask what they saw, heard, smelled, or felt during the incident?
• Ask them what happened first, or what did you do first?
• The standard follow-up question will be, and then what happened next?
• If you do not understand why they did something, ask them why did you do that?
• If they mention anything about their senses, slow down and probe all their senses.

Ask them, what did you think happened?
• Try to be curious and ask them why they think the incident occurred?
• Probe each of their responses.
• Ask them why they think and listen for any evidence that might lead them to their conclusions.
• Ask them something to support their claims.
• Ask them if this incident happened in the past?
• Ask them if there were any signs or warnings before the incident occurred?
• If there were any warning signs, ask them what was done about it?
• Ask them if any measures were indicating that the failure is happening?
• Ask them if they know of someone who also had a similar experience with the incident in the past?

Probe on their Latencies

• Tell the interviewee to forget about this particular incident and try to think about their role, in general.

• Ask them how long they had been working and their overall feeling about their work?

• Ask them if these incidents had happened in the past, and what was done about it?

• Ask them why they think that these incidents happen?

• Ask them if there was anything that can be done to avoid this type of incident?

• Ask them if given a chance, how would you like it to be, or how do you think things should be done here?

• Finally, ask them to think about the incident once again and ask them what they think was done and how they should be done?

4.5: RCFA Logic Tree Diagram

Once all the evidence-gathering teams have completed gathering their pieces of evidence, the principal investigator will call them for a meeting. Each of these evidence-gathering people in charge will present their evidence individually. Once the presentation is completed, the principal investigator together with the evidence gathering team can now construct the RCFA Logic Tree diagram. An RCFA Logic Tree diagram is a tool that uses deductive logic to guide the thought process used to draw correct conclusions. Therefore, a logic tree is a disciplined methodology that prompts the user to answer questions that will eventually identify the root cause of a failed event.

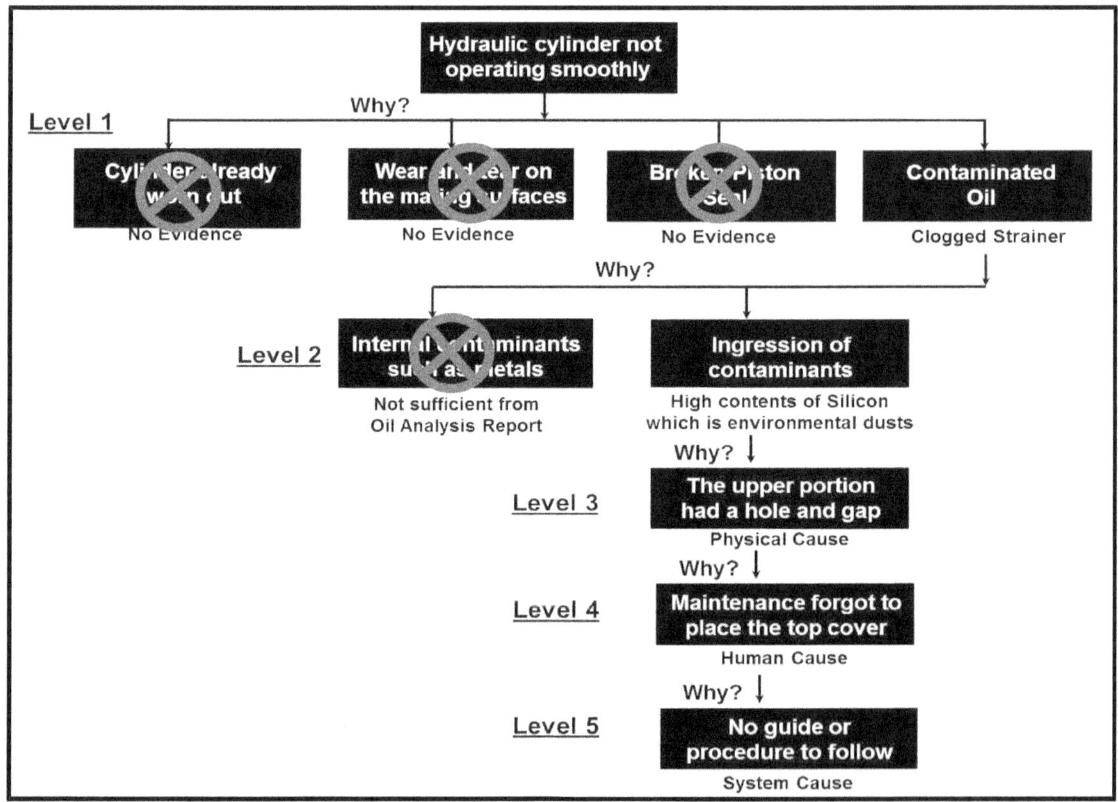

Figure 4.6: RCFA Logic Tree Diagram

The first step in building a logic tree is to properly define the failure event to ensure that the analyst is truly working on the problem and not on its symptoms. The failure event is at the top block and the possible causes or hypotheses will be on the first level of the tree. The analyst will follow each Failure Mode and verify if it actually occurred or not. If it occurred, the principal investigator and the evidence-gathering team will continue to proceed until the next level is revealed. Each of the three evidence gathering teams should summarize their conclusion on what they think is the physical, human, system, and latent cause of the problem before the meeting with the principal investigator happens. Once the physical cause of the problem is reached, each of the evidence gathering teams involved will compare their physical cause based on the evidence they have gathered and conclude if they agree or disagree with the principal investigator's physical evidence. The same process will be done for both humans, the system, and the organizational latent cause of the problem.

There are two ways to construct the RCFA Logic Tree diagram. First is from the top to bottom, where the top will be the problem or incident that took place. The second is from left to right where the left will be the problem or incident. There is a possibility that two or more causes based on the evidence have caused the failure. In this case, I would recommend treating them separately with each having a separate RCFA Logic Tree diagram so as not to cloud the diagram. The RCFA Logic Tree diagram will include the physical, human, and system, and what the principal investigator and evidence gathering think would be the latent causes of the problem. Once a hypothesis is written on the logic tree diagram, each of the evidence gathering teams should confirm for any pieces of evidence gathered. If no evidence exists, then it will be crossed out in the RCFA Logic Tree diagram, and move on to the next hypothesis.

4.6: The Physical, Human, and Latent Cause of the Problem

Before we begin our investigation, we need to ask ourselves why we need to perform a Root Cause Failure Analysis. Do we simply do it to comply with customer's requirements, such as manufacturing plants? Do we perform a Root Cause Failure Analysis because our management wants to know what happened? Do we perform a Root Cause Failure Analysis because we want to know what caused the problem to occur? The reason to perform a root cause is for them to perform the actual investigation for the right reasons, and there is no better reason I can think of than learning from the failure itself. We need to learn from the things that go wrong. Again, do we learn from failure when we punish someone, or do we just expand the gap between our people and ourselves?

I recall a couple of years back when I was teaching Root Cause Failure Analysis for three days in one plant, and after the end of the first day, one of the participants approached me and said, Rolly, I'm afraid I won't be able to attend the last part of your training tomorrow and when I asked why? The person said that he will be serving his suspension starting tomorrow, which will last for one week. My curiosity aroused, so I asked the person who was a maintenance technician what happen. He told me that the operator committed an error, and both of us will have to serve our suspension starting tomorrow. I asked why he was included if it was the operator who committed the mistake, and he said that it was the policy that management wants. And the worst part was, their names, faces, and mistakes had been highlighted, detailing their mistake committed for everyone to see. It is very humiliating, indeed. By the way, this maintenance has already been in that industry for 15 years, and it was his first taste of punishment in the number of years he worked in that industry.

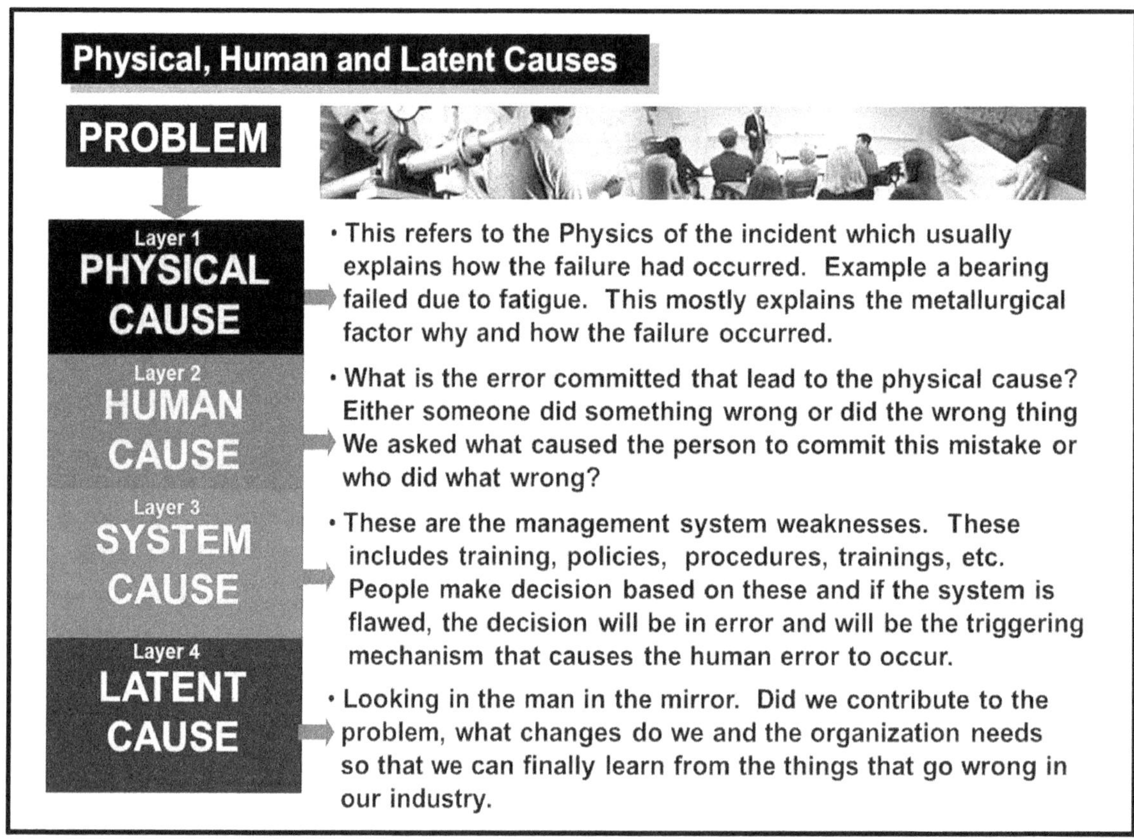

Figure 4.7: Physical, Human, System and Latent Cause

In conducting an RCFA or RCA investigation, the first thing we need to do is uncover the physical evidence. The physical cause refers to the failure analysis or physical reason for why the failure occurred on the equipment. This is also the same as the failure analysis.   The analysis will be up to the component or parts level only. Once the physical cause had been determined, we need to probe further until the human cause had been reached. Once the human cause had been known, we need to probe further until we finally reach the latent cause of the problem.

Physical Cause of the Problem: This is the physical reason why the parts failed. This is the technical explanation of why things broke or failed. This refers to the physics of the incident, which usually explains how the failure had occurred. For example, a bearing failed due to fatigue. This mostly explains the metallurgical factor of why and how the failure occurred. This is the same as Failure Analysis, in which the probe or investigation will end up on the component level or on the part that eventually failed. When we try to fix the equipment, we easily replaced the part that failed and threw it away. The first rule on any evidence-gathering process is collecting and preserving the part that failed and taking photos of the part and equipment. Once an object is identified as evidence, it must be tagged, meaning, it should be placed on a plastic zip lock and label it accordingly. The same goes if there will be metal fragments near the failed part or on the floor. Evidence tagging helps identify the collected item.

Human Cause of the Problem: Both RCA and RCFA believe that all failures are caused by humans. All humans are prone to commit mistakes and errors where either someone did the wrong job or simply did the job wrong. People commit slips and lapses. Believe me, when I tell you that even with the best procedures industry got, people, are likely to commit errors. All people commit mistakes and errors. Errors can be classified into slips and lapses, as discussed in Chapter 6 of this book. Many factors can be attributed to the cause of these errors, such as human fatigue, pressure, environment, inattention to details, which are just to name a few. One thing we must understand is that most human errors are not necessarily the fault of the person who actually committed the error. We need to understand that either the error is caused by external circumstances far beyond their control or by flawed rules and systems that need to be changed. We also need to understand why the error is committed so that we can learn from it. In fact, are we convinced that if we were in the shoes of the person who committed the mistake, or error, would we do it differently under the same circumstances? But the most important lesson is, do we really learn from the failure by blaming and punishing people? Or are we all part of the problem?

The person in charge of people's evidence should summarize the results of the interview before proceeding with the meeting with the principal investigator. Remember that if two or more people have almost identical responses to a particular question, then this is most likely to be considered as a piece of evidence.

System Cause of the Problem: These are the management system weaknesses, including lack of training, outdated policies, and procedures. People make decisions based on these, and if the system or procedure is flawed or outdated, the decision will be in error and the triggering mechanism that causes the mechanical failure to occur. Usually, most system causes are addressed by updating their current or standard operating procedures for the industry, but the root cause should not end here since there is a much and policies deeper cause that eventually leads to the failure to occur unless the latent cause of the problem is not completely exposed and addressed then the problem will still continue to happen and resurface once again. System causes are not yet the latencies but they can be considered as the start of the probe on the latencies.

Latent Cause of the Problem: Underneath every problem lies a deeper cause called Latent Causes. These are the concealed and hidden causes that eventually cause human error to be committed. The only way to address these Latent Causes is to expose them, and we can only expose them if we truly understand what it is all about. Latent Cause is not just about system flaws and procedures that eventually led someone to commit the mistake. It is not just about organizational management weaknesses. It is not just about flawed management decisions, but rather Latent Causes means understanding that we are also part of the problem. Physical, human, and system causes will require corrective action, while latent causes will not require any corrective action. The only way to address a latent cause is for people to change. Learning from the things that go wrong is not as easy as we think it is. The only way we can learn from the things that go wrong is if we can see ourselves in the mirror and admit to the fact that we, too, are also part of the problem. Collectively, we all have our share that eventually caused a person to commit a mistake. The sad thing about this is when this happens, we isolate the person who commits the mistake and put all our fingers on that person and blame them. We always wanted a fall guy. Before we can truly understand the latency behind the problem, we need to ask ourselves the following:

• What is it about the way we are as an organization that contributed to the problem?
• What is it about the way I am that contributed to the problem?

The first question refers to the organizational latencies, while the last question refers to our personal latencies that we need to face and admit. If we are part of the problem, then we must be responsible for being part of the solution too. Engineers, technical people, maintenance can easily arrive at the physical cause of the problem. If some mechanical component such as a bearing failed, these people could dig up evidence that will eventually lead to its physical cause, but digging deeper into the latent cause is the most difficult part as we need to look ourselves in the mirror and admit to the fact that we too are also part of the problem. It means that we need to accept that we are also accountable for why this problem occurred. If you are the management that made that decision that allows the physical and human cause to happen, the question is what prompts or forced us to make this decision. There must be a reason behind it.

Ask ourselves, can each and every one of us be this kind of person? Is it difficult? Latency is not about system causes but rather understanding how we people in a way contribute to the problem since we are taught that we are serving in the industry's best interest. I consider latency to be a higher form of human cause. These latencies serve as a reminder that something needs to be changed within us. There are only two options in this case, either we change or remain the same. Note that dealing with the latencies does not only address the particular physical and human cause but is also possible to address other problems too. Again, exposing these Latent Causes is the only way to understand a true and meaningful Root Cause Failure Analysis. Hence, to answer the question, where do we end our probe on Root Cause Failure Analysis? The answer is when we reach the Latent Cause of the problem.

4.7: RCFA Joke – The Evidence Tells It All

In a faraway distant place in South Africa, there exists a tribe in which the people are free to worship anything. Although these tribes live peacefully by themselves, they seemed to worship anything. Some of these Pagans worship the ants, the stone, the moon, the trees, or anything they can think of. Knowing the situation, the United States (US) decided to send a missionary whose mission is to teach Christianity in this village. His name was Father Joseph. He was warmly welcomed by the tribe and their leader. Let us call him, Chief Korokok. And the story goes that one day the leader of the tribe, Chief Korokok was very upset and looking for Father Joseph. And so the story goes.

Figure 4.8: RCFA Joke – The Evidence Shows It All

4.8: Take Quiz on Physical, Human, System, and Latent Cause of the Problem

1. Failure Analysis indicate strong evidence of fatigue on the shaft
   a) Physical Cause
   b) Human Cause
   c) System Cause
   d) Latent Cause

2. Lack of training causing the operator to commit an error
   a) Physical Cause
   b) Human Cause
   c) System Cause
   d) Latent Cause

3. Management cutting budget on training on their people
   a) Physical Cause
   b) Human Cause
   c) System Cause
   d) Latent Cause

4. Revision of a procedure that had been outdated
   a) Physical Cause
   b) Human Cause
   c) System Cause
   d) Latent Cause

5. Too much torque exerted on the bolt
   a) Physical Cause
   b) Human Cause
   c) System Cause
   d) Latent Cause

6. Control valve fails to open due to solenoid failure
   a) Physical Cause
   b) Human Cause
   c) System Cause
   d) Latent Cause

7. Compromising safety to earn more revenue for the company
   a) Physical Cause
   b) Human Cause
   c) System Cause
   d) Latent Cause

8. Flawed management decision caused the lives of the seven astronauts of the Challenger Disaster in 1986.
   a) Physical Cause

b) Human Cause
c) System Cause
d) Latent Cause

9. Abrasion occurring on a spur gear meshing together
   a) Physical Cause
   b) Human Cause
   c) System Cause
   d) Latent Cause

10. Lack of communication between management and their subordinates
    a) Physical Cause
    b) Human Cause
    c) System Cause
    d) Latent Cause

11. Wrong lubricant being poured into the gearbox
    a) Physical Cause
    b) Human Cause
    c) System Cause
    d) Latent Cause

12. Shaft misalignment caused by soft foot on the motor
    a) Physical Cause
    b) Human Cause
    c) System Cause
    d) Latent Cause

13. Infant mortality failures are usually caused by
    a) Physical Cause
    b) Human Cause
    c) System Cause
    d) Latent Cause

14. The RMS Titanic sunk because they put the engine in full reverse, causing the ship to turn at a slower pace.
    a) Physical Cause
    b) Human Cause
    c) System Cause
    d) Latent Cause

15. Strong evidence of pitting caused by gear overloading
    a) Physical Cause
    b) Human Cause
    c) System Cause
    d) Latent Cause

16. According to an interview with Richard Feynman regarding the Challenger Disaster, he told the media that the engineers and managers are not communicating well.
    a) Physical Cause
    b) Human Cause
    c) System Cause
    d) Latent Cause

17. Cost-cutting schemes and neglect on maintenance is the primary cause of the disaster at Union Carbide at Bhopal India.
    a) Physical Cause
    b) Human Cause
    c) System Cause
    d) Latent Cause

18. The lubrication procedure was revised after maintenance attended training on lubrication making their procedure more detailed.
    a) Physical Cause
    b) Human Cause
    c) System Cause
    d) Latent Cause

19. A flock of birds strikes one of the turbine engines of a 747 commercial plane that caused the life of 350 people and the plane to crash.
    a) Physical Cause
    b) Human Cause
    c) System Cause
    d) Latent Cause

20. 470 people can still be saved if the officers and crew of the RMS Titanic knew the capacity of their lifeboat.
    a) Physical Cause
    b) Human Cause
    c) System Cause
    d) Latent Cause

Refer to Appendix A for the Answers

Chapter 5

Understanding the Physical Cause of Failures

There is no question of having highly skilled people in our maintenance organization. Technically speaking, the goal of MTTR is to reduce repair time, but the real goal of maintenance is to analyze failures if they continue to repeat themselves. Therefore, when the failure keeps on repeating, we need to slow down and start addressing the root cause of the problem.

5.1: Understanding the Process of Wear

It is common to hear our maintenance people say they have replaced the part they assumed to have worn out in their equipment. If the boss asks why the part failed? The most common and saturated root cause the maintenance will provide is wear and tear. First, let us define the process of what wear is all about. Wear is defined as damage to any solid surface caused by the removal or displacement of material caused by some means of mechanical action of either a contacting solid, liquid, or gas, causing significant surface damage. The damage is usually thought of as gradual deterioration. It means that it can be solid to solid mechanical item that will cause wear or a continuous bombarded of either a gas or fluid to a solid object that will cause the object to directly wear out for as long as the bombardment is continuous, just like in the case of a continuous bombardment of a fluid on a pump impeller. Wear may also mean the undesired removal of material from any contacting surfaces through some means of movement or mechanical action. This means that several parts and spares have some movements. These mechanical parts can either slide up and down, left and right cycles or rotate inside our equipment. Any part, item, or spare that moves inside our equipment, machines, or assets will be subject to wear and degradation after a definite amount of period.

Although TPM (Total Productive Maintenance) books claim that there are two types of wear, including accelerated deterioration and natural deterioration. Accelerated deterioration means that the part, item, or spare had failed and has not reached its lifespan. We can state that we have a case of premature wear. The latter means that the part has reached its lifespan and eventually gradually wears out. Suppose we purchase a bearing from a given vendor; in that case, the manufacturer has a projected life associated with the bearing, which is given as L_{10}. Let us assume the L_{10} life of the bearing is 5 years. This means that the reliability of the bearing is given at 90% under a given load and speed with 10% chances that the bearing will fail under normal load. However, this L_{10} life of the bearing is theoretical in which it is assumed that everything is in perfect condition. This means that if the load, RPM, speed, lubrication, and environment satisfy the requirements of the bearing, then the bearing will only fail on one condition, which will be a factor of fatigue. This means that we are speaking about an ideal condition of the bearing. Still, in reality, that is not the case in industries. So the end result is that the bearing will fail prematurely before reaching its dictated life because there will be more than a dozen ways that a bearing can actually fail.

Every movement that a particular item, part, or spare will cause stress on the part after some time. If I explain this wear process in the simplest way possible, I can speak about a boxing match where the opponent is 1 foot taller and his reach is much longer. A tactic that will be used by the taller boxer will be to stick and jab his opponent. As the round progresses from the middle rounds 5, 6, 7, and so on, the opponent's face will begin to puff and become swollen since the jabs continuously hit the face. If we talk about mechanical parts inside our equipment and assets, the number of jabs that hit the face represents the number of cycles or movements. Each of these cycles or movements contributes to a small incremental amount of stress. There can only be an n amount of stress that the mechanical part can tolerate before the stress finally wins and exceeds the strength of the material. In any failure analysis involving metal fracture, the subject of both metal strength and metal stress must be considered because they cannot be separated or divorced from one another (I think the Philippines is the only country in Asia that does not allow divorce). This means that as long as there is stress, wear will occur until the material's strength is exceeded. Once this happens, the process of wear has been completed. By definition, stress is the force per unit area, often considered the force acting through a small area within a plane. Strength in the metallurgical sense is the metal part's property to resist the stress imposed upon the part.

Therefore, all fractures will be caused by stress, a version of the weakest link theory, which applies that a fracture will occur if the local stress finally exceeds its strength of materials.

I have several books on the subject of metal fracture and metallurgy in my home. Each time I intend to absorb this message, it tries to confuse me more than I can comprehend regarding the subject of wear. I seek help from metallurgical experts on the subject matter, yet their explanation was so technical and profound that it was simply too complicated for me to comprehend. I almost gave up on this subject on how the wear process took place until I visited my old-time friend and sensei Mang Tibo. I asked him his thoughts about the wear process. I have so many questions to ask. He told me to calm down and gave me a glass of water. I asked him how the wear process occurred in the equipment parts and components and what this message means that all fractures are caused by stress. If the local stress exceeds the material's strength, then wear takes place. Mang Tibo is like a father to me. He was my earliest mentor on maintenance. He explains things in a very simple way, where even a small kid can comprehend. Mang Tibo lives in a poor neighborhood and is not certified in maintenance, but oh boy, he knows his stuff.

Mang Tibo has 8 children where all of them have already grown up. All his children are already married and have their own families to live with. He once told me that there were many times in his life that his children convinced him to move up with them so that he could live a better life and live the comforts of life for the remainder of his days, but the old man was simply too stubborn and refused them all the time. He preferred to stay in their small home, which he and his family once stayed in for more than 50 years. He had set aside his ambition for a good life, tried to live with his principles, and would rather settle down in their old home with his loving wife. Mang Tibo invited me inside his house. His house had no furniture and just some wooden stool that he made a long time ago. His house had no ceiling, and you can see so many holes in the roof when you look up. I can just imagine his life and be toasted mostly during summertime in our country, the Philippines. Mang Tibo never finished school because when his parents died when he was 15 years old, he was left responsible for taking care of his seven brothers, so he started working early. He was the bread earner during that time. He invited me to sit down (just like in the days of the La Cosa Nostra and the Mafia Bosses) and gave me some biscuits as we discussed his humble version of the wear process. As we sat down, he told me that he had stayed in this house for more than 50 years, as far as he can reminisce. He told me to look at the floor and asked me what I saw. I said that I could see a lot of small dents and fractures on the floor. The floor in his house was just cemented with no tiles. He told me that it used to contain no fractures when their house was new and asked me what I thought would have caused these tiny bits of fracture on the floor. As I tried to think of an answer, I looked up to think, which is a common gesture when people think. They look up in the sky, not to look for birds or superman but just to think. As I looked up, I saw these tiny holes in the roof since his house had no ceiling at all, and each of the holes in the roof was located perpendicular to where the fracture was located on the floor, and it gave me an idea as to what caused the damage on the floor. It was definitely the leak caused by the rain throughout the years. You see, in the Philippines, there are only two weather seasons: the summer months and the rainy season. I told him it was the leak caused by the rain that damaged the floor. He smiled and said that I was correct. He said that when this floor was new, cemented, and shiny fifty years ago, although he did not place tiles on the floor. The floor has its own strength of materials. Each drop of rain that fell on the floor represents one stress, and each stress continues to drop drip by drip into the floor for fifty years during the rainy seasons. A few drops definitely will do nothing and will not damage the floor. Then as the days passed by, a hundred drops, a thousand drops, and as the year passed by, a million drops, a billion drops, he said that although he never counted the drip, which was understandably foolish. All he knew was that the drop of rainwater continued until the stress had ultimately exceeded the floor's strength that eventually caused the fracture and damage on the floor. That was when I finally understood how the wear process occurs. Mang Tibo does not have any Master's, Ph.D. degrees, or any maintenance certification. Neither is he a metallurgist, nor did he finish his schooling, but his experience tells it all, and that is all I have to know about the process of wear. You don't need to be a rocket scientist to know that; all you need is just a modicum of common sense.

Our equipment is not a plug-and-play appliance, just like a TV set that, when you plug it in, it will run even when you fall asleep while watching your favorite Netflix movie or sitcom show without any sort of problem. Inside our equipment are many parts composing of electrical parts, mechanical parts, electronic parts, and electromechanical parts. Some of these parts, especially mechanical parts, have movements, and every single movement that these parts make is equivalent to one stress, which is also one complete cycle. A piston, for example, in a diesel generator set motion, will move on an upward and downward movement. This up and down movement will stress the piston out after (n) several cycles. This simply means that as the number of improvements increases, there will be tiny bits of metal fragments that will chip off on the metal. It will not last forever. It can last longer once we increase the strength of the material to a stronger one, but again, a new problem will arise as the stronger the material's strength, the more the material is prone to brittleness. A piston that moves from an upward and downward motion would be subject to stress after an (n) number of movements. When the equipment is not loaded or is at rest, the piston remains on the centerline. When we start the equipment, the piston will move from the centerline to an upward motion, back to the centerline, then from a downward motion, and once again back to the centerline, which is equivalent to one stress or one cycle. In vibration analysis, this is simply one sign wave.

One movement is called one cycle or one stress. After hundreds of thousands of upward and downward movements, tiny metal debris is removed from the piston and the ring, and the engine's blowby increases. This is now the start of the wear process. Once the wear process has reached its limit, the equipment can no longer continue its operation where the maintenance people intervened and finally replaced the part that had actually worn out. This is the completion of the wear-out process. The process of wear does not happen overnight or even in a week. It is a gradual process, just like when we were kids, up to this point in time where we have grown up and get wrinkles on our face.

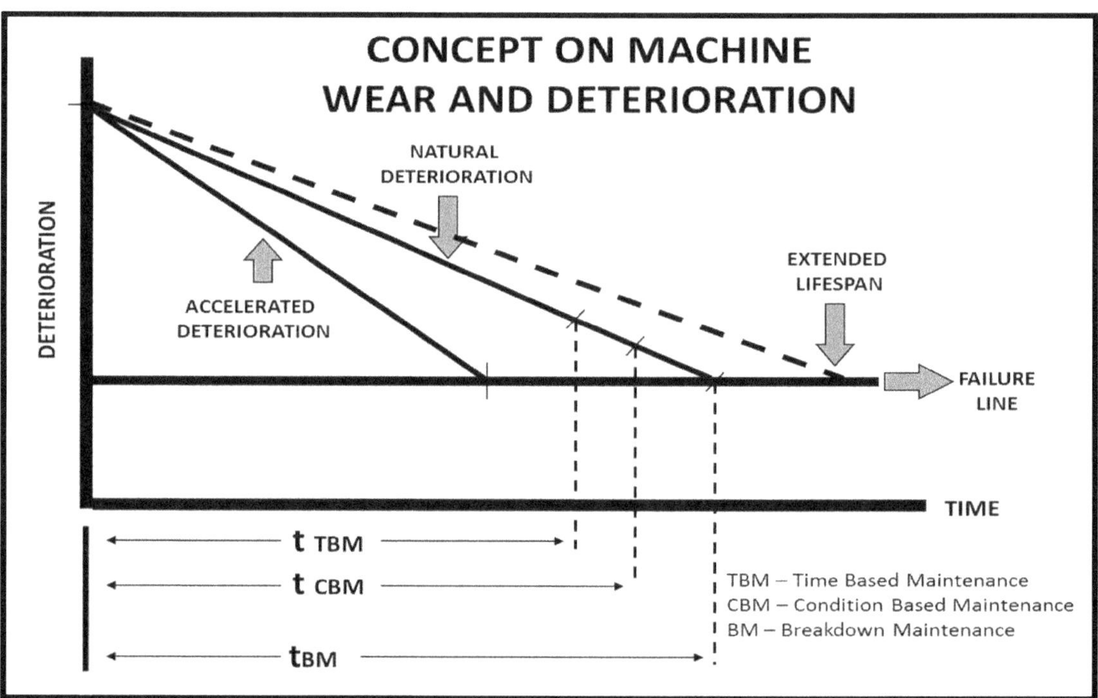

Figure 5.1: Concept of Machine Wear and Deterioration

Wear is a big problem for mankind; on the other hand, wear is also good because this has provided employment to billions and billions of people worldwide. If we try to speculate for a while and think. What if wear does not exist at all? What if God comes down to earth and asks us if we want to remove the process of wear on everything? What will you answer? I think there will be chaos and disorder, and many people will stop breathing and die. If wear no longer takes place, we don't need to buy any pair of shoes because our shoes will last a lifetime, and the entire shoe industry in the world will close down as a result, and its entire people will be unemployed. Because wear will no longer take place, then your hair will no longer grow, and your barber will no longer have work. You will no longer suffer any tooth decay, and that goes for your dentist too. They will close down their clinic. These people will lose their work and result in other means just to survive. All factories that provide replacement such as tires, clothing, brakes, clutches, bricks, food, drugs, and much more will close down, and billions and billions of people from all around the globe will be laid out of work and become unemployed, not because of this **covid 19** pandemic, but because the wear process was eliminated. Therefore, let us thank God that there is wear in this world and so to conclude, always remember that wherever you are or wherever you go, wear will always be with you

Many experts say that everything that is man-made will eventually wear out; well, I tend to disagree with that. There is one thing that will never wear out, and that is the music of the Rolling Stones. Just would like to mention, Rest in Peace, Charlie Watts (Rolling Stones drummer) who passed away last August 24, 2021, at the age of 80. Oh Yeah. ☺.

5.2: Most Common Types of Mechanical Wear Explained

When a part or spare is replaced in our equipment, and the boss asks for the reason why maintenance replaced it, maintenance most common response will be the part worn-out, wear and tear, or the part had deteriorated, but if the boss asks us a bit further such as what type of wear do you think it is, most maintenance will simply scratch their head. I believe that it is important for every maintenance in industries to associate themselves and be familiar with the different types of wear and how each of these wear actually looks like so that they can have some idea on how the wear process is actually taking place and what is really happening inside our assets, equipment, and machines.

According to Dr. Ernest Rabinowicz, equipment may lose its usefulness because of three factors. First, the equipment becomes obsolete, where the equipment or the product it is producing will no longer be used, and the equipment will therefore be retired or decommissioned. This consists of around 15% of why equipment can no longer be used. The second factor is accidents where the equipment encounters a major breakdown or failure. The cost of rebuilding or repairing the equipment will almost equal to purchasing new equipment, which constitutes around 15% of why equipment loses its usefulness. But the bigger percentage of why equipment loses its usefulness is due to surface degradation, which constitutes around 70% of why equipment loses its usefulness. And under surface degradation, 20% will be due to corrosion, and about 50% will be caused by mechanical wear. In this regard, mechanical wear will be caused by abrasion, adhesion, erosion, or fatigue. Dr. E. Rabinowicz states that surface degradation cannot be eliminated and is inevitable, but the good news is that the process of surface degradation can be controlled, or it can be prolonged. What is important for our dear maintenance function is distinguishing between the different types of wear, how they look on parts, their markings, and, most importantly, what causes them to occur on the mechanical part. With this kind of knowledge and information, our maintenance can understand how the wear process happens, how to prolong it, and what should be done to control it.

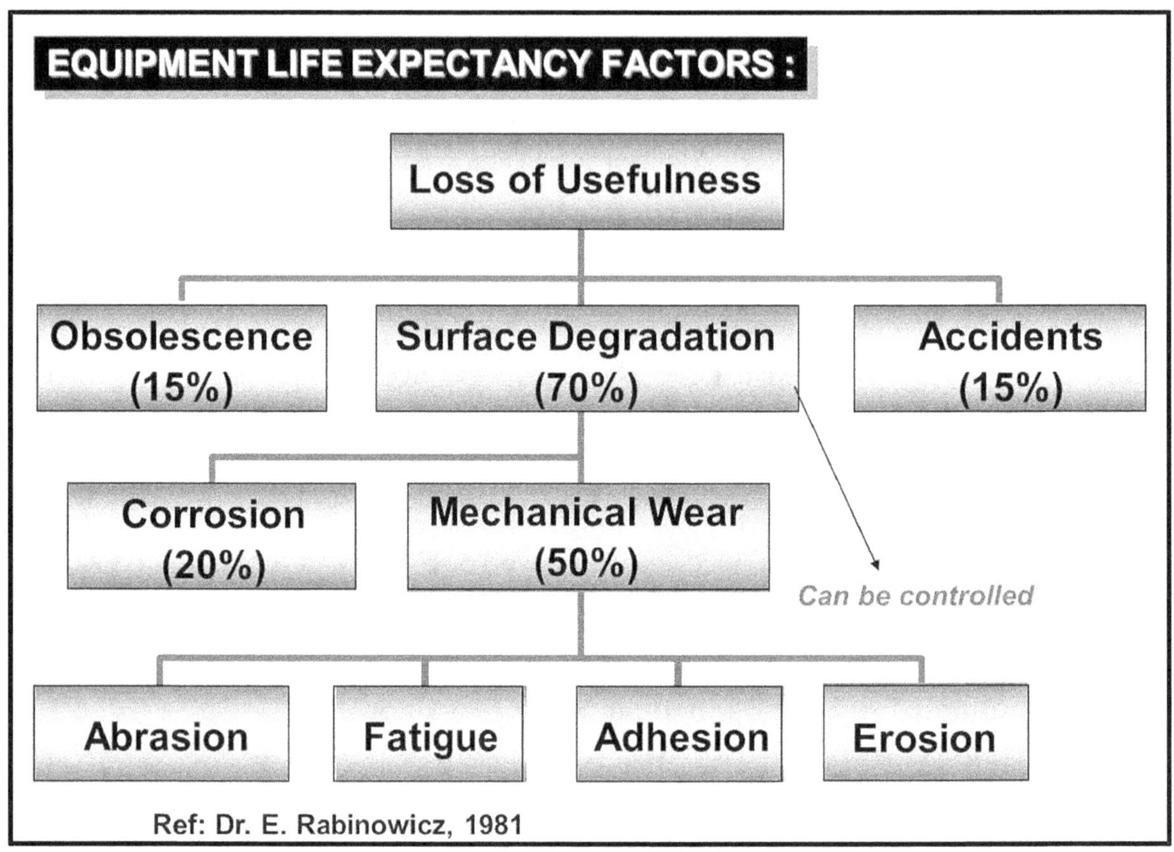

Figure 5.2: Why Equipment Can No Longer Be Used

Our equipment is not like a television set or a plug-and-play instrument. When you plug it, then it will run continuously, or when you click on the remote control of your TV set and watch your favorite sitcom, you fall asleep, and when you wake up, the TV is still turn on, although the channels may have already run out and what you see is a snowy screen. Remember that your equipment that runs in your plant contains mechanical parts such as a gear meshing around with one another, a piston moving up and down, mechanical components that run either horizontally or vertically, separated by a very small clearance in mils or microns. Each movement, whether it is moving vertically, which is in an upward, then back to the centerline, then a downward motion, and then back again to the centerline, will be equivalent to one stress or one complete cycle. A 360-degree revolution of a bearing or rotating equipment will also be equivalent to one stress or cycle. This means that after reaching (n) times the number of stress, the material's strength will deteriorate, and the stress, in the end, wins over its strength.

Let us say that we have a given die punch that moves on an up and down movement and comes in direct contact with the product; the die punch may be made of a hard alloy. Each up and down movement will represent one stress, and so on. Hence, after reaching 1000 movements or cycles, the strength of the material of the punch remains the same and unchallenged, but this small incremental stress or movement will take its toll once the punch reaches 100,000 movements. Stress will now take its toll on the punch, and whenever the punch contacts the product every time it moves, then it will start to wear out, which means that small fragments of metal from the punch tear up. After perhaps 200,000 up and down movements of the punch, then it can no longer provide the correct measurement on the product and will now have an effect on the quality of the product, hence; therefore, it is now time for the maintenance to stop the equipment and replace the punch because it already had worn out but in the metallurgical and technical sense, the punch stress out after reaching its nth number of cycles. The stress in the end, which is just merely an incremental movement, had finally won the battle against the punch's material's strength. I hope you can follow my explanation since I cannot explain it to you simpler than that. The following are the most common types of wear.

[4]**Abrasive Wear:** This type of wear can be categorized by a single keyword, which is cutting. This type of wear occurs when hard particles are suspended in a fluid or projections from one surface roll or slide under pressure against another surface, cutting the opposite surface. This wear mechanism is basically the same as machining, grinding, polishing, or lapping for shaping materials. When you go to a machine shop, and a lathe machine is in operation, cutting a shaft produces long metal debris pieces as the cutting tool comes in contact with the metal; this is actually how abrasion occurs.

If we get a mechanical part and placed it on a Scanning Electron Microscope (SEM), and enhance the magnification to 1000 times. We noticed that the surface of the part is not actually perfectly smooth. Meaning that there are micro or very tiny peaks and valley the surrounds the mechanical parts. These peaks and valleys are usually called asperities. When the peaks and valleys of two opposite mechanical parts come in contact with one another, they will fracture or tear off. These small fragments of metal removed will flow together with the lubricating oil and end up in tight clearances and start the process of abrasion. In layman's terms, if you are involved in Preventive Maintenance overhauls and replacements, you see several spares or mechanical parts and items with hairline scratches, then they are caused by abrasion.

There are two types of abrasion: **two-body abrasion** occurs when the lower portion of the mechanical component's peak or asperity comes in contact with the upper component's peak or asperity since they move in opposite directions. The weaker of the two asperities will break off, creating a third party, which later on can become the other type of abrasion, which is a **three-body abrasion**. In a three-body type of abrasion, a contaminant or particle usually as hard as or harder than the two mechanical components gets trapped in the two mechanical components' clearance. Suppose the contaminant gets trapped in the upper portion of a mechanical part; it will cut or rub the lower portion of the mechanical part. Signs or distinguishing marks from abrasion will be some hairline scratches on the component or part. There are two ways to control abrasion. The first is to harden the mechanical part since the principle for abrasion to occur is that the contaminant or particle trapped should be as hard or harder than the mechanical part itself. But making the material harder will create a secondary problem, which will make the material more prone to a brittle fracture. A brittle fracture is usually a fracture with little or no deformation. Another way of controlling abrasion is to remove the contaminants themselves through the oil by applying high efficiency and beta rating type of oil filtration in which the filter can guarantee to remove more than 99% of the contaminants present in the oil. The higher the beta rating of filters in lubricating hydraulic or engine systems, the more solid contaminants it can remove. The cleaner the oil, the longer the wear process to occur in the

---

[4] Pall Industrial Hydraulics Co., *Contamination Control and Fundamentals*, (New York, 1994), Page 6-9

mechanical part.

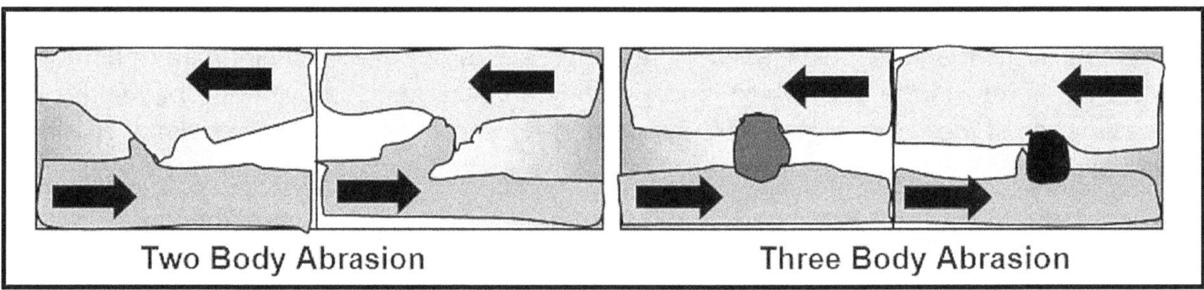

Figure 5.3: Two-Body and Three-Body Abrasive Wear

Adhesive Wear: This type of wear occurs when a peak or unevenness from one surface comes in contact with the other surface's peak or asperity. Adhesive wear has been commonly identified by the terms galling or seizing. There may be instantaneous micro-welding due to the friction or heat involved due to the loss of film on the lubricant. For adhesive wear to occur, there must be a metal-to-metal contact. When the metal comes in contact with each other, friction occurs, and heat is generated. There will be some form of discoloration that occurs in the process. Micro-welding occurs, and the final stage will be a metal breakage or fracture on the mechanical component. Adhesive wear is produced by the formation and subsequent shearing of welded junctions between two sliding surfaces. For adhesive wear to occur, surfaces must be in intimate contact with each other. One way of controlling adhesive wear is by applying the oil's correct viscosity on the mechanical component. Surfaces held apart by very thin lubricating and oxide films will reduce the tendency for adhesion to occur. Adhesive wear starts out on a small scale but rapidly escalates as the two sides alternate weld and tear metal from each other's surfaces. That is what adhesive wear is all about, microscopic welding. The heat produced at the contact interface is very high which is near the two metals' melting points touching each other. Although it started as adhesive wear, these small metal fragments that break off can also become abrasive wear once these metal fragments get trapped in tight clearances between mechanical parts.

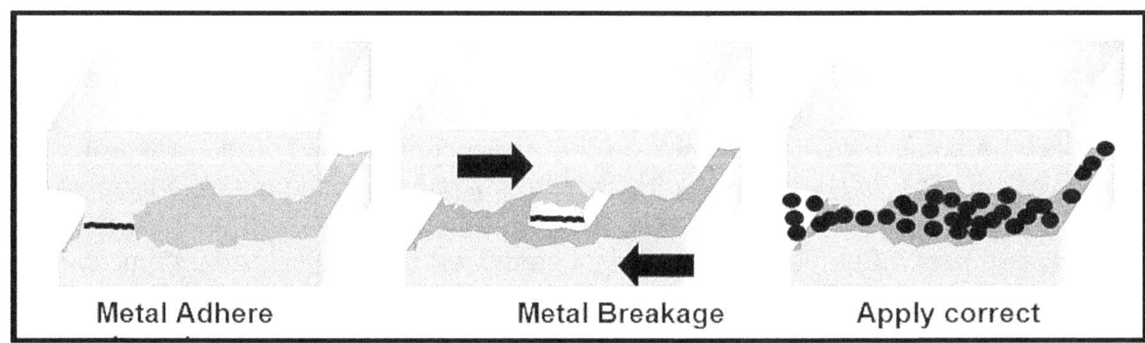

Figure 5.4: How Metal Adhesion Takes Place

Erosive Wear: This type of wear happens due to mechanical interaction between a surface and a fluid, a multi-component fluid, or impinging liquid or solid particles. Erosive wear occurs by the impact of particles upon a surface with a resultant material loss due to fracture. This can occur both on metals and ceramics. The continuous bombardment of particles, either solid, fluid, or air, contacting the surface cuts a tiny particle from the component over time. This tiny particle that is removed is insignificant at the moment but, over some time, will have its impact. Erosive wear can be expected to occur in metal parts and assemblies where the above conditions are present. Common problem areas found are pumps and impellers, fans, steam lines, nozzles, inside of sharp bends in pipes and tubes, and sand shot blasting equipment. Sudden sharp curves or bends cause more erosion problems than gentle curves in tubes and pipes. One way to control erosion is to strengthen the material's hardness, which can prolong the erosion process, but again can subject the metal more prone to a brittle fracture. When a fluid continuously bombards the pump's impeller, the pump's flow rate will be reduced after a given period. This is because the impeller tip will start to erode, which results in an increase in the clearance size of the impeller and the volute casing inside the pump. When this happens, internal leakage occurs where not all of the fluid is pumped out by the impeller but rather returns to the pump. The end result is that the flow rate will drop. When the reduction in the flow rate is no longer acceptable to production, maintenance will schedule to replace the impeller.

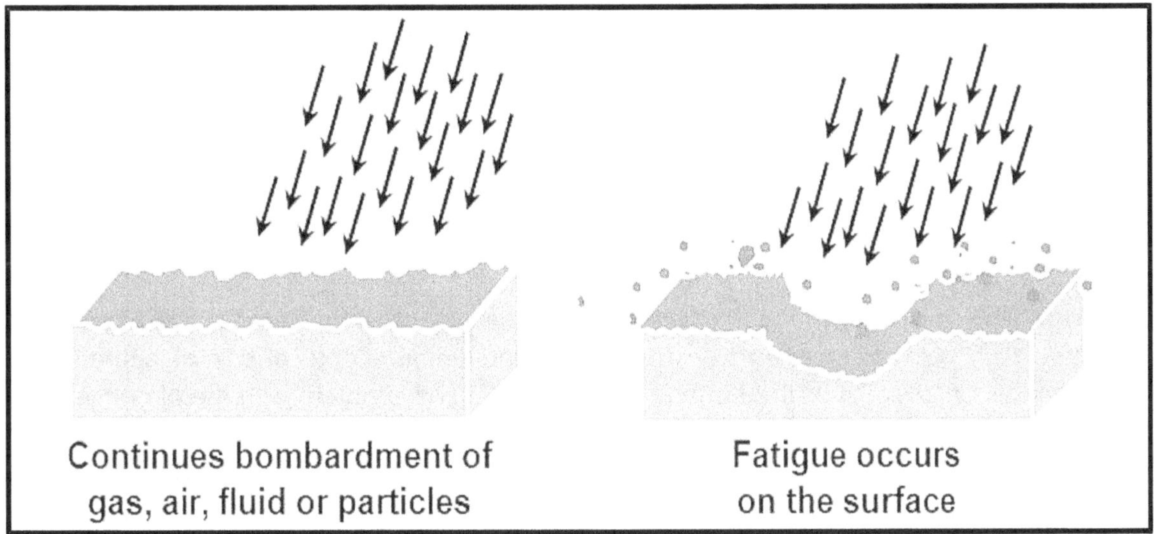

Figure 5.5: The Process of Erosion

Fatigue Wear: Fatigue wear results from continuous cyclic slip under repetitive load applications for hundreds of thousands or millions of load cycles. These removed fragments caused abrasive wear and other damage when carried by the lubricant to other parts of the mechanism. Some fractures will start as microscopic cavities and may stay microscopic if the equipment is no longer loaded. Still, it will start off as a microscopic fracture most of the time but gradually become larger after continuous load. The last stage is a large fracture or crack. Fatigue wear is also a phenomenon leading to fracture under repeated or fluctuating stress having a maximum value less than the material's tensile strength. Fatigue fractures are progressive, which can start as a micro-crack and propagates and grow into larger ones that cause the destruction of mechanical components. For fatigue wear to happen, a contaminant or particle is caught between clearance, and the surfaces are dented, and crack will be initiated. After N number of cycles, the particle trapped is released, and the crack starts to propagate. The last stage will be a fracture of the component. Again fatigue wear can also result in abrasive wear if these micro fractures will get trapped between clearances on mechanical parts.

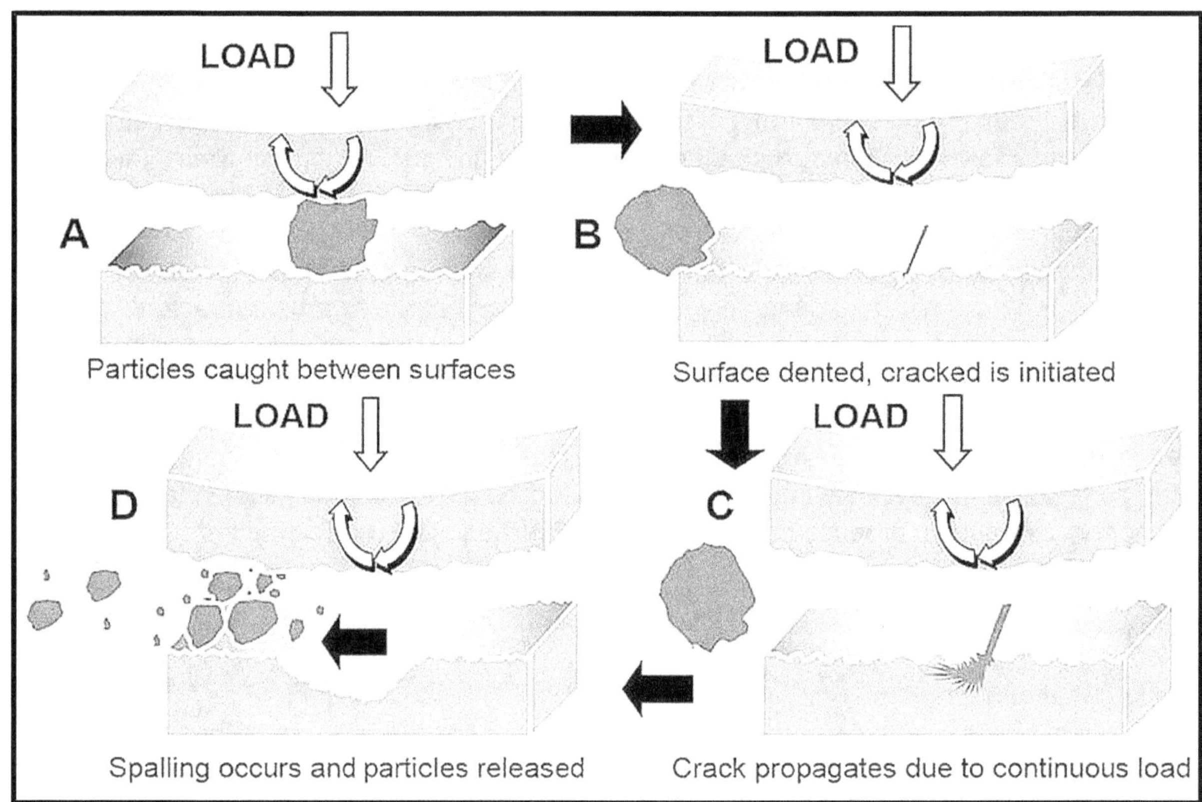

Figure 5.6: The Process of Fatigue Cycle

Corrosion: This is a wear process of materials caused by chemical, electrochemical, or other reactions. [5]It is also defined as the destructive and unintentional attack of a metal. It is electrochemical and begins at the surface of the metal. The problem with metal corrosion is of significant proportions. It is estimated that around 5% of the industry's income is spent on preventing corrosion as well as the maintenance and replacements due to corrosive reactions. There is an actual material loss in metals through dissolution, which is corrosion by forming a non-metallic scale or film. The main difference between corrosion and rust is that corrosion occurs due to the chemical reaction on metal surfaces, while rust only affects iron metals that are exposed to air or moisture.

Corrosion is when the metal is converted to a more stable form, such as its oxide, hydroxide, or sulfide state, leading to its gradual deterioration. This usually happens when the metal is exposed to oxygen and moisture, creating red iron oxides, commonly called rust. For parts under lubrication, corrosion is often caused by the depletion of the additive called corrosion inhibitors or rust inhibitors, either due to the extended use of the oil or the oil containing too many contaminants and moisture content. Although corrosion is not catastrophic from a safety standpoint, it can be disastrous if it fractures metal components, especially in vessels that contain hazardous chemicals. Corrosion is visible as surface deterioration caused by the chemical action of active ingredients in the lubricant. It includes acid, moisture, foreign materials, and extreme-pressure additives. During operation, the oil breaks down and allows

---

[5] Callister Jr., William, **Materials Science and Engineering, An Introduction**, (John Wiley and Sons Inc, 1985), Page 563

corrosive oil elements to attack, as in the case of gear contact surfaces. When a material is subject to corrosion, it loses its material strength, hardness, and toughness. A corrosive attack on a metallic structure can cause the formation of small cracks, dents, and pitting, which further leads to deeper cracks over time due to tensile stress acting on the edges of the cracks. These situations can cause catastrophic failure without significant deformation of the structure.

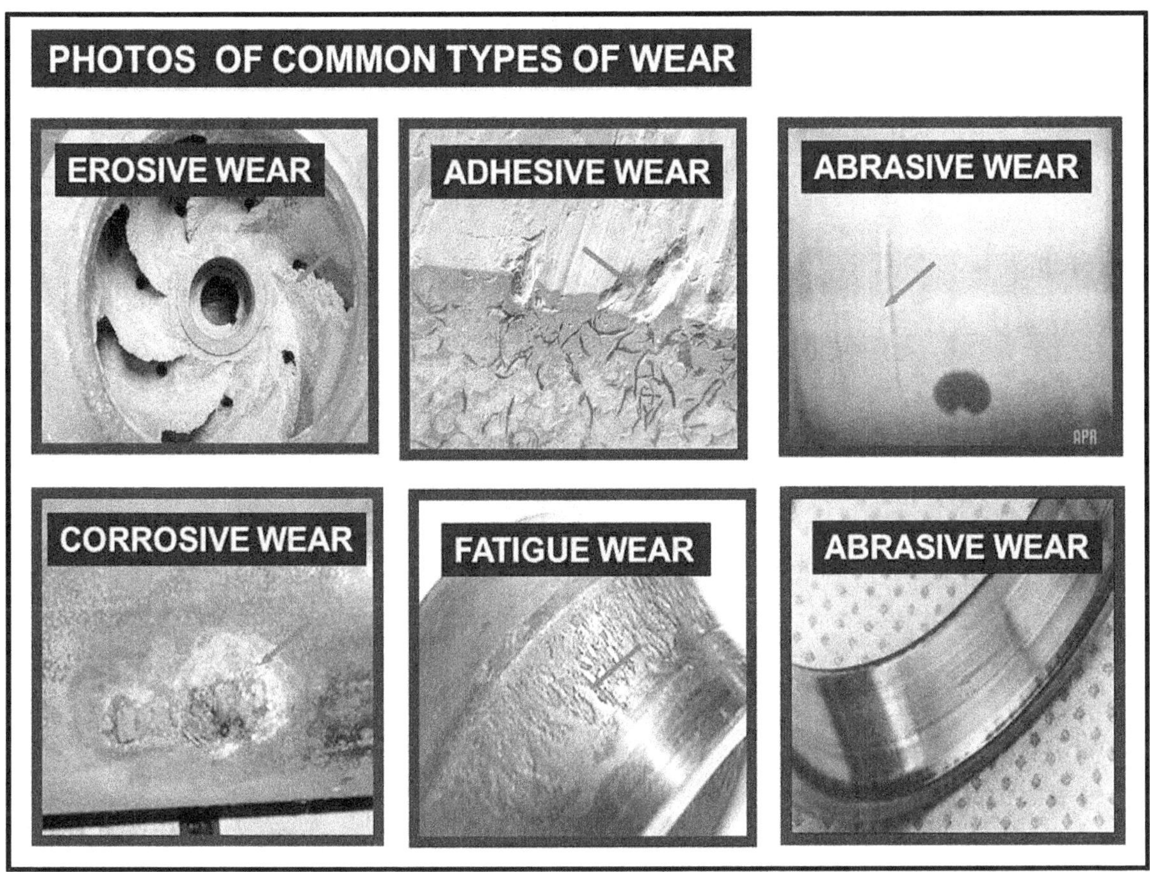

Figure 5.7: Photos of the Different Types of Wear

Friction is simply defined as the resistance to its relative motion. It is the force that resists motion. Wear is the loss of materials caused by friction. Lubrication is the use of fluid in solid materials to minimize or reduce friction and delay the process of wear. The subject of tribology simply tells us that the energy loss due to friction cannot be recovered and is considered lost forever; therefore, the study and application of tribology can prepare us for designing and maintaining better equipment to maximize the longevity and lifespan of its parts and components. In the report by Peter Jost, the founder of Tribology, friction wear, corrosion, and lubrication cost the United Kingdom over 500 million pounds per year or roughly around 639,749,500.00 USD. He reasoned out that significant education and research are required to reduce these losses by 30% of its primary energy consumed by friction throughout the world, while 60% of the machine failures are due to premature wear.

In conducting failure analysis, if we can just identify the type of wear, then we have already addressed 50% of the process. The next 50% will be answering the why and how it occurred on the mechanical component so we can conclude the failure analysis on the failed component or part that wear. To continue the Root Cause Failure Analysis investigation is to probe further until we reached the human and the problem's latent cause.

SUMMARIZING WEAR :

WEAR	PRIMARY CAUSE	HOW TO CONTROL	CAN OCCUR ON
Abrasive	• Particles between two adjacent moving surface Also known as cutting wear, contamination	• Increase hardness of the material • Upgrade filtration by using high efficiency and beta rating	• Rod cylinder wear • Pump Gear, vane, piston • 2 metals with clearances • Bearing raceways
Adhesive	• Loss of surface film • Surface to surface contact due to lack of lubricant which cause metal bonding.	• Application of correct oil viscosity film. • Improve metal surface smoothness or apply coating	• Any two metal bonding together due to the loss of film from the lubrication
Erosive	• Continuous bombard-ment of particles and at high fluid velocity.	• Use of clearance protection protection filter • Minimizing 90 degree pipe to reduce erosion	• High velocity flow on servo, proportional valves, control valves, pump impellers
Fatigue	• Particles damage since surface is subjected to repeated stress and ends with a fracture.	• Control contamination • Correct loading • Reducing machine vibration	• Bearing surfaces • Gear wear • Tooling, dies and punches

Figure 5.8: Summary of Wear

5.3: Different Types of Mechanical Stress

As discussed in Section 5.1, stress is the force per unit area, often considered the force acting through a small area within a plane. Strength in the metallurgical sense is the metal part's property to resist the stress imposed upon the part. These stresses are the forces that can be applied to the mechanical part. Different types of forces can be applied to a mechanical object, which can either be tensile, compressive, torsion, shear, or bending forces. In layman's terms, just like in boxing, the boxer can give his opponents different punches like a jab, straight punch, hook, and uppercut. The mathematical calculations for each of these mechanical stress will be beyond the coverage of this book. The author intends to explain the different forces that can happen to a mechanical component when subject to stress. The result of the forces applied on the mechanical part is called the strain. Strain is the quantity that describes the amount of deformation that can occur within a material body whenever a load is applied. The stress-strain relationship varies depending on the material used, and mechanical stress applied. A moment represents the effectiveness of either a rotating, bending or twisting force upon an object. All materials have their respective hardness. The hardness of the material is the measure of the material's resistance to plastic deformation. The most common method to determine the hardness of the material is the Rockwell Hardness Test, which can test the hardness of metals and alloys. With this method, a hardness number is determined by the difference in the penetration resulting from applying an initial minor load followed by a major larger load. Stress (S) = Force (F) / Area (A)

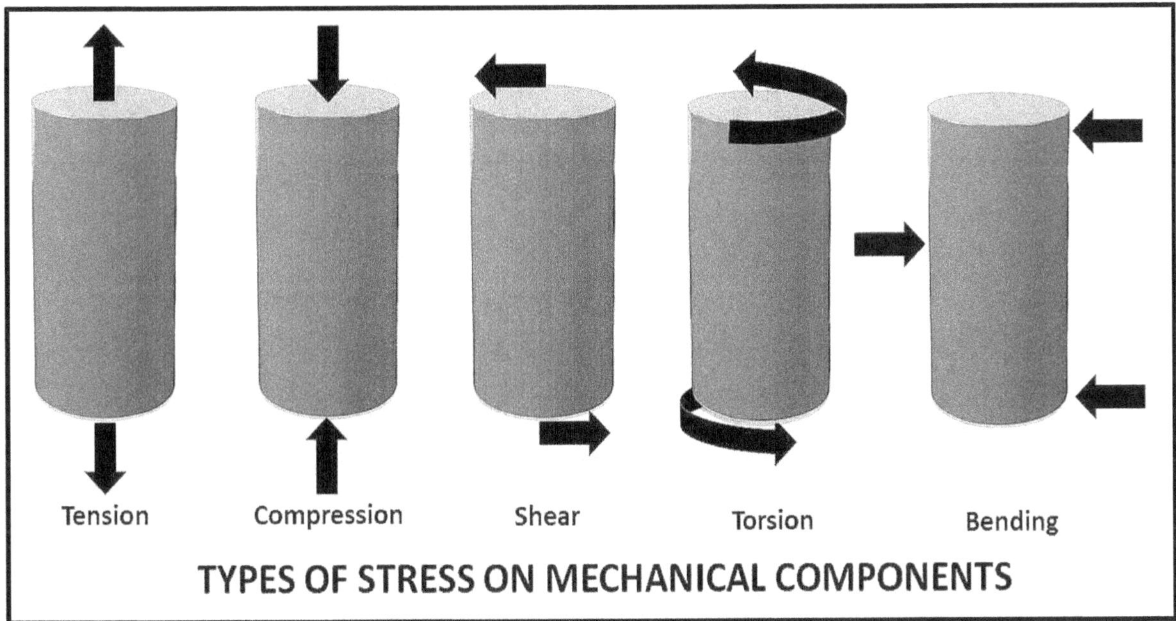

Figure 5.9: Different Types of Stress for Mechanical Components

Tensile Stress: When you have an object in a vertical position such as a cylinder, and an upward force is applied to an object, there will always be an opposite force which will be equal to the downward force. Both the upward and downward forces will cause the object to be under tensile stress. Both the downward and upward forces will extend toward the object. This means that when we pull or stretch an object, that object will be under tensile stress. The strain will be the effect of the change of the object, which can lead to a ductile fracture if an elongation happens on the object due to heat, or it can create a brittle fracture if excessive force is applied that results in no deformation of an object. A good example of tensile stress can be seen when you tune a guitar in which you are stretching the strings until it reaches the correct note. If the tensile strength of the guitar string had been exceeded, the string would break. The tensile stress (or tension) applied can lead to an expansion of the object. We can also experience tensile force if we stretch a rubber band. If we get a clay and form a shape of a cylinder, place it horizontally on the surface, and stretch it gently with both our hands, the clay elongates; as we apply more force, it will elongate more until it reaches its limit and breaks in half.

Compression: When we applied a downward force on an object, the object will have an equal upward force, which is equivalent to the normal force. Once we increase the downward force, then we also increase the upward force. These two opposing forces will cause the object to compress. A good example is placing a hardbound book (let us say the book contains 500 pages) on a table surface. If we place two additional books on top of each other, a downward force is applied to the initial book. This means that the book on the bottom of the surface is under compressive stress. The more books we add, the more compressive stress is applied. Generally, it will take much more force to compress an object compared to pulling an object.

When we talked about tensile and compressive stress, we refer to the maximum stress applied to a material. Once we exceed the maximum stress, the result will be a fracture or breakage that will ultimately lead to the failure of the mechanical part. The strain will be the change of shape of the object after compressive or tensile stress is applied, which can also be equal to the Modulus of Elasticity (E), which is equal to the stress divided by the strain of the object. For tensile and compressive stress, the force applied is perpendicular or at 90 degrees to the area.

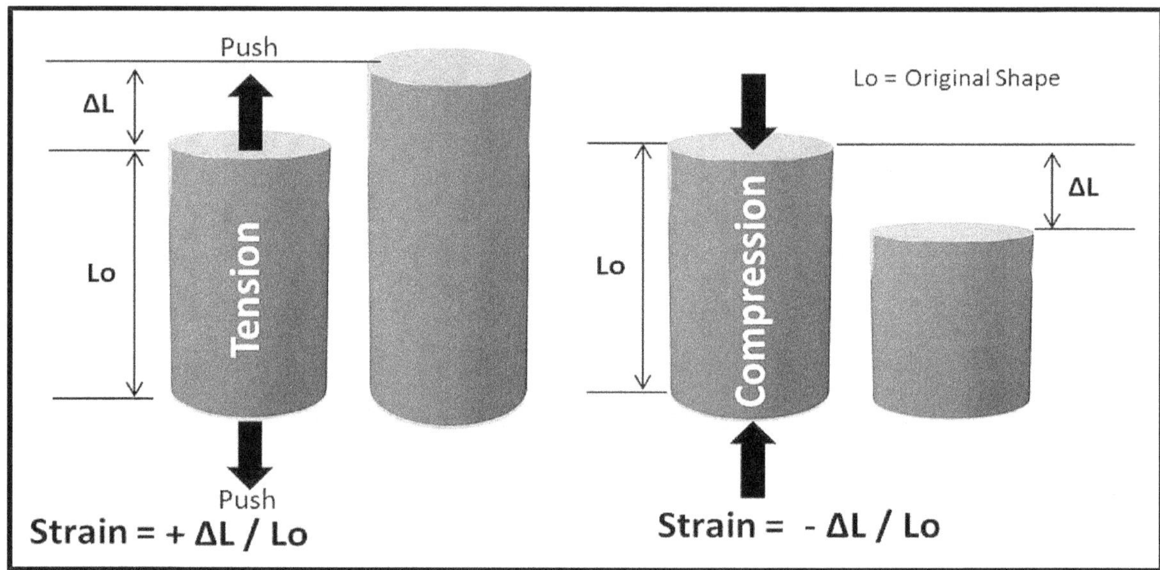

Figure 5.10: Compressive and Tensile Stress

Shear Stress: Also called tangential stress. Unlike both tensile and compressive stress, the force applied is not perpendicular but parallel to the area. Once a force is applied in parallel in opposite directions, it can cause the object to deform. Shear stress is a force acting parallel to a surface or to a planar cross-section of an object. The shear force applied can be from an upward to a downward motion, like when we cut a loaf of bread, or from the left and right motion. Sliding is also a form of shear force. The word shear means that a force can either cut or shear through the surface or object under strain. Shear force is the sum of the effect of shear stress over a surface that results in a shear strain. The important thing to consider is that the force acting on an object is parallel. A ball or roller element sliding on the outer raceway of the bearing is a good example of where shear stress occurs.

Torsional Stress is the twisting of an object caused by force acting on the object's longitudinal axis. The twisting effect is known as the torsion. Torsional stress can lead to deformation. The twisting is called torque which is also called the twisting moment. The most common examples of torsional stress can be seen on shafts when it rotates. Non-cylindrical objects subject to torsional stress will produce warping. When we apply torsional stress on the shaft. The longitudinal axis will be displaced, which creates an angle of twist. Depending on the material, each type of material will respond respectively to torsion; some may deform or fracture depending on the type of material. When we wash our clothes manually and squeeze out the water, we hold the clothing with both our hands. Our right hand will move on a clockwise rotation. Our left hand will move on a counterclockwise rotation, thereby squeezing the water from the clothes. The more force we apply, the more water is squeezed and removed. The more water will be squeezed out. In this case, we are applying torsional stress to the clothes.

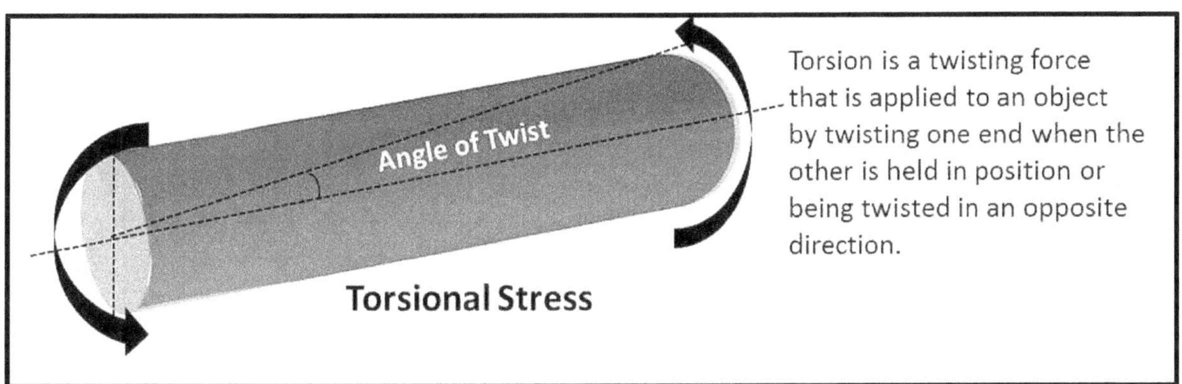

Figure 5.11: Torsional Stress

Bending Stress This is a force an object encounters when subject to a load at a given point which causes the object to bend or warp. This type of stress usually occurs in objects when subjected to a tensile load. This is the amount of energy it takes to compromise the item from its natural shape or condition. Usually, a bending force combines both tension and compression. This can be most experienced in structural beams.

5.4: Ductile and Brittle Fracture

Fracture refers to the separation of a solid body into two or more pieces imposed upon by stress, which is usually static and at temperatures relative to the material's melting temperature. The applied stress can be considered as tensile, compressive, shear, or torsional. A fracture will involve two processes: crack formation and propagation up to its point of rupture. The mode of fracture is dependent on the mechanism of crack propagation. The fracture can either be brittle or ductile. Any fracture will involve two steps: the formation of a crack and its propagation until its final rupture.

Fractography refers to the science and art of evaluating broken things. People who study fractures are called fractographers. The study of fractography can provide many clues to the person who wants to understand why the component failed or breakdown. Fractography is often thought of as a means of getting information about how and why the part cracked by examining the fracture surface and its forces involved. More advanced fractography can also be used to shed light on many types of wear mechanisms. Fractography is getting information about how and why the part cracked by examining the fracture assembly of the failed component or part. What is important in the study of fractography is to determine the stress or forces, whether this is a tensile, compressive, shear, torsional, or bending force that caused the object or material to deform or fracture.

A **brittle fracture** is a fracture that involves little or no permanent deformation. Many non-metals lack ductility and are subject to brittle fracture. A brittle fracture occurs when a part is overloaded and breaks with no visible distortion. In a brittle overload failure, separation of the two halves is instantaneous but proceeds at a tremendous rate where the crack begins at the point of maximum stress. A **ductile fracture** occurs by plastic deformation if the yield strength of the material is exceeded. Ductile materials exhibit a substantial amount of plastic deformation with high energy absorption before they fracture. This causes a permanent change

131

in the shape of the part, which becomes apparent if we attempt to reassemble broken components. There is a great deal of distortion on the failed part. There are instances when a brittle fracture appears in normally ductile materials, which indicates that the load was applied rapidly and vice versa. When you heat up the material, it becomes ductile and deforms. Usually, a ductile fracture will be a slow process, while a brittle fracture will happen instantaneously and very fast. A ductile fracture will be much preferred than a brittle fracture since a brittle fracture can occur without any warning; on the other end, a ductile fracture will involve the process of plastic deformation and provide some warning that a fracture is imminent, allowing maintenance time to replace the items before it ends up into a catastrophic failure. Ductile fractures have also been termed the cup and cone fracture.

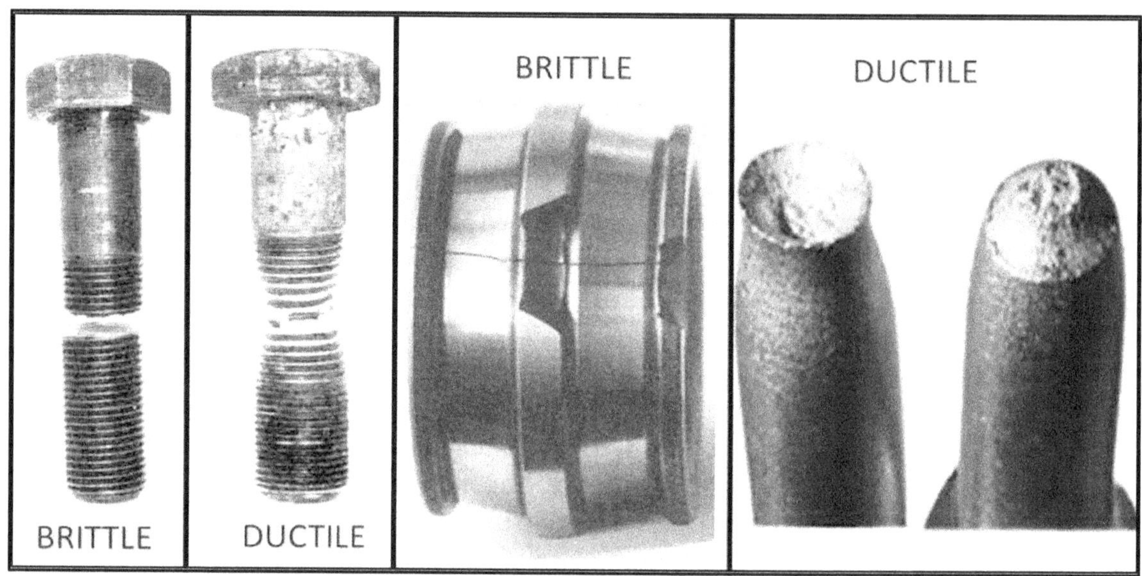

Figure 5.12: Ductile and Brittle Fractures

Although failures are inevitable and can never be eliminated, there will always be the subject of parts wearing out, and when parts wear out, they actually had achieved their useful life, and therefore it failed in their technical sense. The role of maintenance is to control the timing of failure or eventually prolong the duration of the failure itself. Not all failures are created equal, which is important in determining which parts will undergo a thorough RCFA. The investigator must depend upon the consequences of the failure itself. Therefore, maintenance aims to control the timing of failure to select or perform a task before it happens. Making our equipment more reliable is about extending the life and the Mean Time Between Failures (MTBF). TPM says that breakdowns can be eliminated, which is unlikely to happen. What TPM is declaring is that it is simply attempting to increase the Mean Time Between Failure (MTBF) of a particular part or item. Mechanical components such as shafts, fasteners, and structures aren't supposed to fail, but when they do fail, they can tell us exactly why. Experts say that the causes for more than 90% of all plant failures can be detected with a careful physical examination using a low power magnification of some basic physical testing. Every failure will leave a clue as to why it happens. A group of people can use some failure analysis to diagnose the mechanical cause of the failure itself. Inspecting the failure can tell us the forces involved and whether the load is applied cyclically or not. The direction of the critical load and influence of outside forces such as residual stress and corrosion. By knowing the physical roots, we can now proceed with the human, system, and latent cause of the failure.

5.5: Understanding Bearing Failures

Typically, when a bearing fails, what do we do? We simply replace it like any other part; we change the bearings every time we disassemble the equipment. Most often, a bearing does not attain its useful or calculated life for various reasons such as heavier loading that had been anticipated, inadequate lubrication, careless handling, ineffective sealing, or fits that are too tight or too loose, or how the bearing was even stored. A failed bearing will leave some clues for maintenance to figure out what cause it to fail. The bearing life is determined by the number of hours it will take for the metal to fatigue, and that is a function of the load of the bearing, the number of rotations, and the amount of lubricant a bearing receives.

There are several causes why a bearing fails, but according to experts, the majority of cases for bearing failures can be grouped into four causes.

UNDERSTANDING BEARING FAILURE

Survey shows that the most common type of bearing failures are :

Poor Lubrication (36%)
- Using wrong lubricant type
- Mixing lubricants
- Improper lubrication amounts
- Lack of lubrication
- Loss or improper additives

Fatigue (34%)
- Bearings are overloaded
- Due to misalignment unbalance
- Premature fatigue cause by lubrication problems

Poor Installation (16%)
- Improper installation can lead to failures such as load imbalance, misalignment, tight or loose fits

Contamination (14%)
- Typical failures due to contamination may include excessive wear, abnormal surface stress, corrosion such as water and liquid contamination

Figure 5.13: Common Causes of Bearing Failures

Lubricant Failure: In many cases, a bearing subjected to abnormal operating conditions exhibits signs of lubrication failure. The abnormal operating condition often produces excessive heat, increasing the fluid film thickness, allowing metal-to-metal contact between the raceways and the balls. Poor lubrication can lead to other problems such as overheating, contamination, fatigue, and corrosion. Dried and discolored (blue/brown) raceway and balls are symptoms of lubricant failure. Excessive wear of balls, rings, and cages will follow, resulting in overheating and premature catastrophic failures. The bearings depend on the continuous presence of a very thin film between the ball and the races. Once lubrication is inadequate or restricted, it will give rise to excessive temperature and degrade or destroy the lubricant oil properties and deplete the additives one by one. Oil starvation can be attributed to several causes. Wiping, blue discoloration, and adhesion wear to shaft or pin are common indicators. Restricted lubricant flow or excessive temperatures that degrade the lubricant's properties typically cause failures. Lubricant failure will lead to excessive wear, overheating, and subsequent bearing failure. Temperatures above 400 ° F (204 ° C) can anneal the ring and ball materials causing the metal to become ductile and deform. The resulting loss in hardness reduces the bearing's capacity causing premature failures. Rust will form if moisture or corrosive agents reach the inside of the bearing in such quantities that the lubricant cannot protect the steel surfaces. This process will soon lead to deep-seated rust. Another type of corrosion is fretting corrosion. Evidence of lubricant failure includes dried raceway. Extreme cases result in the deformation of balls, rollers, and rings. Extensive fretting corrosion on the outer raceway of a deep groove ball bearing can occur if lubricant is inadequate.

Heat is another cause of premature bearing failure. Heat will cause the lubricant to increase its viscosity, causing more heat as it loses its ability to support the load and form a varnish residue; this will destroy the ability of the grease or oil to lubricate the bearing and introduce particles in the lubricant. Leading bearing manufacturer states that the life of bearing oil is directly related to heat. Non-contaminated oil cannot wear out and has a useful life of about 30 years at 30 °C (86 ° F) and states that the life of the bearing oil is cut in half for each 10 ° C rise in the temperature of the oil.

Fatigue Failure: Fatigue is a phenomenon leading to fracture under repeated or fluctuating stress having a maximum value less than the material's tensile strength. Fatigue fracture is progressive, which begins as a micro or mini cracks that grow under repeated fluctuating stress. When stress exceeds the strength of the material, then fatigue occurs. Fatigue fractures are generally considered the most serious type of failure in machinery parts since they normally occur without excessive over-loads and under normal operating procedures.

To Understand Fatigue Fracture
1) Try to straighten out a standard paper clip
2) Flex it a little, and then let it go. You will notice that it returns to the straightened position. You may repeat this cycle many times (many years actually) without breaking (fatiguing) the metal because you are cycling it in its elastic range.
3) Now, let's bend the paper clip a lot further, and you will note that it did not return to the straightened position. This means that you have stressed the metal way beyond its elastic range.
4) If you bend the metal back and forth in its elastic range, it will fracture and break in less than 20 cycles. The metal fatigue more quickly because it work-hardened, and it became brittle. Hence fatigue is a function of stress and cycles.

When a bearing is pressed on a rotating shaft, the load passes from the inner race through the balls to the bearing's outer raceway, and the ball carries a portion of the stress as the ball rolls under the load. Other terms used include spalling, a fracture of the running surface, and subsequent small, discrete material particles removal. Spalling can occur in the inner ring, outer ring, or balls. Fatigues are progressive, which start as micro cracks until their final rupture and a marked increase in the machine's vibration. The bearing life is determined by the number of hours it will take the metal to fatigue, and that is a function of the load on the bearing, the number of rotations, and the amount of lubrication the bearing receives. Although showing heavier ball wear paths, symptoms are normal fatigue, greater evidence of overheating, and a more widespread and deeper spalling. By examining the path patterns on the raceway in a dismantled bearing that had been in service, it is possible to gain a good idea of the conditions under which the bearing had operated, and we may know if the bearing had been running under normal or abnormal working loads. Excessive loads usually cause premature fatigue. Tight fits, brinelling, and improper pre-loading can also bring a premature fatigue failure.

4 Points to Understand Fatigue
• Without stress fluctuations, fatigue cannot happen.
• Fatigue happens at stress levels well below the tensile strength of the material.
• Where corrosion is present, the strength of metals continuously decreases.
• The cracks take measurable time to progress across the fractured face, where the end result will be a
   fracture.

Contamination can be said as one of the leading causes of premature bearing failure. Contamination symptoms include dents or scratches embedded in the bearing raceways, balls, rollers, and cages resulting in undue bearing vibration and wear. Contaminants may include airborne dust, dirt, or any abrasive substance that gets into the raceway of the bearing. Principal sources are dirty tools, contaminated work areas, dirty hands, and foreign matter in lubricant. Contamination includes ingress of solids or fluids into the bearing. Solid contamination can enter a bearings cavity under an extremely dirty and dusty environment, even with the most effective seals. Poor lubrication and contamination can account for many bearing failures as well as secondary damages and can be prevented by applying Total Contamination Control Program besides performing routine lubrication schedules performed during Preventive Maintenance. What is important for maintenance to note is that there will always be a direct relationship between the amount of contaminants present and the degradation of the oil, as well as the number of failures we experience in our equipment, machines, and assets. In fact, a more accurate way of determining when to change the oil should be based on the amount of contaminant the oil have instead of the number of running hours the oil had been used. Here are some cases concluding that there is a direct relationship between the number of contaminants present in the lubricant and premature failures.

• GM Corporation test shows that 82% of internal wear comes from contaminants less than 40 microns.
• SKF states that improper lubrication accounted for 54% of bearing failures.
• Ford states that 80% of hydraulic failures can be treated to particulate contamination.
• Cummins Technical Center indicates that wear can be reduced by as much as 91% using a by-pass
   filter combined with a full-flow filter.
• SAE states that contamination in the lubricant of engines, transmissions, and hydraulic systems causes
   70% of equipment failures.

- Canadian National Research Council studies indicate that contamination was found out to be the leading cause of wear in a variety of industries investigated. In fact, 82% of all wear was found to be particle-induced
- Nippon Steel reduced bearing failures by 50% through aggressive contamination control.
- International Paper's Pine Bluff Mill reported a 90% reduction in bearing failures through an aggressive contamination control program.
- Caterpillar claims indicate that dirt and contamination are the number one cause of hydraulic system failures. This means that with regards to hydraulic systems, they must be kept clean at all times.
- British Hydromechanics Research Association, which covered a 3-year study on 117 hydraulic machines composed mainly of injection molding, machine tools, mobile, construction, marine, metals, concluded a dramatic relationship between fluid contamination and its service life. Improve fluid cleanliness can improve MTBF from 10 to 50 % of its life.
- General Motors AC Delco Division tested their DDA engines and found an 8 fold improvement in wear rates and engine life with lower lube oil contaminant levels. GM reports compared to a 40-micron filter, engine wear was reduced by 50% with 30-micron filtration, and wear was reduced by 70% with 15-micron filtration.
- Albertson's Inc., a supermarket chain, conducted a study on a series of over-the-road Cummins tractor engines that found reduced wear rates with a greater lube oil cleanliness. After analyzing 6 engines having 600,000 miles, engine crankshaft journals showed only 0.0005 inches of wear. The rod and main bearings had not even worn through to the copper layer. The compression ring and oil ring shows no wear.
- U.S. Department of Defense states that approximately 30% of all engine failures are caused by metal contamination in lubricating oil systems.
- Alumax of South Carolina reported a per machine reduction in component replacement cost from $15,000 to $500 per year.
- Kawasaki Steel implemented a similar contamination control program and achieved an almost unbelievable 97% reduction in hydraulic component failures.
- Oklahoma State University reports that when a fluid is maintained 10x cleaner hydraulic pump, life can be extended by 50 xs.
- Machine Design Magazine reports that less than 10% of all rolling element bearings reached the fatigue limit since contamination usually causes wear or spalling failure earlier.
- MIT states that 6 to 7% of the gross national product ($ 240 billion) is required just to repair damage caused by mechanical wear, which is a result of contamination

All these cases mean just one thing, the more contaminants present in the oil, the more failures can be experienced on the equipment. This means that the cleaner the oil, then the longer the life of the equipment.

Poor or Improper Installation: Misalignment failure can be detected on the raceway of the non-rotating ring by a rolling element wear path that is not parallel to the raceway edges. Excessive misalignment can cause abnormal temperature rise or heavy wear in the cage pockets. The most prevalent misalignment causes are bent shafts, burrs, soft foot, or dirt on the shaft or housing shoulders shaft threads not square with the shaft seats locking nuts with faces that are not square to the thread axis. Misalignment occurs when the centerlines of rotation of two machinery shafts are not in line with one another. There are two types of misalignment: parallel and angular.

In most cases, machine misalignment is actually caused by a combination of these two types. Misaligned shafts can cause increased bearing load, reduction in bearing life, increased seal wear, increased vibrations, increased noise, and increased energy consumption. Correcting shaft alignment regularly with a laser alignment can minimize these effects.

False Brinelling: Fretting damage happens if a bearing is subject to vibration when it is stationary, giving rise to what is known as false brinelling. Standby rotating machines inside the storeroom should also be rotated slightly to change the ball's points and the bearing's outer raceway, especially if vibration forces are felt. If there are standby rotating machines, it is good to run the machines alternately to reduce the risks of fretting. This is done for motors and other rotating assemblies with anti-friction bearings. The shaft of the motors should be rotated at least 1 and ¼ turns regularly. I would prefer a weekly or bi-weekly rotation for standby components. False brinelling can occur on standby units and rotating components inside the storeroom where vibration is felt. When the decision is made to stock the spare parts in the storeroom, it can be held in the inventory for many years and becomes a non-moving item. Therefore, it is important to have a program to ensure that these components remain in serviceable condition. Rotating spare parts like pumps, motors, gearboxes, blowers, and other rotary components can be damaged by the environment and vibration. Over five years, motors and other rotating components inside the storeroom inventory should be tested to ensure they are still in working condition.

When a bearing is loaded, which means that if the bearing is rotating, the balls are separated from the outer and inner raceway through a film produced by the lubricant. The load will always be heaviest between 4 to 8:00 if we speak about the clock. However, if the bearing is not loaded as in the case of a standby component. The balls are resting on the outer raceway of the bearing. The vibration produced by the running component is not only confined to the duty but will slightly be passed and absorbed on the standby component. This means that the balls of the bearing are vibrating respectively to its inner and outer raceways. The reason for rotating the bearing 360 degrees + 60 degrees or 720 degrees + 60 degrees is to change the location of the balls with respect to the outer raceway of the bearing since if the balls are vibrating on the same exact spot, it will create stress marks on the raceway.

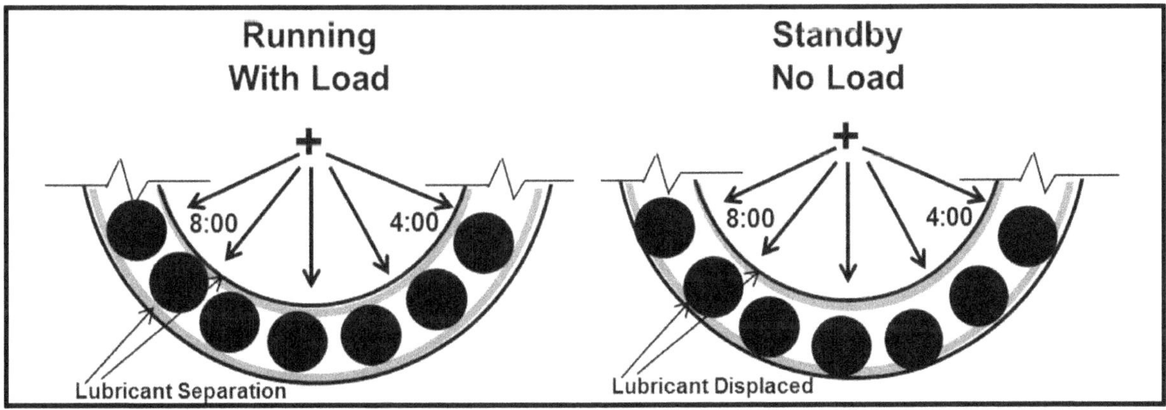

Figure 5.14: False Brinneling Explained

When the bearing does not reach its useful life, maintenance people must take the time to perform a Root Cause Failure Analysis to determine the cause of the failure. The evidence of the failure will always be visible in the raceway of the bearing. Usually, if lubrication causes the problem, both the inner and outer raceway will be dried and have varnishes or discoloration, indicating that the lubricant had dried up; this will serve as evidence for the physical cause of the bearing failure.

When trying to select a bearing, determine the maximum load for the bearing. This is important for both axial and radial loads. We also need to determine the minimum and maximum running speed or RPM that the bearing will be subject to. This will help us to determine the correct viscosity of the lubricant for the bearing. Next, determine all possible environmental conditions to which the bearing will be exposed. Extremely hot or cold environments often require valid bearing specifications, affecting the type of lubricant requirements. It is also important to ask the vendor when the bearing was manufactured and how it was stored before the purchase. Ask the bearing vendor about storage and handling procedures.

5.6: Why Lubricating Oil Fail?

Although for this section, we shall only cover lubricating oil. I have written a separate book about lubrication, volume number 6 of the World Class Maintenance Management, The 12 Disciplines series if the reader wants to know more about the subject. The primary purpose of lubricating oil is to provide a film or layer to create a separation to avoid metal-to-metal contact. When we have two mechanical components such as a ball and the raceway, two gears meshing one another, two mechanical blocks moving in the opposite direction or a piston moving up and down from the wall, there is a tiny clearance between these two mechanical parts. These two metals cannot come in contact with one another as heat will be generated, that is why lubricating oil is applied. If we rub both our hands, you can feel some warmth in the palm of your hands, but if you place something in between then we delay the process. This is the main function of the lubricant in our equipment and assets. Although depending on the oil's application, there are other functions of lubricating oil; for example, for transformer, the oil is not used for lubrication but for insulation purposes.

Lubricating oil is made up of base stocks and additives. The base stocks can be made from petroleum, synthetic, or para-synthetic. The base oils are then treated with chemicals and additives to form the different grades of lubricating oils for various applications being used in industries. Lubricants are specifically designed to reduce the friction and wear of contacting surfaces in machine components. Various additives are placed on the lubricating oil, depending on the type. Motor or engine oil will consume the most amount of additives. Usually, the additive will represent around 1 to 30% of the total volume of the lubricant depending on the type of lubricant. There are three main reasons why lubricating oil fails, and they are connected in every way. These include oil contamination, increase in temperature, and additive depletion.

Oil Contamination: This may be in the form of solid, liquid, or even gases that should not be present in the lubricant. Contamination is also the reason for oil degradation. Hence, the need to change the oil on a time-dominated (running hours.) frequency is essential. This means that more contamination means more chances of failures and frequent oil changes; therefore, it is important to analyze oil on the number of contaminants and what elements are present and not by the frequency of changing oil itself. By knowing this information, maintenance can strategize measures to improve fluid cleanliness and lengthen its drain interval. Solid contaminants can get trapped into tight clearances that can produce abrasion. Once abrasion occurs, a new set of contaminants is added to the oil, which is a continuous cycle, which will deplete the additives one by one.

Dynamic Clearances	Size in Microns
Servo Valves	1 to 4 microns
Servo Valve Flapper/Nozzle Spacing	40 to 80 microns
Servo Valve Spool to Sleeve, Radial	1 to 4 microns
Proportional Valves	1 to 6 microns
Directional Control Valves	2 to 8 microns
Journal Bearings	0.5 to 100 microns
Roller Element Bearings	0.1 to 3 microns
Hydrostatic Bearings	10 to 25 microns
Anti Friction Bearings Radial	0.5 to 1 micron
Gear Pump (Gear to Side Plate)	0.5 to 5 microns
Gear Pump (Tip to Case)	0.5 to 5 microns
Vane Pump (Tip to Vane, Estimate)	0.5 to 1 micron
Vane Pump (Sides of Vane)	5 to 13 microns
Piston Pump (Piston to Bore, Radial)	5 to 40 microns
Piston Pump (Valve to Plate Cylinder)	0.5 to 5 microns
Control Valve Orifice	130 to 10,000 microns
Control Valve Spool to Sleeve (Radial)	1 to 23 microns
Control Valve Poppet Type	13 to 40 microns
Control Valve Disc Type (Estimate)	0.5 to 1 microns

Figure 5.15: Critical Clearances for Mechanical Parts

Increase in temperature: Heat is the number one reason that destroys oil. An indicator of heat tolerance is Flash Point. Excessive contamination can result in an increased temperature in the lubricating oil, which can cause the oil to oxidize. All lubricants, whether grease or oil, have a corresponding temperature in which they can tolerate and can be used. Increasing temperature in lubricating oil can cause many problems such as oil oxidation, building up of varnishes, gums, and sludge. Excessive moisture in oil can cause increased viscosity, hydraulic pump cavitation, oxidation, hydrolysis, aeration, acid formation, rusts, corrosion, and other problems on our equipment, machines, and assets. This means that the lower the Flash Point, the greater chance of vaporization loss at high temperature and oil to burn off on hot cylinder walls and pistons, leading to oil thickening and deposit build-up on critical engine components. The minimum Flash Point should be at 400 ° F or 204 ° C.

Depletion of Additives: Oil comprises of two things: first is the base-stock, which can be petroleum-based, extracted from Mother Nature, or synthetic oil, which is man-made oil, and then we have the additives. The base oil is treated with chemicals and additives to form the different oil types. Once the additives deplete, there is no reason to use the oil, and, therefore, it should be changed since it will no longer serve its purpose. Additives deplete due to contamination present in the oil. Contamination does not only come in the form of solid or foreign particles but also through moisture and air, which causes foaming in the oil. Excessive contamination and increased temperature can cause the oil additives to deplete, which means that the base is no longer protected.

If I explain lubrication in the simplest way I can, it's just a battle between the good and the bad. If you love to watch action movies. The base can be referred to as the President of the United States, and the additives are the secret service agents protecting the President. There are several additives in lubricating oil, and each of these additives has a specific function. But the overall function of these additives is to protect the base from these contaminants. Simple as we may think but read again, here's the catch.

Unknowingly, maintenance people become collaborators with these terrorists, I mean contaminants. It will start with how we stored our lubricants. Do not assume that new oil is also clean oil since it is not. New oil already contains contaminants or terrorist cells, but they are in a small group and cannot do anything at the moment. Still, since maintenance collaborates with these terrorists, we store them anywhere, mix them, over-lubricate, under-lubricate, and then the process of oxidation happens. The terrorist cell becomes bigger and stronger, killing these agents one by one; that is why 95% of the time, the terrorist wins, and the agents lose and are killed. Then we change the oil, and the same thing repeatedly happens for the rest of our lives, but the truth of the matter is that the base never wears out. The additives are self-sacrificing, just like the secret service people willing to risk their lives to protect the President. Everything will start on how we store our lubricants and providing education not only to maintenance but everyone on how oil is being contaminated. Creating an Oil Contamination Awareness will be the first step in reducing equipment-related failures attributed to lubrication. This simply means that if lubricating oil is clean, then the lesser the contaminants in the oil which means that the additives can last longer or even indefinitely. Imagine if you take your dinner and leave the food on the table, then the food will be spoiled the following day. If we place the food inside the refrigerator, then it can last for days. Still, if we place the food in the freezer, then it can last for weeks. But suppose we subject the food under cryogenics in which the food will be completely frozen at -150 ° C or lower; in that case, the food can be preserved at a much more longer span of period, perhaps years. This means that if the oil can be maintained clean, then the additives can last much longer.

Lubricating oil contains two things, the base and the additives. Each of these additives serves a particular Function, but the primary function of these additives is to protect the base from any unwanted contaminants. Once the additives depletes, the base is no longer protected and the president dies, but in reality, the base (President) is pretty much alive and well.

Figure 5.16: Lubricants Base and Additives Explained

<u>5.7: Understanding the Physical Cause of Failures</u>

In several cases, there is no single cause and no single sequence of events that lead to a failure. There will always be several causes. The physical failure for example fracture, explosion, fatigue, corrosion, or premature wear will be obvious through some physical examination, however, there are other levels of failures that allow the physical cause to occur. Performing failure analysis on the failed part, item or component will be part of a deeper probe on the investigation whose intent is to prevent a recurrence of the same causes of the problem. Many people think that conducting a failure analysis is already addressing the root cause when what they are referring to is only the physical cause. The key to extracting the physical cause is to preserve the evidence. To determine the root cause, the human, system, and latent causes must be uncovered. All the discussions we covered in this chapter have something to do with the failure's physical cause. This will be the first of a series of causes that the Principal Investigator and the crew of evidence gathering team should uncover before moving forward to determine the problem's human, system, and latent cause. The physical cause of the failure is similar to conducting failure analysis on broken components for RCFA. This is the physical reason why and how the parts failed. It is also the technical explanation of why and how things broke or failed.

Both the how and why will be answered in dealing with the physical cause of the failure. The question of how usually will provide the method or process that took place. It will answer the physics of the failure. The question of why will always begin with the word "because," since we are interested in knowing the reason that particular event happened. Let us provide the following example. Why did the bearing fail? The bearing failed because it fatigued. How can fatigue occur on the bearing? To answer this question is to provide the process of how fatigue happens in the raceway. The ball forces lubricant into the crack, which caused the metal to stand proud of the surface. Due to this continuous process, the metal sheared off, forming a solid metal particle. The small crater formed slightly changes the vibration characteristics of the bearing. As the ball passes continuously into the crater, it makes it bigger, creating more metal fragments. Soon, the balls themselves get damage since they are no longer rolling on a smooth surface. At some point, the bearing becomes audibly noisy and starts to get hot.

In performing Root Cause Failure Analysis, we are interested in knowing the real cause of a particular failure by verifying each hypothesis based on the evidence uncovered until we reach the final cause, which is the latent cause of the failure. There are also possibilities that there will be two or more physical causes based on the evidence collected; in this case, treat each cause separately. This means that a separate RCFA Logic Tree Diagram will be done for each cause of the failure.

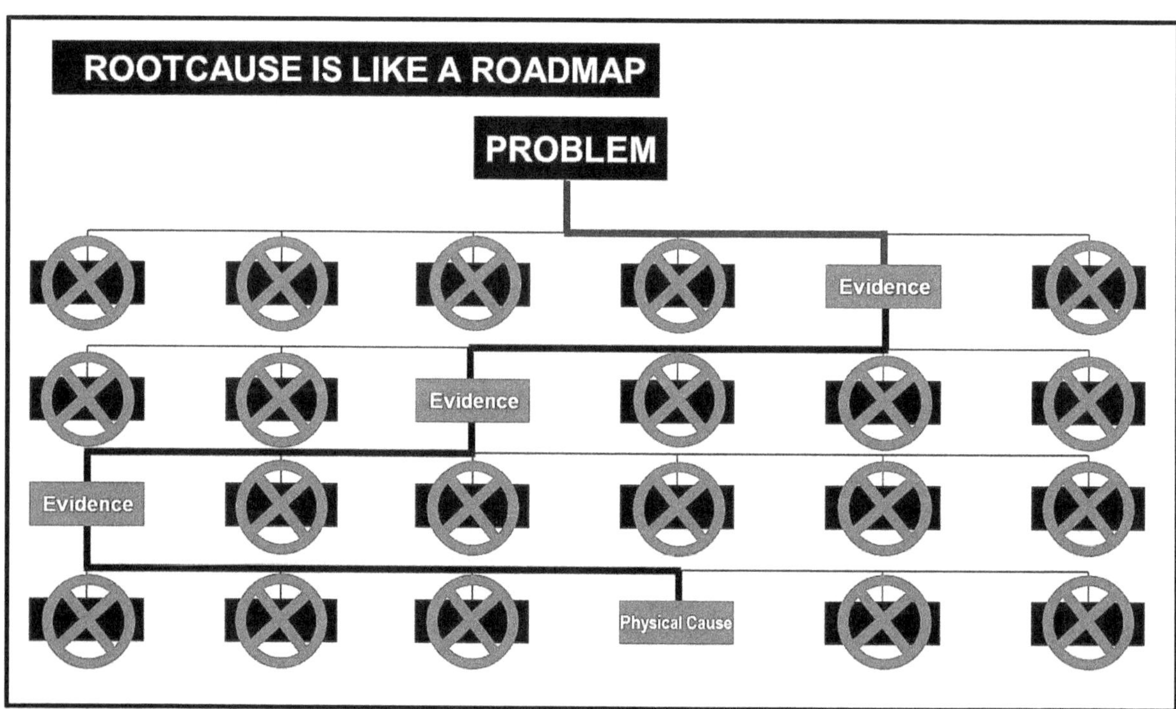

Figure 5.17: In Root Cause We Follow the Evidence

When conducting failure analysis on the equipment or asset, the failed part or item needs to be extracted with care. Maintenance people responsible for repairing should be trained on how to extract the failed item so that the integrity of the evidence can be preserved for later investigation. The following steps should be followed if the physical evidence will be done by a third-party metallurgical lab.

• The physical evidence should be extracted with care and must not be tampered with. Use gloves in handling the failed mechanism
• Ensure that the failure analysis is performed by a qualified person. If instruments such as a scanning electron microscope are needed, it is best to send the failed mechanism to an independent metallurgical laboratory, or an experienced 3rd party vendor.
• Provide all necessary details to the vendor to ensure that the person in charge of conducting the failure analysis understands the situation.
• If the failed part will be shipped to an independent courier, use bubble wrap on the failed part when shipping so that it will not be damaged during shipment. Provide necessary instructions if needed.
• Ensure that the physical evidence must be well preserved and should not be tampered with.

Typically, conducting failure analysis on the mechanical part that failed will require a scope with magnification to examine the problem such as fracture, and how it propagated. This means that the person conducting this should have some knowledge about the stress involved, the different types of wear, and basic knowledge about fractography. An alternative option is to send the part to a third-party independent metallurgical laboratory and they will be the ones to conduct the failure analysis of the failed part.

5.8: Electronic Failure Modes and Causes

The physical cause will not only be limited to mechanical failures but also to electrical and electronic failures. These electronic components have a wide range of failure modes. Electronic failures can be caused by any of the following; increases in temperature, excessive current or voltage, electrostatic discharge (ESD), ionizing radiation, handling and storage, moisture contamination, excessive machine vibration, failure of individual components on a printed wiring board (PCB), mechanical shock, mechanical stress, poor solder contact, faulty circuit design, change in the operating environment, thermal stress, humidity, and many other causes. In semiconductor devices, problems in the device package may cause failures due to contamination, mechanical stress of the device, or open or short circuits. One of the major sources of contamination that threaten electronic components on circuit boards Is moisture. Moisture can be acquired during the component of the manufacturing or assembly process. Condensation can occur on the printed wiring board and within component packages, especially during cold temperatures hence room temperature must always be regulated and monitored.

Chapter 6

The Study of Human Errors

> *Not all human error is necessarily the fault of the person. In many cases, human error is either forced by external circumstances or by obsolete rules and policies. Therefore, if blame is to be allocated for any error, care must be taken to identify the real cause. Remember, industries will never learn from their failures if we blame people for their mistakes and errors.*

6.1: Understanding Human Errors

It is much easier to blame people for their mistakes and errors than to learn from the things that go wrong, whether in industries or about lessons of life itself. If you ask why? It is because of human nature. Let us admit it, myself included. This is how people react. On the other hand, people that committed the error would always have tons of excuse for not doing the right job rather than taking their time to analyze the root cause of their errors. If possible, they will always look for ways to pass the error to someone else rather than admit them. If one of the C-level people or decision-makers made an error or mistake, then no one will admit to the fault as if nothing happens. As a result, most industries have some policies and rules for people that commit mistakes. If an operator commits an error during operation or when a technician poured the wrong oil into the equipment, we punish or reprimand them instead of understanding the reason behind their failures. Why? Because it is much easier, faster, and most importantly, this is how we do things in our plant since the beginning of time. In short, it had been part of our culture for many years. If you are an engineer and your management people commit an error, we have limitations on what we can do. Instead of correcting them, we just avoid them as we do not want to be involved. Roger Biosjoly of Morton Thiokol warned his management of the O-ring problem on the solid rocket booster they supplied to NASA and recommended to delay the launching, but he was ignored, and the shuttle exploded, killing the six astronauts and the teacher Christa McAullife. What would you do if you were in the management's shoes?

By definition, human error can be defined as an action planned but not executed according to the plan. Errors are committed daily by physicians, doctors, engineers, designers, carpenters, nurses, and you name it. For people involved in the medical field, some errors can be fatal and can compromise the lives of their patients. You have road accidents reported almost daily by the news, errors from management decisions, flaws in systems, procedures, and policies can have a detrimental effect on its product, flaws and cost-cutting schemes that caused a bridge or structure to collapse, operations deferring Preventive Maintenance to cope with production, outdated procedures, industrial disasters, and you name it. But the most important factor in human error is learning from our failures to become a better person. If people create problems, then people must also be capable of correcting them, but this can only happen if we look ourselves in the mirror and admit that we, too, are also part of the problem.

Humans play an important role for industries during the design, installation, production, and maintenance phases. In maintenance, human error may be defined as the failure to perform a specified task that could disrupt scheduled operations or result in damage to property and equipment. While human error had existed since the beginning of mankind, only in the last 50 years has it been the subject of scientific inquiry. There are various reasons for human errors such as stress, fatigue, working too long without adequate break time, incorrect tools, forgetfulness, complacency, pressure, distractions, insufficient knowledge, lack of training, lack of communication, misinterpretation, ignorance, lack of standards, lack of awareness, and the list goes on. However, with every error we make, there is typically an associated change or something out of the ordinary occurring in our environment. The difference between humans and machines is that people can sense, change, hear, smell, see, feel, or taste something different and take the necessary actions to correct the anomaly. In industries, human error exists either because we tolerate them or we are just ignorant. Let me state some cases.

Case 1: Try to loosen some bolts in your equipment and place a toolbox near it. Let us say that the toolbox will contain a hammer, pipe wrench, adjustable wrench, pliers, socket wrench, double-ended open type wrench, and a torque wrench. Ask one of your technicians in the plant to tighten the loose bolts and observe how he tightens them. After he has tightened the bolts, ask him to loosen them again. Try to call 4 to 5 more technicians, one at a time, and give them the same instructions to tighten the bolt and write down how they perform the tightening process. Perhaps you will observe that some have similar ways of tightening the bolt and have used the same tools, while others simply use other methods to tighten the bolts. Why? Because you just asked them to tighten the bolts, but if we are explicit in our instruction and procedure, we can instruct them to use a torque wrench with this amount of torque, or perhaps a warning to indicate not to use an adjustable wrench, hammer, or pipe wrench so that each one of them will do the exactly the same procedure on tightening the bolts.

Case 2: Every single PM Checklists has a form of lubrication activity that is being performed in a timely fashion, which states something like this:

• What: Perform lubrication on bearing
• When: Monthly
• Who: Preventive Maintenance Group

• As to what specific brand and type of grease or oil to be used are unknown?
• What specific NLGI number and thickener will be used for the bearing?
• If grease is to be used, what specific temperature of grease is required for this application?
• How many shots or pump is needed or when do I know when to stop pumping the grease gun?

• How many grams of grease is required for this application, and how do we measure it?
• How many pumps or strokes are required before stopping?
• What is the correct interval to perform lubrication?

These are just a few questions to raise because lack of lubrication or over-lubricating will not be good for the bearing. Frequently we blame it on the lubricant, but most lubricant failures are caused not by the lubricant we use but by the way we perform our lubricating practices in our plant. Applying the wrong lubricant or mixing lubricants of different brands can cause incompatibility problems with the base and additives. This would likewise shorten the lifespan of the part it lubricates, not to mention the life of the lubricant itself.

Categories of Human Errors

When considering people and machines' interactions, human errors can be classified into four categories: In conducting Root Cause, not all human errors are necessarily the fault of the person who made the error.   In many cases, the error is either forced by external circumstances or other factors; therefore, if blame is allocated for any error, care must be taken to identify the real root cause of the problem. Here are the different categories of human error.

A) Anthropometric Factors: These factors rely on the person's size, shape, and strength. Errors occur because the person simply cannot fit into the space provided, cannot reach something, or is not strong enough to move or lift something. Industries aiming for multiskilling should also consider anthropometric factors to avoid human errors. The human resource department hires people with special built, height, or perfect eyesight vision because it is mainly required in their work. Hence when a taller person is required to operate certain equipment, we simply cannot replace him with a shorter person because the height requirement is needed to perform their job, just like a flight attendant from an airline industry. Many industries that practice multiskilling should also note if there are limitations and one of them is anthropometric factors. Perhaps a good example I can provide here is a scene from the movie Schindler's Lists where the children who were also factory workers were not allowed to board the train, as Schindler and his factory workers will be transferring to another location. The German soldiers detained the children, and they were not allowed to board the train. Schindler interfered and told the German soldiers, these children are my workers. Schindler grabbed a child and told the soldier to look at their fingers. How am I supposed to polish the inside of 45-millimeter casings? And the German soldiers released the children.

B) Human Sensory Perceptions: These are factors that people can see, hear, touch, feel, and even smell what is going on around them. Some failures will give some sort of symptom or warning that they are in the process of occurring, which can be captured by the human senses, such as excessive vibration, loud noise, heat, and foul smell. Therefore if errors are occurring or thought to be likely to occur for any of these reasons, then the human error is not the root cause of the failure, and we need to dig deeper into the real cause of the problem. When maintenance cannot smell a burnt motor because he has a terrible cold and was not allowed to go home by his boss, he should not be blamed for the failure.

C) Physiological Factors: The term physiological factors refer to the environmental stresses which affect human performance. These stresses include high or low temperatures, loud or irritating noise, excessive humidity, high vibration, exposure to toxic chemicals, radiation, or working too long without adequate break time. Exposure to these stresses leads to reduced sensory capacity, fatigue, and reduced mental alertness. These are all manifestations of human fatigue, and all greatly increase the chances that the people concerned will either make a slip, lapse, or mistake. If errors occur or are thought to occur for any of these reasons, again, human error is not the root cause of the problem. For example, reducing temperature or providing hearing protection should minimize exposure to noise and stress of the workers.

D) Psychological Factors: Psychological factors can be grouped according to those that are unintended and intended. Unintended actions can be grouped into slips and lapses, while intended action is grouped into mistakes and violations.

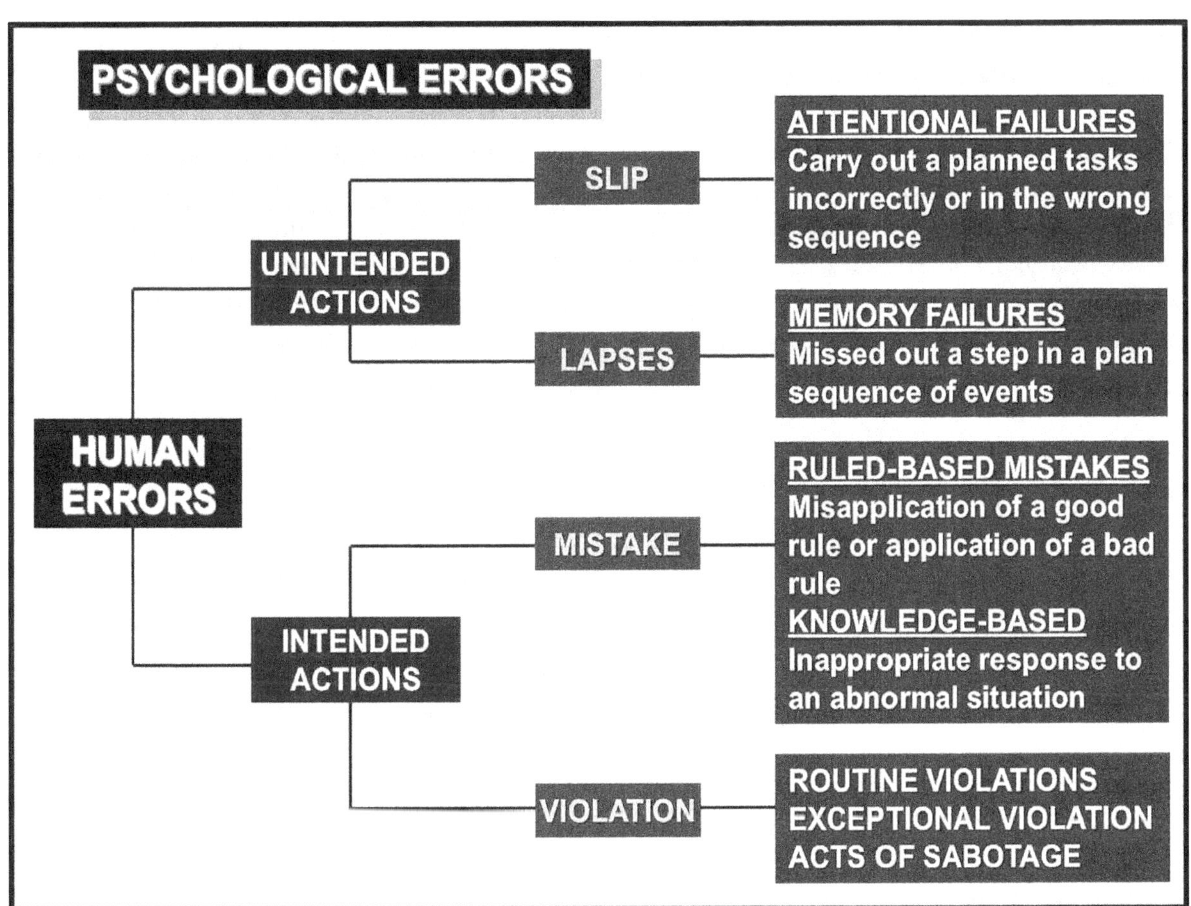

Figure 6.1: Understanding Human Errors

Slip occurs when somebody does something incorrectly or does something in the wrong sequence. If a person performs something in sequences such as step 1, step 2, step 3, step 4, and step 5, and the person does step 1, step 2, step 4, step 3, and step 5, the person committed a slip. These are human actions that take place which was not intended. The occurrence of a lack of one's train of thought, which is derived from unconscious behavior.

Lapses occur when someone misses out on a key step in a sequence of events or activities. For example, a mechanic leaves a tool behind after working in a machine or simply forgets to fit a key component while reassembling it. If a person will do step 1, step 2, step 3, step 4, and step 5, and the person missed out on step 4 in the process, he committed a lapse. As humans age, our memory will not be as sharp during our high school or college days. People tend to forget many things as they grow old. If you get in your car on your way to work and while driving you forgot your wallet or id and return to your house, then what you have committed is a lapse. Both slip and lapses happen in doing Preventive Maintenance, especially during overhauls and replacements.

Mistake is a planning failure, where actions go as planned, but the plan was simply bad. These are errors in judgment. Mistakes are the real challenges in the analysis of human errors. If someone knowingly or intentionally chooses to do something wrong, then it is considered a violation, which means that one has deviated from safe practices, procedures, standards, or regulations. Mistakes are influenced by external factors and can be group as knowledge-based, ruled-based, and violations.

Knowledge-Based Mistakes are types of mistakes that occur when someone is confronted with a situation that had not occurred before and had not been anticipated. In other words, there are no rules or procedures to follow. In situations like this, the person has to decide quickly about an appropriate course of action, and a mistake occurs due to the wrong decision. This suggests that the first and foremost way to avoid knowledge-based mistakes is to improve people's knowledge to make these decisions.

Ruled-Based Mistakes usually occur when people believe that they are following the correct course of action when doing the tasks based on a specific rule, but the course of action is not appropriate.   An example will be replacing parts in equipment to comply with their PM procedure even if the pattern of failure is random as Preventive Maintenance is only feasible for age-related failures.

Violation occurs when someone knowingly and deliberately commits an error intentionally. Violations can fall into three categories, routine violation, exceptional violation, and acts of sabotage. When an operator does not wear safety goggles during his duty, then this is a routine violation. When a company car is exceeding the plant's speed limit of 10 kph, since he is carrying an injured person to the nearest hospital, then this will be a case of exceptional violation. Sabotage is the deliberate destruction, disruption, or damage of equipment, or company properties. An example of sabotage is using weapons or other materials to destroy or damage company properties or introducing a virus into a company's software.

Therefore whether we like it or not, humans will continue to make errors and mistakes. That is why the need to perform a thorough Root Cause Failure Analysis is a must. Failures are not really bad after all if we can learn from them. Failures allow us to begin more intelligently. We can only benefit from failure if we can learn from it. Perhaps we cannot change the human condition, but we can change how humans perform their work. When people commit errors, we must steer away from blaming or punishing the person. We must understand the reason behind the error or failure. Changing policies, training, procedures, and systems is not enough. It also requires a change in oneself. The person who commits the error and the people accusing them should also change because the more we blame, the less we understand the failure. When human errors occur, we blame or point fingers at someone; it is more important to understand the reason behind the failure. All people commit mistakes and errors. We must realize that most of these errors happen beyond the person's control. Remember that the most successful people in this world failed miserably in their lifetime before they become successful. One thing these people have in common is that they admit their human error, learn from their mistakes and become better, and I think that's all I have to say about that.

6.2: Human Fatigue – The Circadian Rhythm

[6]Human fatigue includes the feeling of tiredness, physiological, in the execution of work concerning time. The body temperature fluctuates throughout the 24 hours. This means that our mind and body's energy is not constant. The lowest point or the time humans feel fatigue will be around 0200 to 0300 hours and the highest is at 1200 to 1400 hours. The difference is 0.5 C. This means that the body will fatigue more in the early hours of the morning. This diminishes and the body increases its level of energy until just past noon and decline in the afternoon. These fluctuations are examples of the Circadian Rhythms which means around the day. It comes from the Latin words circa which means around, and the word dies which means a day. These circadian variations are governed by a biological clock located in the brain. Skill-based errors exhibited a significant circadian rhythm, being most prevalent in the early hours of the morning. However, the circadian rhythm can be disturbed by several factors such as a change in time zone, especially when we go to a country with a different time zone or from a shift change. It is much faster to adjust from a day shift to a night shift, rather from a night shift to a day shift. During the night shift, the body temperature decreases. Alertness begins to reduce. This means that there are much greater chances of human errors to occur during the night shift as it includes working in the early hours of the morning. It is much easier to sleep during the night for workers than for night shift people to sleep during the day. I have experienced this personally. There are a lot of distractions sleeping during the day compared at night.

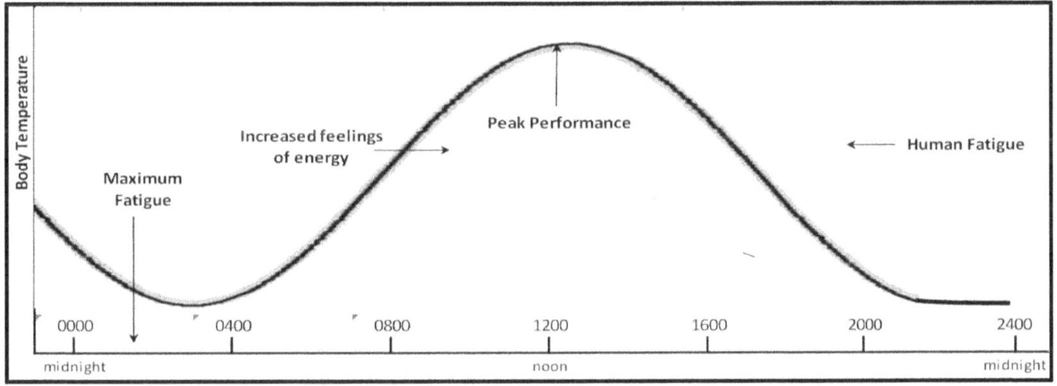

Figure 6.2: The Circadian Clock

[66] Reasons, James and Hobbs Anthony, **Managing Maintenance Errors**, Ashgate Publishing Limited, 2003, page 31 to 32

Investigating Equipment Failures Through Root Cause Failure Analysis

Several industrial disasters such as Chernobyl, BAC1-11 window blow out, Three-Mile Island, Bhopal Tragedy, and so on occurred during the early hours of the morning. According to James Reasons, human causes can be caused by any of the following. First, the person approach focuses on the premise that an error is caused by a single individual, whether it be due to forgetfulness, inattention, negligence, and other personal lapses. Second, the system approach focuses on the premise that humans are fallible and errors are to be expected, hence, the errors are a result of system failures and the root cause of errors ultimately falls on the system rather than the individual.

Fatigue workers during the night shift can become irritable and may tend to lose concentration during their work especially if the work is routine. Research shows lack of sleep or sleep deprivation experienced by these workers is very similar to those produced by drinking alcohol. Both mental and physical performance on several tasks will be affected by a worker who is awake for 18 hours straight. This is like having a Blood Alcohol Concentration (BAC) of 0.05%. It is highly recommended that shift rotation should not go beyond one month for industries since night shifts are more prone to human errors. A worker should have at least 11 rest hours between shifts. It is also important for workers to have at least 2 rest days a week. This will allow the worker to start fresh when reporting for work.

6.3: Maintenance Induced and Non-Maintenance Induced Errors

Maintenance Induced Failures are failures that are directly caused by the maintenance people themselves. Maintenance Induced Failures include infant mortality failures or intrusive maintenance, overhauling, and inducing early failures right after endorsing the equipment back to operators. Perhaps the maintenance reassembled the equipment incorrectly. Non-Maintenance Induced Failures are failures on the equipment that are not caused by maintenance, but other factors such as how the equipment was operated, design errors, commissioning errors, cutting costs, administrative errors, and many more.

From the book of Keith Mobley, An Introduction to Predictive Maintenance, he quotes: From studies of equipment reliability problems conducted over the past 30 years, maintenance is responsible for about 17 percent of production interruptions and quality problems. The remaining 83 percent is totally outside of the traditional maintenance function's responsibility. Inappropriate operating practices, poor design, non-specification parts, and many other non-maintenance reasons are the primary contributors to production, product quality problems, and not maintenance.

Although the percentage of human errors induced by maintenance is not that big, the irony still lies that the worst industrial disasters that occurred globally resulted from maintenance and, of course, from human errors. Human errors will have a wide range of consequences, and the problem is that we only act if the consequences of the failure already occurred. Even if we already knew the problem for as long as nothing happens, we always result in the same old ways. Here are some possible situations where the failure are caused by maintenance itself or Maintenance Induced Errors:

Infant Mortality Failures: Early failures can occur on the equipment right after a major Preventive or scheduled maintenance is done. Operators often say sarcastically that if maintenance did not perform their regular overhauling and replacement on this piece of equipment, I bet it will be running without any problems. But isn't it that the main purpose of a scheduled Preventive Maintenance is to ensure that the equipment performs as intended, but the opposite happens whenever maintenance does something. Something goes wrong. Is it profound? No! Honestly, we are just a victim of what we called Infant Mortality Failures. In almost all cases, Infant Mortality Failures are caused by human errors and intrusive or forced maintenance. In fact, many things can go wrong when we try to dismantle the equipment for overhauling purposes. Human errors such as slips and lapses can occur on the part of the maintenance performing the overhauls. Infant Mortality Failures are failures that occur at the beginning of life; others refer to them as commissioning failures, start-up failures, or debugging failures. Many factors affect Infant Mortality Failures, including poor equipment design, poor quality manufactured, incorrect installation, incorrect commissioning, incorrect operation, unnecessary or intrusive maintenance, human errors, or simply bad workmanship.

The case of infant mortality failure, pattern F of the six failure patterns, starts off with a high incidence of early failures, which eventually drops to a constant or very slow, increasing conditional probability of failure and ending in a no wear-out zone. This means that the failure can occur at the beginning of operation. If Infant Mortality Failure exists frequently, the question raised is, can we eliminate them? The answer is no, we cannot eliminate infant mortality failure completely, but it can be reduced. Reducing unnecessary PM activities we perform on the equipment can reduce infant mortality failures. In layman's terms, the more we touch, the more it fails; the less we touch, the less it fails. Suppose we review the lists of every activity performed on the equipment; we can find out that there are maintenance tasks that duplicate themselves. It is already performed by regular maintenance yet still being performed by other contractors. There may also be maintenance tasks that are deemed unnecessary and do not address any failure modes. These should be removed from the PM lists of activities. There might also be activities that should be done but do not exist in the Preventive Maintenance lists of tasks. We can also reduce infant mortality failures through the application of Precision Maintenance. There are also cases where the interval is too short. Precision Maintenance involves performing maintenance work in a consistent, precise, and industry-accepted way. If properly implemented, this means that maintenance should yield the exact same results, no matter who is performing the tasks. Precision Maintenance is much more than merely having procedures on PM. It involves performing maintenance work accurately and precisely, which means that the maintenance should provide the exact same results no matter who is performing the work, whether the work is done by the most or the least experienced maintenance craftsperson in the plant. Precision Maintenance must be the correct culture of the organization.

Installation and Reassembly: Human intervention is involved during a major overhauling activity on the equipment. Although there is little or no human error when dismantling equipment, human error happens mostly when reassembling or putting the equipment back in one piece. The people involved in these tasks should be fully focused on completing the tasks correctly. A guided and detailed procedure on the correct way to overhaul can minimize the chances of human error. This means that if this is how the equipment is overhauled by the most experienced person in the plant, then this should also be the case of the least experienced person thereby producing the same results. There should be no shortcuts on dismantling, reassembling, and using the right tools to dismantle the equipment. Maintenance should also know that there are *do not disturb* portions of the equipment left untouched. Maintenance should have the right knowledge, skills, tools, and guidelines to install, disassemble, and reassemble the equipment.

Faking Maintenance Documents to Comply with PM Specs: If we can refer to the Top Ten Problems on Preventive Maintenance I wrote in my other books, this problem ranks third among the lists. What maintenance is thinking is that it is better to fake the documents than to be incomplete, and besides, this is just a minor thing in which no one will know, or nothing happens after all if I place a check on this activity. Besides, I will be audited if I did not fake this document, or my shift is ending, and I will be late as the company shuttle bus will be leaving in 5 minutes. This is one minor problem that can lead to a dangerous situation. This is one of the major issues if maintenance will be done by these people 100 percent of the time. Operators are never involved in maintenance because the operations manager just wants these operators to operate and produce output. You see, if Autonomous Maintenance is established in an industry, there will be inspection and other tasks that will be done not only for maintenance but also for operators themselves.

Not Performing a Functionality Check on Protective Devices: A protective device is placed in the equipment to serve a particular function, which is to protect something. That something is called the protected function. Protective devices should be inspected at the correct frequency because these failures are considered hidden. The only time we know that these protective devices had failed is if the protected function also failed. When both protective and protected functions are both in a fail state, then, we experienced a multiple failures. This is what we are totally avoiding. Try to think of your protective device as a parachute. If you jump off the plane with your parachute and it malfunctions, or it did not open when you press the button, do not panic as there is always a reserve parachute in place. When it is time to open the reserve parachute, and still it did not open, then all I can say is that it will be the end of your life. These protective devices, such as led, beacons, alarms, sensors, emergency stops, are strategically placed in the equipment as a defense to alert operators and maintenance of upcoming problems, deviations, and abnormal conditions.

Here is a classic case problem. April 20, 2010, Deep-Water Horizon Oil Spill in the Gulf of Mexico. According to a federal investigation officer, vital warning systems on the Deep-Water Horizon oil rig were switched off at the time of the explosion to spare workers from awakening up by false alarms. The revelation that the alarm systems on the rig at the center of the disaster were disabled and that key safety mechanisms had also intentionally been switched off came in from testimony by a chief technician working for Transocean, the drilling company that owned the rig. Mike Williams, who maintained the rig's electronic systems, gave evidence to the federal panel in New Orleans investigating the disasters, which killed 11 people. Mike Williams also told the hearing that no alarms went off on the day of the explosion because they had been inhibited or switched off. Sensors monitoring conditions on the rig and in the Macondo oil well beneath it were still working, but the computer had been instructed not to trigger any alarms in case of any adverse readings since it always gives a false alarm which wakes up the people mostly early in the morning. Both visual and sound alarms should have gone off for the sensors detecting fire or dangerous levels of combustible or toxic gases.

Making Maintenance a Jack of All Trades: You are the mechanical technician today for dynamic industries that always keep reorganizing their people; next week, you will be transferred and become an electrician. New top management was hired, and after a week of observing the plant, he decided to reorganize his staff, and you were a part of it. I have been employed in a dynamic industry where there are changes in the organization almost every month. Is this a good thing? Honestly speaking, I just can't tell. But I see a lot of wasted projects and improvements that were abandoned due to the reorganization. All I can say is that if you plan to make maintenance a jack of all trades, try also to consider the anthropometric factors.

While Non-Maintenance Induce Errors are errors that were not manifested by the maintenance function and cause by others beyond maintenance. This is a much bigger problem because maintenance may have little or no control over the situation.

MRO Spare Parts is Not Managed by Maintenance: MRO Spare Parts are not controlled or managed by maintenance in many organizations today. This is a big problem as the best people to manage the storeroom are the maintenance people because they are the parts' users. They know these parts better than any other people in the organization. While in some industries, they can be managed by Supply Chain, Warehouse, Purchasing, Finance, Admin., or even others. You see, the storekeeper's role is not only to issue and receive parts from the vendor. The role of the storekeeper is also to maintain the parts inside the storeroom. For example, belts should not be placed in a cold area, as resiliency property will be lost. Bearings that are in boxes should not be placed vertically but horizontally. Large motors and rotating components inside the storeroom should be rotated once or twice a week to avoid any chances of false brinelling, especially if there is some form of vibration felt in the storeroom. It should be regreased every 6 months. There will be a wide array of problems that will happen if maintenance does not own the storeroom. I would strongly recommend that maintenance be the ones to own and manage the storeroom as they are the users of these spare parts.

Purchasing Going for the Lowest Bidder: In most industries, purchasing will source for at least three bidders for a particular part, but the end result is that they will award the part to the lowest bidder. So the Purchasing Manager together with their people, are celebrating because they have saved a few cents on the part but what they do not realize is that they make the life of maintenance much more difficult and miserable because the part breaks easily and cost thousands or even hundreds of thousands in US dollars a year not to mention the cost of downtime and repair. There were 3 vendors for this type of oil, and the purchasing awarded the oil to the lowest bidder. What purchasing do not know is that the Flash Point of the oil of this vendor is at 300 ° F or 149 ° C, which caused a lot of evaporation and tapping up of oil on the maintenance side, and now they question maintenance why this equipment consumes too much oil. There is always a temptation to purchase equipment or spares based on the lowest or cheapest vendor. This might not be a good idea since purchasing based on the initial costs only tells us one side of the story. The true costs can be seen based on its performance and total life cycle costs and not on the initial costs. In my experience, this is the problem with most procurement and purchasing departments since they decide to purchase parts based on the lowest bidder or lowest possible costs. The savings that these departments claim are insignificant since both operations and maintenance can encounter many failures. They always look at the part's initial cost and not the cost of problems the part may give the user in the long run which is part of the running costs. This is the problem with purchasing cheaper parts since the equipment will fail easily and will likely yield a much larger cost in a period compared to the slightly higher cost of equipment or spare. All I can say is that buying cheap parts may be expensive, and buying expensive parts may be cheap in the long run.

OEM Design Errors: Since God did not create the equipment, no equipment is perfect by nature. It will always contain flaws and inherent design weaknesses. Although this can be addressed by the maintenance by deploying the Planned Maintenance pillar of TPM. Planned Maintenance Phase 2 will address this issue. What is important for maintenance is to observe the equipment, list all the parts that fail and break easily, and challenge the team on how to extend the lifespan of these parts through modification or redesign.

When a person commits an error or mistake, you will be the talk of the town. Why are people inclined to blame other people for their mistakes? Our boss gets irritated with the error maker and wonders how maintenance can be irresponsible, careless, silly, incompetent, or stupid, especially when the error has damaging consequences. To battle this issue, the boss gets mad and yells at the person that committed the error. They are punished, some disciplinary action was given, and the boss will ask someone to write a new procedure. We blame or shame the culprit, or on the soft side, the person is sent back to training. Many organizations focus their limited resources on preventing the error to the level of the human maintenance that committed the error. In summary, most organizations will change the human condition instead of the conditions and environment humans work in. Management and decision-makers should treat errors as an expected part of maintenance work since human errors are part of being human.

100 years ago, machines were not fully automated, so the consequences of human error may not be as devastating compared to nowadays. Today the implications and consequences of maintenance human error can cause far greater harm, damage, and devastating consequences, leading to loss of lives and environmental damage. For as long as we rely on human hands and minds to maintain our assets, we face the irony that maintenance is a significant, or many say the major cause of failures. Every industry has its procedures, SOP, specs, or whatever you name it, and most of the people that wrote them rarely, if ever, have really carried out the activities under real-life situations or conditions, or some had already retired with the industry. Although human error cannot be eliminated, the good news is that human error can be managed. If man created the problem, then let the man find ways to solve the problem.

As Michael Jackson in his song says, blaming is part of human nature, but it is not a good thing about humans. Errors and mistakes are part of our life, and we will continue to do them. According to Roosevelt, the only man who makes no mistakes is the man who never does anything. Remember that the more we blame, the less we understand and learn from the failure. It is easy to say that mistakes and errors make us stronger, yet we still practice our human nature of blaming others. In fact, the truth of the matter is that we can only learn from our mistakes if we look ourselves in the mirror and accept the fact that we too are part of the problem. There is no shame in failing. I failed many times in my life, in my class, in my work, in my teachings, in writing my books, and in so many things, and I admit to it, correct it, and become better. There is no shame to fail and in failing, but what we need to understand is that before we can move on, we need to accept it wholeheartedly.

6.4: Classic Case of Human Error - The Sinking of the RMS Titanic

One of the classic cases of human errors I can think of is the RMS Titanic sinking, which I covered in my Root Cause Failure Analysis Master Class. I think most of us know this classic story. On April 10, 1912, RMS Titanic set sail from Southampton, England, on her maiden voyage to New York. At that time, she was the largest and most luxurious ship ever built. At 11:40 PM on April 14, 1912, she hit an iceberg about 400 miles off Newfoundland, Canada, despite several warnings from other ships. Although the crew had been warned about icebergs several times that evening by other ships navigating throughout that region, she traveled at a near top speed of about 20.5 knots when one grazed her side and met her faith. There were 2,223 people on board, and out of those, 1,517 people died, and there were around 706 to 708 people who survived. The overall survival rate for men was 20%. The majority that survived belong to the upper class or rich people. For women, it was 74%, and for children, it was 52%. Most of the people that died on the Titanic did not drown but were caused by hypothermia. People can only survive for a few minutes at a cold, freezing temperature. Many of those who didn't get into the lifeboats were alive when they went into the water, but the water was so cold, which was at a freezing temperature, that they could not survive very long after being immersed at this temperature. Before we proceed with the human errors committed on the RMS Titanic, let us discuss the physical cause of why the RMS Titanic sunk:

Figure 6.3: RMS Titanic A Classic Case of Human Error

Figure 6.4: RMS Titanic the Elegance Inside the Ship

The RMS Titanic was the biggest ship ever built during that time. It was not only grandeur in size but also luxurious and elegant too, especially for the first-class passengers. There were decks for the first-class, second-class, and third-class passengers. Only the elite and rich people can afford the first-class which has almost all the comforts of a 5-star hotel. The rooms for the first-class people vary in size depending on how much you can pay for it. The food was served according to its class. The second-class accommodation and facilities were spacious compared to first-class amenities in other ships of the time. 284 passengers boarded the second-class which had a total capacity to accommodate 410 second-class passengers. The third-class accommodation was also comfortable. A dining room provided the passengers with simple and provided three meals a day compared to other ships in which you need to bring your own food during the voyage.

Physical Cause on Why the RMS Titanic Sunk: In 1994, metallurgists studied a piece of the Titanic's hull retrieved from the wreck site. The plate was 1" thick and still showed remnants of the original paint. Scientists used a Charpy test to determine the brittleness of the steel. The sample is held in place in that test while a pendulum is released and swings down as it hits the sample. The forces in the impact area of the pendulum are electronically monitored. The sample of steel used in modern shipbuilding and a sample from the Titanic were cooled to about 28 °F or -2.2 ° C, the same temperature of the water at the time of the collision with the iceberg. Modern steel was tested first. When the pendulum hit the sample, the test piece bent into a "V" shape. When the Charpy test was repeated with the sample from the Titanic, the pendulum broke the sample in two, which means that the Titanic's hull steel was brittle when it hit the iceberg. Many human errors occurred that led to the sinking of the RCM Titanic that caused the lives of 1,517 people. Let us discussed this one by one.

Divers prepare a piece of *Titanic's* hull (the "Big Piece") to be hauled aboard the research vessel 'Abeille'

The "Big Piece" being recovered by the research vessel 'Abeille'

Charpy test

Figure 6.5: Failure Analysis on the RMS Titanic

Human Error Number 1: The Officers Do Not Know the Capacity of the Lifeboats on RMS Titanic: It was a fact that had the Officers filled the lifeboats according to their specification, an additional 470 people could have been saved. Most lifeboats were not full and not even half because the officers on board were unsure how many people each lifeboat could carry. This was very unlikely for an officer of a ship. Should the officers know exactly how many each lifeboat can carry, an additional 470 people can still be saved. Just imagine that human error. Titanic carried 20 lifeboats: 14 wooden boats, each with a capacity to carry 65 persons; 2 wooden cutters, each able to carry 40 persons, and 4 collapsible boats with a 47-person capacity. Altogether, the 20 lifeboats were built to carry 1,178 people, but around 705 to 708 people were saved and could make it to the lifeboats. Officers on board a ship should know the emergency procedures and precautions, especially if the passengers' lives are at stake.

Investigating Equipment Failures Through Root Cause Failure Analysis

Although the RMS Titanic faith was already doomed that night, a total of 470 people can still be saved if each of these officers on board this ship exactly knew each lifeboat's capacity. You might also be wondering why the RMS Titanic carried a few lifeboats instead of having enough lifeboats to accommodate all the passengers on the ship; well, that will be another set of human errors discussed as you try to read along what went wrong with the RMS Titanic.

- Total passengers on Titanic   2,223
- Maximum Lifeboat Capacity   1,178
- Actual Loaded   708
- Passengers that can be saved   470

HUMAN ERROR 1: TOTAL NUMBER OF PEOPLE IN THE LIFEBOAT

TIME LOADED	LOCATION	OFFICER	LIFEBOAT #	CAPACITY	LOADED	PERCENT	NOT SAVED
12:45 AM	Starboard	William Murdoch	7	65	28	43.08%	37
12:55 AM	Starboard	Harold Lowe	5	65	36	55.38%	29
12:55 AM	Port	Charles Lightoller	6	65	28	43.08%	37
1:00 AM	Starboard	William Murdoch	3	65	32	49.23%	33
1:00 AM	Starboard	William Murdoch	1	40	12	30.00%	28
1:10 AM	Port	Charles Lightoller	8	65	28	43.08%	37
1:20 AM	Port	Charles Lightoller	10	65	35	53.85%	30
1:20 AM	Starboard	William Murdoch	9	65	56	86.15%	9
1:25 AM	Port	Charles Lightoller	12	65	30	46.15%	35
1:30 AM	Port	Harold Lowe	14	65	58	89.23%	7
1:30 AM	Starboard	James Moody	13	65	65	100.00%	0
1:35 AM	Port	James Moody	16	65	40	61.54%	25
1:35 AM	Starboard	William Murdoch	15	65	65	100.00%	0
1:40 AM	Starboard	William Murdoch	Coll - C	47	44	93.62%	3
1:45 AM	Port	William Murdoch	2	40	25	62.50%	15
1:45 AM	Starboard	William Murdoch	11	65	70	107.69%	-5
1:55 AM	Port	Charles Lightoller	4	65	32	49.23%	33
2:05 AM	Port	Charles Lightoller	Coll - D	47	24	51.06%	23
2:20 AM	Starboard		Coll - A	47	???	#VALUE!	#VALUE!
2:20 AM	Port	Charles Lightoller	Coll - B	47	???	#VALUE!	#VALUE!
TOTAL				1178	708	60.10%	470

Figure 6.6 Actual Number of Passengers Loaded on the Lifeboat

Human Error Number 2: The Law Itself was the Problem: When the lifeboat needs were finalized, the general feeling was that the modern ship was engineered and built so well that even if a ship was in a situation where it might sink, there would be plenty of time for other ships' to rescue the ship in distress. Why were there so few lifeboats on RMS Titanic? Well, believe it or not, the Titanic actually had exceeded the number of lifeboats required by the Board of Trade at that time. The regulations, ratified in 1894, applied to ships of 10,000 gross tons or larger, indicate as ships increased in size over the years, the lifeboat requirements stayed the same; this law was only ratified when the RMS Titanic sunk.

The Titanic was designed to carry 48 lifeboats, but the White Star Liner, which was the owner of the RMS Titanic, decided that passenger comfort was more important. They believed that an increase in the number of lifeboats (beyond 20) would have cluttered the decks and taken up valuable space. Besides, the RMS Titanic already complied with what was required by the law. The law was only changed after the sinking of the RMS Titanic.

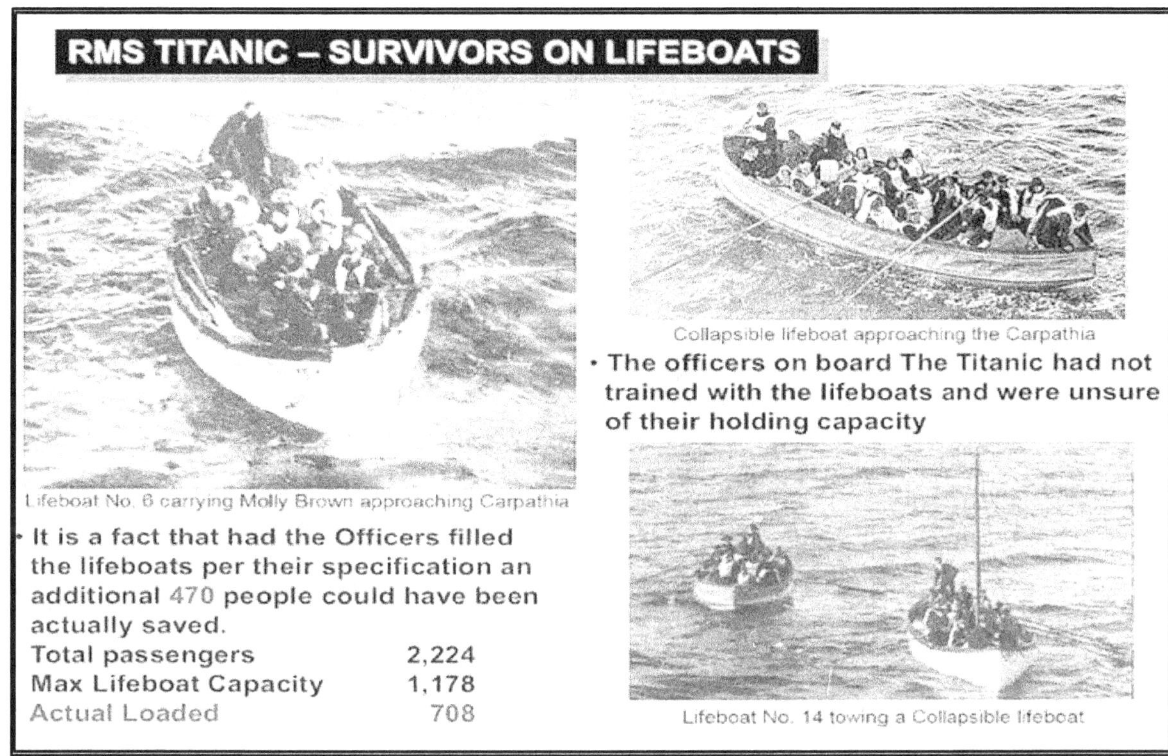

Figure 6.7: Limited Lifeboats on RMS Titanic as it is not required by Law

Human Error Number 3: The Lookout Did Not Have Any Binoculars: Having binoculars might have prevented the RMS Titanic from hitting the iceberg. Fredrick Fleet was the person on duty as the lookout. He had no binoculars during that night. If the lookout had binoculars with them, they could spot the iceberg from a much further distance. Remember that it was a cold night on April 14, 1912, with a freezing temperature. Looking with your naked eye might be quite a painful experience. In a test done to determine the stopping distance of the RMS Titanic, the ship was accelerated to 20 knots, and then the engines were reversed at full power. Remember that ships have no brakes and the only way to stop a ship is to put the engines in a reverse position where the propellers will rotate at a counterclockwise rotation. The distance required to stop the Titanic was about half a mile. When the iceberg was spotted by the naked eye, the distance between the ship and the iceberg was less than a mile.

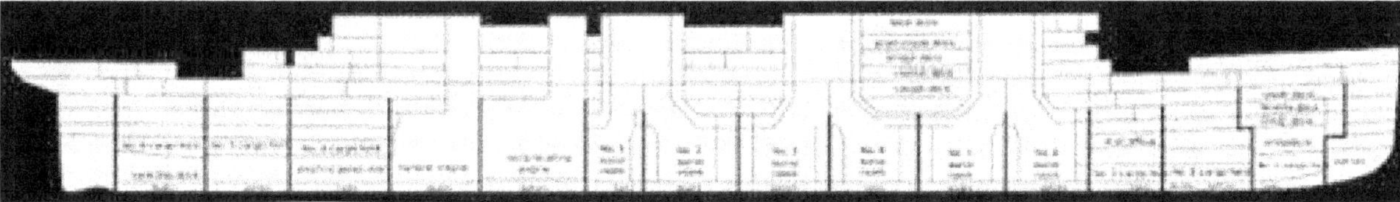

Figure 6.8: RMS Titanic 16 Compartments

Human Error Number 4: Design Error on the 16 Compartments: The Titanic design was that the door for each compartment was watertight, but each compartment was not completely watertight. For the compartment to be watertight, each of the six sides must be completely sealed. The RMS Titanic design problem was that only five out of the six compartments were watertight and completely sealed. The Titanic design extends to the lower decks, making the bulkhead or the upper ceiling not watertight. Making it watertight was either to remove some decks for passengers or make the passengers walk on stairs to the other side, which is not a practical idea. The RMS Titanic had 16 watertight compartments, and the ship could stay afloat with up to four of these compartments flooded. After hitting the iceberg, water began flooding the Titanic's forward's six compartments.

Figure 6.9: Titanic Radio Operator Jack Philips

Human Error Number 5: Radio Operator Ignored Warning of Icebergs: On April 14, the day it hit the iceberg, the RMS Titanic received seven heavy ice warnings, including one from the Californian less than an hour before the fateful collision with the iceberg. The message said we were stopped and surrounded by ice. The RMS Titanic radio operator sent back a message that said, shut up, we are busy. It was at 7:15 pm where the first iceberg warning from the Baltic was posted on the bridge. First Officer Murdoch then ordered a sailor to secure a hatch, from which a light glow might have been obscuring the view on the crow's nest. At 7:30 pm, the Californian radioed Titanic and reported ice. An additional ice warning from the Mesaba ship came in at 9:40 pm but was never taken to the bridge. Should the message have reached the Captain or the officers on deck, they can be more prepared and have slowed down the RMS Titanic, traveling at 20 knots per hour. It was also known that Jack Phillips prioritized delivering the messages from first-class people instead of attending to the iceberg warnings from other ships.

Human Error Number 6: Command Error by First Mate William Murdock: The RMS Titanic sank 2 hours and 40 minutes after hitting the iceberg. It probably took the RMS Titanic about 15 minutes to sink the ship to her final resting place beneath the ocean floor. That means that the Titanic sank at a rate of 10 miles per hour or 16 km/hr. The RMS Titanic hit the iceberg on the starboard or right side of the bow. Immediately right after, the lookout spotted the iceberg and informed the deck. First Mate William Murdock immediately commanded an astern or reversed the engine. Reversing the engine will cause the ship to turn at a slower phase. Should the officer did not order a reversal of the engine, then the ship can turn to the port side or left side at a much faster rate and might have avoided the collision of the iceberg by a few meters. It has also been speculated that the Titanic may have suffered only from minor damage and minimal loss of life and will not sink if it had hit the iceberg head-on with the engines at full reverse. It has also been suggested that the Titanic may have completely avoided colliding with the iceberg had the bridge not requested that the engines be reversed (Full Astern) before steering the ship to the left (Hard-a-port). This action would have decreased the forward momentum of the Titanic, causing it to turn at a slower rate. William Murdock committed two human errors, first, the lifeboats have not been maximized because he did not know exactly how many people each lifeboat can carry, which was very unlikely for the ship's officer where he was the second in command. He was believed to have committed suicide at the time of the sinking.

Figure 6.10: Last Remaining Survivors of RMS Titanic

Figure 6.11 are photos of relics on the ill-fated RMS Titanic that was auctioned. Since the RMS Titanic sank in international waters, there is a debate among the US who want to auction the items and the British Museum to preserve the relics to honor those who perished in the tragedy. As of this point of writing, all survivors from the ill-fated RMS Titanic had died. The last known survivor was Milvina Dean, who died recently last May 2009 at 97 years old. She was the youngest survivor of the Titanic. Milvina was only a baby when she boarded the RMS Titanic. May their souls finally Rest in Peace. But perhaps the biggest human error was thinking that RMS Titanic was unsinkable because of its size and grandeur and defying the laws of Mother Nature. Although technology is advancing as time goes by, nature shall remain undefeated. Let us not be too arrogant with the current technology we have and preserve our nature. Perhaps the greatest lesson we can all learn from the RMS Titanic Tragedy is not to be so arrogant and claim the unthinkable. They thought the Titanic was so big that it was unsinkable just took 15 minutes to finally sink in this final resting place. The wreck of the RMS Titanic was discovered lying upright in two pieces on the ocean floor separated 100 meters apart at a depth of about 4,000 m (about 13,000 feet), which is located at 41° 46' N 50° 14' W. Whatever technology we have right now, we still cannot defy and challenge Mother Nature itself.

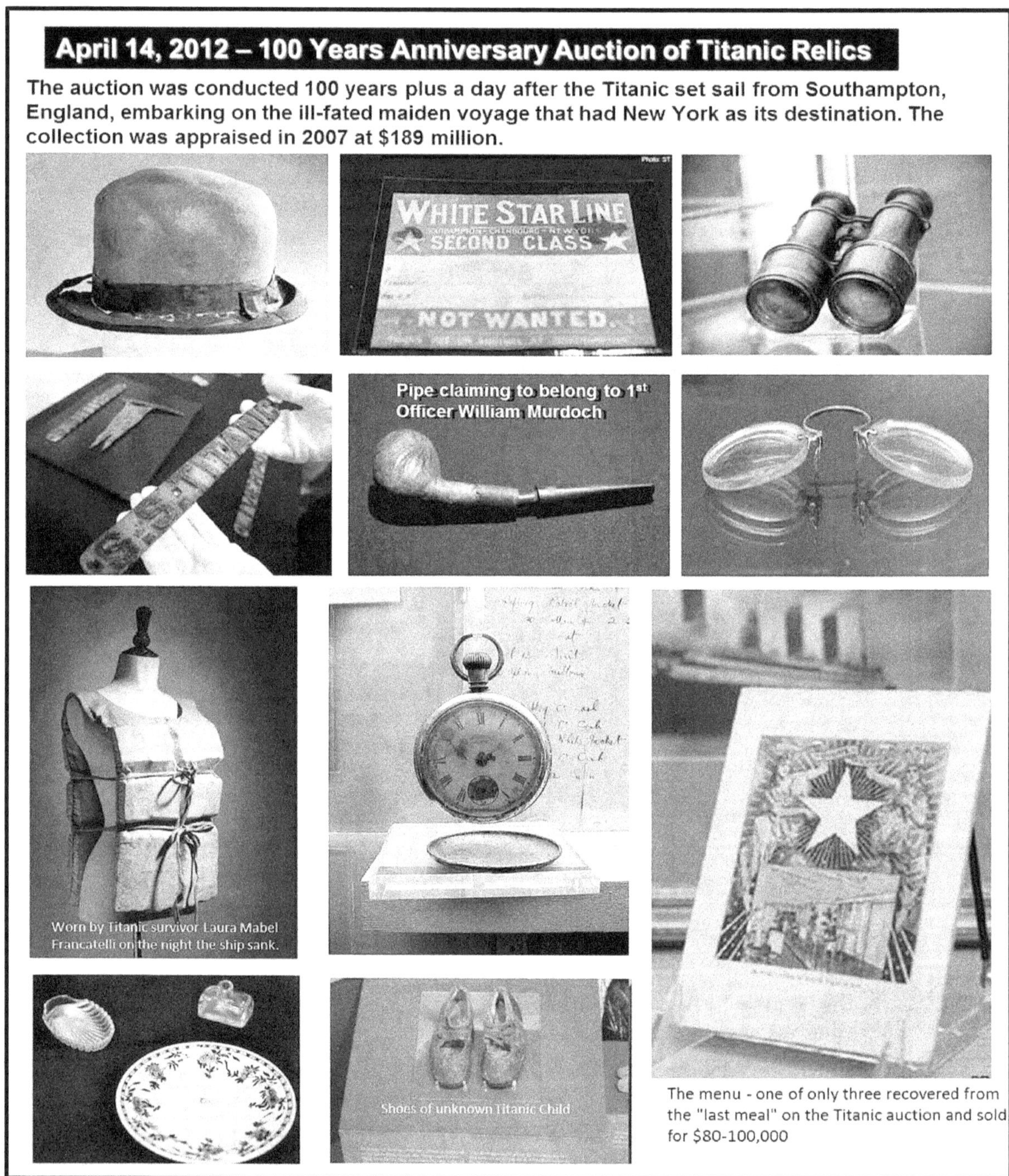

Figure 6.11: Relics Retrieved from RMS Titanic and Auctioned

Human Error Number 7: Bruce Ismay Pressuring the Captain: Bruce Ismay was the Managing Director of the White Star Liner that owns the RMS Titanic. He was also on the RMS Titanic's maiden voyage. It was believed that Bruce Ismay was pressuring Captain Smith to light the remaining boilers so that the ship can run faster in the icebergs' vicinity. The reason for pressuring Captain Smith was due to the stiff competition in the shipping business. Bruce Ismay does not just want the RMS Titanic to be the biggest and most luxurious ship during that time, but he also wanted it to be the fastest ship of his size. Another human error Bruce Ismay did was that he forgot or intentionally did not give the message of the radio warnings to the officers on duty. The message of iceberg warnings remained in his pocket. Bruce Ismay was despised for saving himself while 1,517 people perished when the RMS Titanic sank. Ismay survived by rescuing himself into one of the lifeboats even if the instructions given by the Captain was woman and children first. He was ridiculed by the media and called "The Coward of the Titanic." Bruce Ismay was exiled and spent the rest of his life living out of the public circulation in Costello, Ireland, before returning to London until he died in 1937.

Human Error Number 8: Mistake by Captain Smith: Captain Smith allowed the ship to run at top speed on icy waters. He also allowed himself to be pressured by Bruce Ismay to light up all the boilers so that the ship could run much faster to reach its destination much earlier than expected. There was very fierce competition during that time in the shipping industry. No practice drills for emergency preparation were done. Captain Smith skipped an emergency drill. As a result, there was chaos after the ship hit the iceberg and nobody knew what to do or how to use the lifeboats. Captain Smith ordered the ship at full speed. This was despite the reports of icebergs and a clear dark sky that evening. No proof had been found that Mr. Ismay had previous discussions with the Captain to have the ship go full speed. Titanic Passengers and crew died of cold or hypothermia. The temperature was below freezing. Suppose Captain Smith ordered to return the ship back to the iceberg and docked the passengers on the iceberg; there could only be fewer fatalities. It can take hours for people to freeze, but it would just be minutes before they can die when submerge in an icy cold freezing temperature on the water.

6.5: The Real Heroes on RMS Titanic

Are there really maintenance heroes? There are actually two that I can think of, and they were the maintenance crew of the ill-fated RMS Titanic. The real heroes from the ill-fated RMS Titanic were not Rose nor Jack, but these brave maintenance and engineering crew who never left their post even if it cost them their lives. Joseph Bell, the chief engineer, asked his crew to make a choice, he told his crew that they can leave or continue to serve. All agree to stay with the Titanic and serve their purpose even at the expense of their lives so that the passengers can be saved. All those who serve perished, and there were no survivors. Their names were never remembered.

The engineering, electrical, and maintenance crew stayed manned their post so that the pumps can bail out the water on the compartment to slow down the flooding of the compartments to provide more time for the passengers to board their lifeboats. During the evacuation, the maintenance crew operated the generators to provide power and allow the passengers to proceed on the lifeboats and power the wireless radio system to keep sending distress calls. They also powered the generator to have lights so people could be placed on the lifeboats until the very end. These people bravely kept at their post to save 708 people even though at the expense of their own lives. At 2:20 am, the RMS Titanic sank together with these heroes. No one survived.

Figure 6.12: The Real Heroes on the RMS Titanic

Another group of extraordinary heroes belongs to the musical orchestra of the RMS Titanic. When the Titanic collided with an iceberg, Wallace Henry Hartley gathered his orchestra, who bravely made their way to the main deck to serenade and calm the passengers to avoid panic. The band's last song was titled, Nearer, My God, to Thee, as quoted by Charlotte Collyer, a survivor of the Titanic. The musicians of the RMS Titanic all perished when the ship sank in 1912. They played music, intending to calm the passengers, for as long as possible, and all these musicians went down with the ship. Their names and age were Theodore Ronald Brailey 24, Pianist, Roger Marie Bricoux 20, Cellist, John Frederick Preston Clarke 28 Bassist, Wallace Hartley 33 Bandmaster and Violinist, John Law Hume, 21 Violinist, Georges Alexandre Krins 23 Violinist, Percy Cornelius Taylor, 40 Cellist, and John Wesley Woodward 32 Cellist. All were recognized for their heroism. Kudos to the engineering, maintenance crew, and the musicians of the RMS Titanic, and May you Rest in Peace.

Although everyone knows the story of the RMS Titanic Tragedy, little is known that the Philippines holds the record for the worse maritime of all times. The greatest maritime disaster in peacetime happened on December 20, 1987, when the Philippine inter-island ferry Dona Paz collided with the MT Vector, a small coastal petrol tanker. The impact triggered an explosion in the tanker. The resulting fireball set fire to the ferry and the sea's surface as the spilled petrol spread. Of the Dona Paz's 4317 passengers (only 1586 appeared on the manifest), only 25 survived; all 58 crew and 11 of the Vector's crew of 13 perished. Of the 4386 that lost their lives, including the crew of the coastal petrol tanker, over 1,000 were children.

Figure 6.13: Musicians of the RMS Titanic

Traveling from Leyte Island to the Philippine capital of Manila, the vessel was seriously overcrowded, with at least more than 2,000 passengers who were not listed on the manifest. It has also been claimed that the ship did not have a radio, and the life jackets were locked away. However, official blame was directed at Vector, which was unseaworthy and operating without a license, a lookout, or a qualified master. With an estimated death toll of 4,386 people and only 25 survivors, it remains the deadliest peacetime maritime disaster in history.

Figure 6.14: The Worst Maritime Disaster of All Times

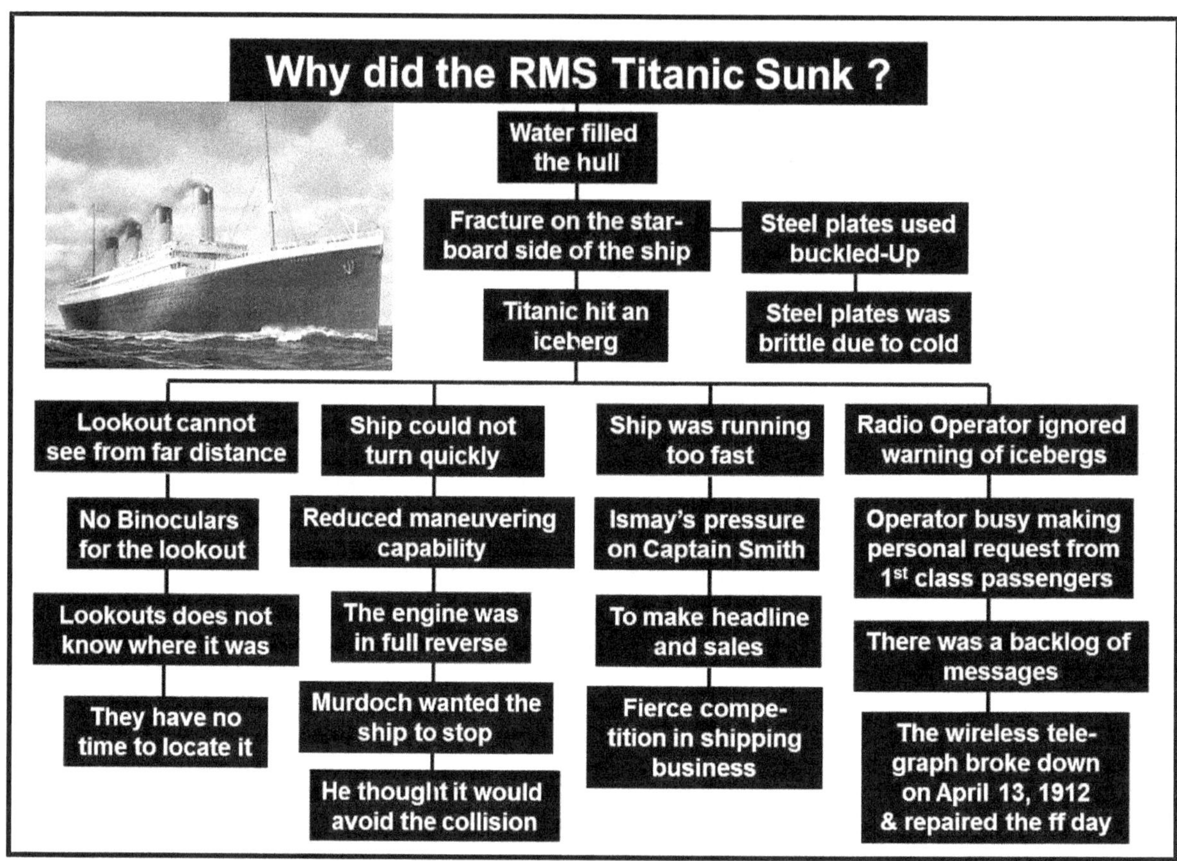

Figure 6.15: Why Did the RMS Titanic Sank

6.6: It's Human Nature to Blame

By definition, human nature is said to be the basic characteristics, feelings, psychology, behaviors shared by people. Although not every human being has the same nature since we all have our different experiences, views, and opinions about being human in our life. This is where the disputes begin. It can also be said that human nature refers to the characteristics of mankind. It has something to do about the way we think, the way we feel and act naturally. Views about human nature are often influenced by cultural experiences. These are the common dispositions, characteristics, and attitudes that are the foundation of our thoughts, and behavior. Life can be hard and cruel at times. By definition, human nature is a concept that denotes the characteristics of human beings such as the way they think, their feelings, the way they feel, and react naturally. Humans react based on their instinct. Instinct can be defined as an inborn impulse or motivation to action typically performed in response to specific external stimuli. It can be said that in certain situations, humans react based on their gut feelings. It can be considered as an intuition, feeling, impulse, or gut feeling. This is how we react to certain situations.

• When a dirty and fouled smell beggar approaches us, we stay away.
• When someone criticizes us, we either deny or retaliate.
• When we hate a post on Social Media, we write negative or derogatory comments or bash them.
• During election campaigns, we vote on people who promise us good things.

- We cry when watching a drama and laugh at something funny.
- We stay away from snakes when we see one.
- We hate banks that follow up on us on our credit cards.
- We blush when we see our crush.
- We avoid people who are rude to us.
- Or, we can be rude or sarcastic to people whom we do not like.
- We resent people who have done us harm in the past.

Figure 6.16: Human Nature by Michal Jackson

We love to talk about people who committed mistakes and errors at work, but nobody wants to talk about their own mistakes. If you ask me why? Because it has something to do with the song of Michael Jackson, "**Human Nature**." As Michael sang in his song "Human Nature," If they say, Why? Why? Tell 'em that is human nature; why, why does he do it that way? If they say, "Why? Why?" Tell 'em that is human nature. Why, why does he do it that way? If you will just listen and understand this song and other Michael Jackson songs, it contains several metaphors. If you can read in figure 6.16, on the second verse, if this town is just an apple, then let me take a bite. My interpretation of this goes way back to the Christian Bible on the origin of the original sin, in which Eve asked Adam to take a bite on the apple since Eve was tempted by the snake. This means that if this is what everyone is doing, then let me just join the crowd. For industries, we want to make changes, or we have recently attended a training that you think can benefit your industry, so you started to implement it, but since, everyone is busy reacting to failures, no one is listening to you and at the end, you just forget about everything and we just mix or join the crowd.

When President Joe Biden pulled out the US Forces in Afghanistan, the Taliban Forces was quick to capture Kabul and now controls Afghanistan. We read a lot of negative remarks from the people, the opposition, and the world. Perhaps the reason why President Joe Biden started pulling out of the US is to finally end the conflict. My question is what if the US Forces were pulled out during the Trump Administration, then the same thing happens and he will be blamed. But what if the US Forces were pulled out during the Obama Administration, then the same thing happens and Obama will be blamed. People love to blame people, we always see the negative things and not the good things. If you are an employee that works hard, has done some improvements, and completed them, nobody will ever talk about that. The good things you have done for your industry will not go viral, but if you committed just even a single error, then you go viral and you will be the talk of the town.

When we dislike someone, we tend to put a touch of Schadenfreude to that person. This means it provides us great pleasure from another person's distress. This is true especially if they perceived the person deserved it because they engaged themselves in a bad deed. If you have a LinkedIn account, try to observe and read the comments of people on whatever topic. Some people just love to criticize, defame, be sarcastic, and rude in their comments. Others will even hit you below the belt. I experience this a couple of times. It's fine to disagree with what we read, but when we comment derogatory or defaming remarks, there's a feeling of satisfaction in what we did. Perhaps we should also consider that the person who posted might know something we do not know. Whatever sarcastic or negative remarks we post to the point of humiliating someone creates a lasting impression on that person and I believe that is not the way to do business. And they say that LinkedIn is for professionals and business liked people. Are they? Again why do people do that? Simple, because it's human nature?

In industries, when someone commits a mistake in which the person was reprimanded, we talked about them. It's human nature to do it. We even go deep down to the character of the human being. As Leo Tolstoy once said, people love to change so many things except for one thing: themselves.

6.7: Is It Possible to Eliminate Human Error in Maintenance?

It is much easier to blame people for their mistakes and errors because it is simply a case of human nature. On the other hand, people that committed the error would always have tons of excuses for not doing the right job rather than taking their time to analyze the root cause of their errors. If possible, they will always look for ways to pass the error to someone else rather than admit it themselves. As a result, most industries have policies, procedures, and rules for people who commit mistakes. If an operator commits an error during operation or when a technician poured the wrong oil into the equipment, we reprimand or, worst, punish them instead of understanding the reason for the failure. Why? Because it is much easier, faster, and most importantly, this is how we do things on our plant since the beginning of time. In short, it had been part of the company's culture for many years. Perhaps if I will rephrase this, I think culture is not the right word. What I meant was it is because of the company's tradition for many years.

If we speak about equipment-related problems, all failures and breakdowns are man-made. From the way, the equipment was designed, fabricated, commissioned, operated, or maintained. There is no such thing as perfect equipment. Every piece of equipment has built-in design flaws and weaknesses that can be addressed by maintenance, reliability, or even the operators themselves.

There are two ways people can go wrong, they can either do something they should not have done or failed to do something they should have done. We get irritated with the person who committed the error and blame the person who committed the mistake. Why? Simple because it is a matter of human nature. We love to blame and see other people's mistakes, but we do not want to see and talk about our own mistakes. We judge people quickly and avoid looking at the Man in the Mirror.

Figure 6.17: Defenses on Human Error on Our Equipment and Assets

If we look at maintenance errors, it is not random, indeed. Most maintenance errors happen during installation, reassembly, overhauling, and repair activities or when maintenance is trying to experiment with how to repair the failure, especially if this is the first time they have experienced this kind of breakdown. We called this experience. These are the things being taught at the University of Hard Knocks.

As a contingency plan and to cope up with these human errors, industries created a defense. There are two kinds of defenses, which include hard defenses and soft defenses. The problem is that each of these defenses has a gap, and even though how many defenses we place, once that gap is breached, a human error occurs. In short, there is no such thing as a perfect defense.

Suppose we study and take a moment to understand human errors; most errors in maintenance are caused by either a slip, lapse, and in most cases, both. Humans make errors. The principle of error management is that even the best people can make the worst mistakes. Often very good people in established organizations keep on making the same blunders. Maintenance errors have been the cause of major industrial accidents and disasters in various industries worldwide. Human error is part of being human. Errors, mistakes, failures are part of being human and part of our life as well. What is important is learning from these errors. If these mistakes and failures we encounter just repeat themselves, industries never learn from their own problem. Whether we like it or not, humans will continue to make errors and mistakes; that's why analyzing the failure is a must. Human errors cannot be eliminated, but they can be managed. Failures allow us to begin more intelligently.

Therefore, whether we like it or not, humans will continue to commit errors and mistakes, and that is why the need to perform a thorough Root Cause Failure Analysis is a must. Failures are not really bad, because it allows us to begin more intelligently. We can only benefit from failure if we can learn from it. When people commit errors, we must steer away from blaming or punishing the person, instead, we need to understand the reason behind the error or failure. Changing policies, procedures, and training is not enough. It also requires a change in oneself. The person who commits the error and the people who accused them should also change because the more we blame, the less we understand the failure. When human errors occur, it is easy to blame or point fingers at someone. It is more important to understand the reason behind the failure.

<u>6.8: Reducing Human Errors in Maintenance</u>

Perhaps if we compare maintenance during World War II, a human error committed may have less impact than today's timeline. The risks, damage, and consequences of failure attributed to human error are simply beyond comparison since the demand for services and products is extremely high compared to during the war. This is where industries earn their revenue and profits. The problem with most industries today is that if they have less revenue, they will result in dangerous cost-cutting schemes tactics. Industries can adopt many strategies to cope with their losses, and cost-cutting is not on my agenda. The problem is industries focus their investment on high-end equipment, software, and technologies, thinking that it will solve their problems. The investment should be in their people since the people will be the ones who will improve their equipment and processes. This cannot be reversed. Here are some recommendations on how to reduce human errors in maintenance.

1. Having a Unified Standard for Quality, Safety, and Reliability: One common thing I see in most industries is that Quality, Safety, and Reliability work in isolation and are independent of each other. But even if Quality, Safety, and Reliability people team up, human errors are still inevitable. The majorities of the World's Worst Industrial Disasters are not only safety-related but maintenance-related and can be attributed to Human Errors. The reduction of human errors can only happen if we stop prioritizing each other and calling ourselves first. Quality, Safety, Reliability, people should not only confine themselves to their department by building stricter rules and policies and trying to police every single employee that creates a very slight deviation from the norm. We are human beings, we love to blame, and that is part of our human nature. We need to change that behavior. Management people believe that operators and maintenance who commit human errors result from carelessness or stupidity. They believe that the only way to eliminate these errors is to outright punish the person who committed the error. Reliability and maintenance people should also educate Safety and Quality people on what reliability is and what it wants to achieve. While these three functions have different goals since Safety will aim for Zero Accidents in the plant, Quality will aim for

Zero Defects and Reliability together with Maintenance people will aim to reduce unplanned breakdowns and improve the reliability of their equipment and assets, each of them share a common vision which is to safeguard the integrity of the plant's existence, stay in business and remain competitive. But the problem is that they work in isolation and are independent of each other instead of having a single vision and mission.

Working together hand in hand instead of isolating each other can dramatically reduce human errors in industries. [7]According to James Reasons and Alan Hobbs, errors are not intrinsically bad. Success and failure spring from the same roots. We are error-guided creatures. Errors marked the boundaries of the path to successful action. What is important is that we learn from the failure as well as the error itself. Human errors will happen, and they cannot be eliminated 100%, but we can manage them more intelligently by learning from the things that go wrong in our industry.

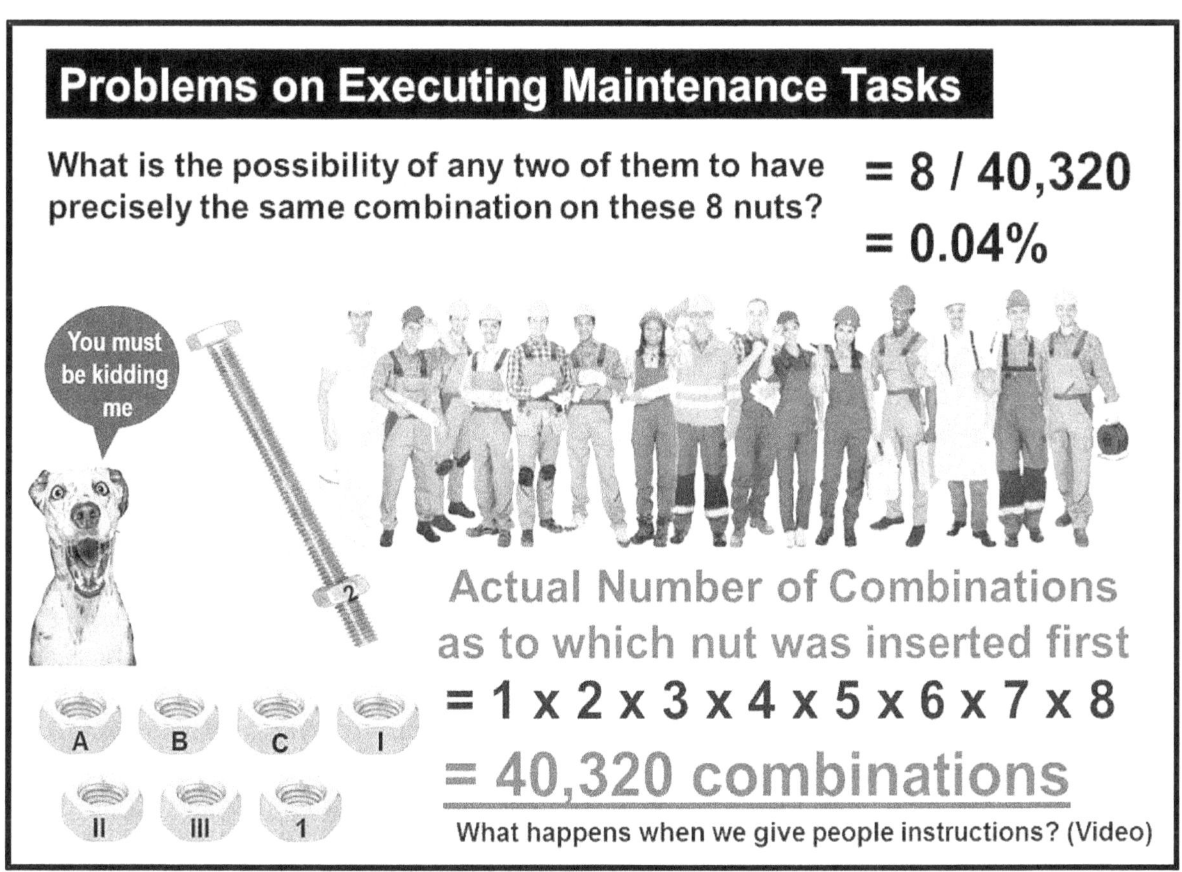

Figure 6.18: How Many Combinations there is for One Bolt and Eight Nuts

2. Application of Precision Maintenance: Let us say that I have one long bolt with eight nuts. I have marked each nut with a code. I have fifteen people with me, and I asked the first person to place the eight nuts on the bolt. I asked them to write their sequence on which bolt came first and so on. After the person completes it, my instruction is to disassemble the nuts and give them to the next person. Once the next person completes, he passed on to the next

---

[7] Reasons, James and Hobbs Anthony, **Managing Maintenance Errors**, Ashgate Publishing Limited, 2003,

person until all the fifteen people completed the assembly and disassembly process. At the end of this exercise, my conclusion is that it is unlikely that any two or more people will have the same precise combination of the first nut they placed until the last nut. Why? The reason is quite simple, if we speak about one bolt and eight nuts, there will be at least around 40,320 combinations here. What I want to drive at this point is that human errors mostly occur during reassembling the equipment, especially when overhauling activities are present. Reassembly is more vulnerable to human error than disassembly. The problem in this case of reassembly is that the chances of detecting the error are very less. You do not have to be a rocket scientist to figure this out; all you need is a modicum of common sense. Application of Precision Maintenance means having guided and detailed instructions in which whoever perform the tasks will yield exactly the same results whether the person is the most or least experience of the craft.

3. Ergonomics: Improve the Working Condition of the People: Ergonomics is the design of equipment, operations, procedures, and work environments compatible with the workers' capabilities, limitations, and needs. It is a vital complement to other engineering disciplines that seek to optimize the hardware performance and minimize capital costs with little or no consideration of how the equipment will be operated and maintained. For example, a salvaged dial thermometer attached to a nozzle just below the third level platform might be a very inexpensive way to accurately measure the temperature in a tower. But from a human point of view, the design is flawed as both operators and maintenance need to climb a ladder, which is around 3 floors, just to get the reading and log it on their records, which needs to be done during their shift. While we always blame people for committing mistakes and errors, we should improve the operators and maintenance working environment and conditions that they are exposed to.

4. Develop a Functionality Inspection for Protective Devices: Equipment Designers placed protective devices on the equipment to alert us of an upcoming problem or error. They may be in the form of sensors, alarms, lighting arresters, overload protection, ground fault protection, overcurrent protection devices, emergency stops, over-speed devices, and so on. They serve as defenses on the equipment to alert operators of an upcoming failure. This means that they should be functional 99% of the time, depending on the criticality of these protective devices. Reliability-Centered Maintenance called this Failure Finding Task or functionality inspection. These will be the only maintenance tasks that can be provided for these devices to ensure that they are still functioning in the equipment, asset, or system that they are placed in. Failure of most of these devices will be hidden. The only time that maintenance or operators know they have failed is when the ones they are protecting also failed, which is called the protected function. Try to assess the risks of damage. Suppose the risks are high enough and unacceptable; these functionality inspections should be regularly carried out. Everyone should be informed of the consequences and the risks involved in both the protective device and protected function failure. Remember that they are placed in the equipment for a reason, and that is to protect something. That is why they are called protective devices. The questions to ask will be what could happen in the event under consideration did occur? Is anyone likely to be hurt or killed as a result? How likely is the failure to occur, and is the risk acceptable?

5. Tidiness and Housekeeping: It is less stressful to work and operate in an environment where good housekeeping is in place rather than working in a workplace where everything is untidy. Provide adequate lighting in the workplace. Implementation of 5's will be a good starting point to address housekeeping issues and reduce the chances of slip and trip hazards. People fatigue easily if the workplace itself is untidy. Slip and trip accidents in the workplace account for many injuries attributed to poor housekeeping. This is less likely to happen if good housekeeping is in place in the organization. It also creates discipline with us, and that good housekeeping starts in our home.

6. Simplify your Procedures: If we look at all the procedures we used in our organization, we may have outdated or even obsolete procedures that are still being used and implemented until this very point in time. Perhaps these procedures were effective when it was written but would no longer hold true for now. Today, there are equipment that had already been disposed of, and some procedures still exist and are still being followed up to this point in time. Procedures should be simple, precise, and should be easily understood by the user. What is important is that if there is a way of simplifying the things being done, procedures should likewise be revisited, reviewed, revised, and simplified whenever necessary. Avoid complicated words. When industries invited me to their plant for in-house training, I can already feel how things are being done before entering the plant. How long security took for me to be allowed inside. Sometimes due to very strict procedures, it will take me 30 minutes or even longer to enter the plant. Although we understand their point, I sometimes think that the bigger the industry, the more problems exist. Sometimes I told my sponsors, especially if this will be an in-house training if they have coordinated my entry to their security ahead of time? I remembered in 2006, I was scheduled to go to this semiconductor plant for two days of training on Maintenance Indices and KPI; at their security, you need to enter a metal detector where you will be scanned just like in an airport. When I entered, it alarmed, and the security told me to remove my belt, so I did; then, after removing my belt, I passed the detector again, then it alarmed for the second time. So the security told me to remove everything in my pocket, which I complied with. Once again, I entered the metal detector scanner, and it alarmed again. I told the security that I have some metal braces on my leg and asked the guard on duty if he wanted me to call my doctor and remove them so that I can enter to conduct my training? The security called his office, and the officer still cannot make a decision. They called the human resources. That was the only time I was able to get inside and conduct my training. The time I lost from security alone was around 45 minutes. If you ask me, what was the name of this plant? Don't bother; it's already closed as they moved to China. When you plan to visit me at my residence, give me a call or message when you come, and I'll instruct my dog that you will be coming. I can guarantee you that you will be inside my house in less than a minute. Why can't industries be like that?

7. Use of Visual Controls in the Workplace: Visual controls are placed in the workplace to expose problems and deviations easily by just looking at them. It makes deviations easily seen by operators by alerting them that something is not right. Visual controls are designed to make the control and management of the plant as simple as possible. They originated from the Japanese. These visual controls are not meant to make us dumb or look stupid. It simply entails making the problems, abnormalities, or deviation from standards much more visible to everyone in the plant. When these deviations are clearly visible and apparent to all, corrective actions can immediately correct these problems and anomalies. Visual Controls allow us to see problems more easily. They are also meant to provide instructions and convey information easily. These are devices or mechanisms designed to manage or control our operations process to meet the following purposes. They make the problems, abnormalities, and deviation from standards easily known to the operator and everyone in the organization. It will easily display the operating or progress status, provide instructions, convey information, and provide immediate feedback to people. Remember that the simplest solution that works will probably be the best. Good visual control management should tell the operator immediately if a process is going beyond the specified parameters.

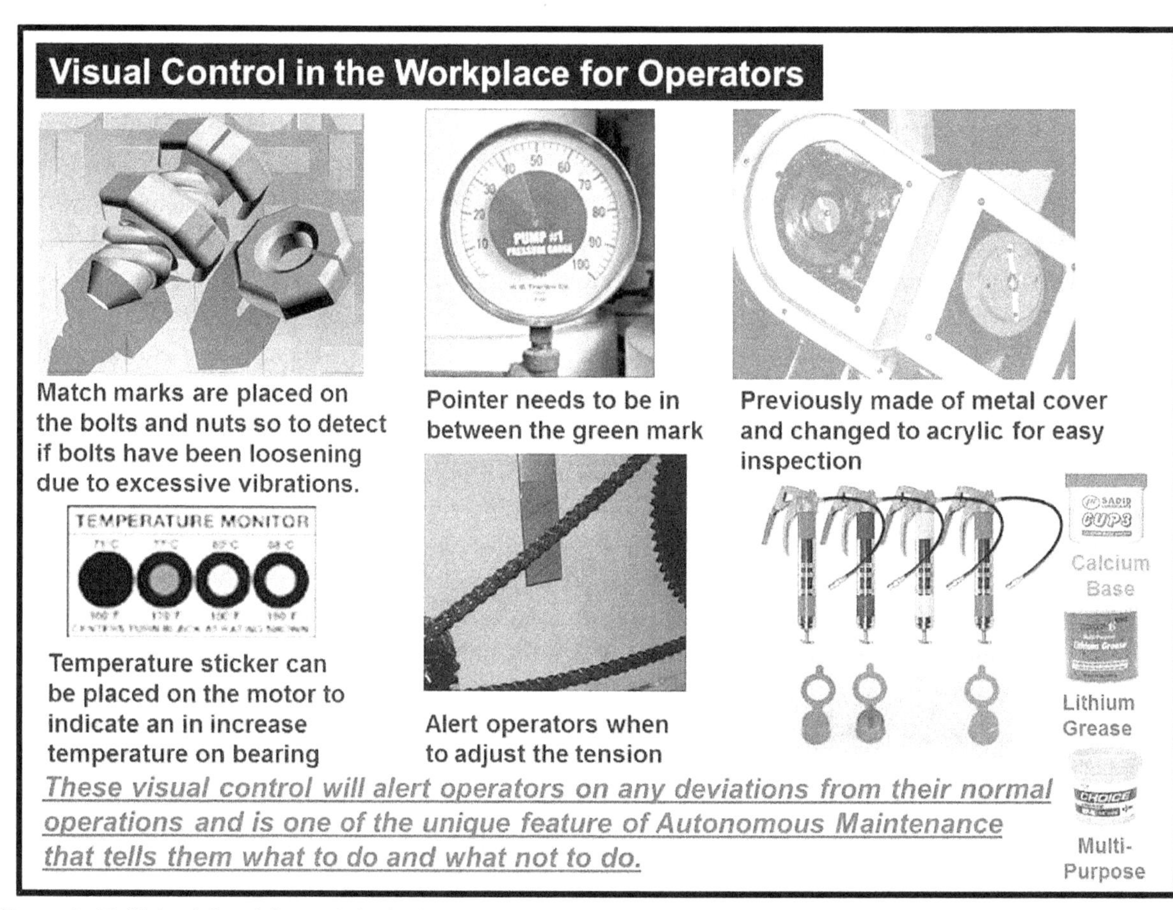

Figure 6.19: Using Visual Controls in the Workplace

Ideally, the process would be stopped automatically. Like when you are inside an airport, signs can guide the passengers to save them a great amount of time instead of asking directions on knowing where you want to go. Visual Control is also sometimes called Visual Communication. In the early 1970s, U.S. executives went on tours to different Japanese industries to see why they were being beaten so badly in quality by their Japanese competitors. They noted one specific thing: the workplaces were filled with pictures and diagrams of what to do and what not to do. With the way things were displayed in the plant, anyone passing by, whether a U.S. executive or a new employee, could sit down and do the assembly themselves without even knowing the Japanese language or asking for instructions. The most important benefit of visual control is that it shows when something is out of place or missing.

8. Perform Root Cause Failure Analysis on Recurring Failures: Although I would not recommend that for every single breakdown or failure, a Root Cause Failure Analysis should be performed as this will be a very tedious process indeed, but when the failure keeps on repeating itself, then I think it is high time to perform a Root Cause Failure Analysis investigation and take things slowly. Note that a Root Cause Failure Analysis is different from implementing problem-solving or analytical tools such as Fishbone, Pareto Analysis, FMEA/FMECA, 8 Disciplines, and the like as explained in the previous chapter. The failure should be fresh when conducting a Root Cause Failure Analysis, which means it just happened. This will require a Principal Investigator and a group of people to gather evidence to shed light to see the bigger picture of why the equipment failed. Root Cause is evidence-driven. The goal of conducting a Root Cause Failure Analysis investigation is to understand why and how the failure occurred so that we can learn from the things that go wrong in our industry. Conducting a Root Cause Failure Analysis will allow the investigating body to determine the Physical cause of the failure, the human cause, the system cause, and finally, their probe and investigation on the latent cause of the problem.

9. Apply Poka-Yoke Solution: Poka-Yoke is a Japanese term that means mistake-proofing. Poka-Yoke examples include elevator alarms, 110/220 volt or auto volt appliances, low battery indicators, fuel indicators in the car, child lock door, and so on. Poka-yoke is any mechanism used by industries, especially manufacturing processes, that helps an equipment operator avoid errors and mistakes. Its purpose is to eliminate product defects by preventing, correcting, or drawing attention to human errors as they occur. It was developed by Shigeo Shingo as part of the Toyota Production System. The term Poka-yoke was applied by Shigeo Shingo in the 1960s to industrial processes in which its primary purpose is to prevent human errors. Poka-yoke aims to design the process to detect and correct mistakes immediately, thereby eliminating defects at their origin.

10. Enhance Maintenance and Operator's Training: The saying that people are the companies' greatest asset is not always true. I believe that the right people are the company's greatest asset, and the wrong people are called liabilities. We can only have the right people if people are equipped with the right knowledge to do their jobs right the first time around. But nowadays, the only training maintenance got is from the University of Hard Knocks. As the great leader Nelson Mandela said, education is the most powerful weapon to change the world. In my first book on World Class Maintenance Management, The 12 Disciplines, I indicated that training is the foundation and backbone of any cultural change. This will be the foundation for setting up the basic maintenance strategy in the plant. Maintenance, as well as operators, should be trained and educated first. But the sad part is that training is one of the primary targets for cost-cutting schemes in most industries. While industries invest in adding hardware resources to expand their business, they forgot to invest in their human resources. Suppose industries deprived their people of training; my advice is to educate yourself, read, surf the internet and join maintenance forums. Training and gaining knowledge can come both formally and informally. For the operator's training, their coach and mentors will be the maintenance people in implementing the seven Steps of Autonomous Maintenance, where the ultimate goal is to empower operators.

My last work was in the mining industry in the Philippines. It was located in Mankayan,

Benguet, 16 hours from Manila by bus or 45 minutes by plane. Our main office was in Makati, but I used to teach in the plant in Benguet. When I teach, everyone was happy because of an abundance of food. They even have some take-away at the end of each training day. One day, the finance manager called me and questioned why I have a big budget for food. My answer was simple, I do not want to teach a person with a hungry stomach. After a month, I went back to the plant for another batch of scheduled training. I noticed that there was no food during the break time, which I learned was the finance manager's directive. When I went back to the plant, I confronted him and told him that you are saving on small pebbles while the big savings will come from the people I teach. That day on, I filed my resignation and never returned to that plant the following day.

11. Develop a Risk Criticality Assessment for Failures: I believe that this is one area where both Quality, Health, Safety, Environment, Reliability, Maintenance, and operations people should sit down, relax and communicate as one team and discuss this matter thoroughly and develop a comprehensive list of all possible critical failures that can occur in their industry and assess the risks involved. Not all failures are critical; the problem sometimes or many times (I just can't tell) on maintenance is that all failures are considered critical according to them. That is why their storeroom is loaded with excessive stocks that were not even used which ended up as a non-moving item. When the equipment failed and the stock was not around, the boss yells at you, making the spare part critical. The criticality of the part will depend upon the consequences of the failure. Remember, for a critical failure, the risk is high, and the aftermath is unfathomable. What I am saying is that critical failures seldom happen but is possible to happen, and when they do, the impact and consequences of the aftermath will definitely hurt your industry's reputation, image, and business. The worst failure you can ever imagine will be those failures that can affect the environment. Failures that can cause harm, injury, or even death are likewise considered critical. For failures in which the boss gets mad. Would you consider this critical? I leave that good judgment to the reader.

12. Remove Departmental Barrier for Better Communication: If you are employed in an industry, that industry will be divided into different departments or functions, and each department will have its own set of cultures, which can be different from other departments. I am not recommending getting a sledgehammer and remove the partition walls that divide every single department. For all I care, what I am saying is that each department should sing just one song. While Finance people love to sing the song of Cyndi Lauper, *Money Changes Everything*. At the same time, Operations, Safety, and Quality want to sing the song from Police, Every Breath you take, Every move you make, Every bond you break, Every step you take, I'll be watching you. Maintenance wants to sing the song from Beatles; when I found myself in times of trouble (maintenance is troubleshooting and having a hard time fixing the repair), *Mother Mary comes to me speaking words of wisdom, let it be.* (It seems it cannot be fixed, so let it be) The problem is they sing their songs simultaneously at the same time, and what we got is simply noise. I am saying that all departments from industries should sing just one song so they can be in harmony. The only time these people communicate is when you are called for in a meeting. Sometimes your time in the plant is consumed by attending so many meetings starting in the morning till evening to end up with the same old problems that occurred again and again. Break down barriers between departments. People in research, design, sales, and production must work as a team to foresee their production problems.

According to Dr. Edward Deming, 14 Points of Management. Just telling people to work together doesn't do much good if the management system drives them to different behaviors. Such support for teamwork is merely a slogan without the necessary management commitment. Management can support but never commit to the initiative. We need to change the management system and the behavior of those in leadership positions in the organization. When we create incentives to optimize parts of the system, the overall system is sub-optimized. To achieve the best overall results, individual parts of the system may suffer and sacrifice. When we evaluate people and provide bonuses and promotions to optimize a portion of the system, this creates pressure against cooperation across departments. When departments have to compete for budget and staff, this builds barriers between departments like the US Republicans and Democrats. When departments have their budgets to protect and spend, it often creates barriers. Often, this even progresses to the point where employees are considered more a part of one department than company employees. If they transfer to other departments, that is seen as a sign of disloyalty. The supposition is prevalent worldwide that there would be no problems in production or services if only the production workers would do their jobs in the way they were taught. In reality, the workers are handicapped by the system, and the system belongs to the management. As I said, there are no proponents for more barriers between departments as a management strategy. And there is often talk about the importance of cooperation and that we are all one team, but when the management system is structured to undermine that notion, it is not much use to tell people to work as if optimizing the overall system being desired. The management system needs to encourage the behavior the organization wants to see. Too many organizations still have difficulty breaking down barriers between departments due to the management systems they have already in place that work against their goals. Likewise, Safety, Quality, and Reliability people cannot eliminate human errors 100%. Stricter rules will only make matters worse. We need time to set aside our priorities, pride, ego politics, bureaucracy, and understand each other's strengths and weaknesses. There is no such thing as first. Learning from one another would make a difference instead of working in isolation.

13. Get Enough Rest: Do not make it a habit of working too late at night just to get some overtime pay unless our work is really needed. As humans, our bodies also need to get enough rest. According to the circadian rhythm, when we work in the morning, our energy rises until 1200 hours and decline slowly in the afternoon. When you have lots of work to do, prioritize them according to their importance and deadline. Try to avoid the nightlife during weekdays where you need to wake up at 0600 hours to prepare for work. Having 8 hours of sleep has many benefits and one of them is improving our memory and alertness which can reduce the chances of committing human errors. According to a study from the AAA Foundation for Traffic Safety, those individuals with 6 to 7 hours of sleep can double the chances of getting you into a car accident, while for those with 5 hours of sleep or less, the chances of getting into a car accident quadrupled. When a person's body has enough sleep, our immune cells and proteins get the rest they need to fight common colds and flu. Another benefit of having enough sleep is that it will lessen the chances of heart problems such as high blood pressure and heart attacks. This is because lack of sleep can cause the heart to release cortisol, a stress hormone that triggers your heart to work harder. Cortisol works with certain parts of our brain to control our mood, motivation, feelings, and fear.

14. Forgive but Never Forget: When someone did something wrong and apologizes to us, humans will always say that it's fine; let us forget about that and move forward. This will not be the case with Root Cause Failure Analysis. Almost all industrial accidents are a result of human errors. These errors will be remembered and written down in history not to ridicule or embarrass the people involved and the industry itself but to remind us what went wrong so that industries can finally learn from the things that go wrong. Humans have flaws, and we are not perfect by nature. We will continue to commit these errors every day as long as we continue to do the same old things. Reminding us of these errors is what makes industries better.

6.9: The E-Experiment

Typically, if a product defect reached the customer and the customer complains about a defect in manufacturing industries, corrective action will be conducted to address the problem. A typical corrective action is to conduct a 100% visual inspection of the product, but what most industries do not know is that this is not a fool-proof solution. In fact, performing a 100 % visual inspection can only guarantee 60 % accuracy. I usually conduct this experiment in my class on World Class Maintenance Management – The 12 Disciplines and Root Cause Failure Analysis. After this experiment, two conclusions are evident. First, not all delegates will get the correct answer, which means that 95% or more will likely get the wrong answer. The second is that not everyone will get the same answer.

Figure 6.20: The E-Experiment

Try this Exercise on the E-Experiment: The instructions here are to count all the letters "e" in figure 6.20. It does not matter whether the letter e is a capital letter or not. Try to do your best to count all the letters e only by using your eyes. The correct answer will be found in appendix A at the back of this book. Count all the letters e, including the title "The E-Experiment." in figure 6.20. The correct answer is in Appendix A of this book. If you got the correct count, then well and good, but 95% or even more who will try this just by using your eyesight will commit a lapse, meaning you missed something

Chapter 7

Understanding the Latent Cause of the Problem

> *All physical failures are triggered by humans, but humans are negatively influenced by latent forces. Therefore, the goal of any RCFA investigation is to expose and identify these latent causes. Root Cause concludes, once the latencies had been exposed and identified.*

7.1: Latent Cause Explained

Organizational Latent Causes is when the whole organization is asking what is it about the way we are that contributed to our problems and what are we going to do about it? Personal Latent Causes is when the whole organization is asking what is it about the way I am that contributed to our problems and what am I going to do about it? This is looking at the man in the mirror. When something goes wrong, we will all try to understand why each person did what they did to such an extent that we're convinced that we would have done the same thing if we were in their shoes. Big things go wrong because we do not act on small things. Small things that go wrong are the most valuable phenomena of life. Waiting to do RCFA on big problems will just assure a continuance of bigger problems.

The word latent means hidden, secret, concealed, invisible, lurking, veiled, inherent, unseen, dormant, immanent, or unrealized. What is it about the way I am that contributes to our problems? What is it about my being and behavior that I need to change? What are my shortcomings? A system cause will always end up being procedural, while a Latent Cause is about asking why I need to wait for a failure before I can change something. What is hindering me from doing it? Why do I need something bad to happen before I react? What do I need to change on the "man in the mirror?" When we continue to address system causes and neglect to correct latent causes, we do not address the real cause of the problem and will likewise assure us a continuance of bigger problems. Root Cause Failure Analysis is performed on three levels. The first level is to determine the physical cause of the failure, but it is also believed that all physical causes were caused by humans. This means that there is always a human involved in the error. In this case, we just cannot punish or blame the human for causing the error since something triggered that human to commit the error. If the error was unintentional, we need to explore the reason behind why the person committed that error until the depths of the latencies have been exposed. Industries must understand that there are so many cases that the human who committed the error is not actually the root cause of the problem. A deeper cause led to the error, and there are people involved, which I call ghosts, with an s, who are also responsible for the human error. But the problem is that even these people do not realize their involvement in the problem, thinking that it is all the fault of the person who committed the mistake or error.

The system causes are not yet the latent cause, but it can be said as the start on the probe on the latent cause. Mostly system causes will refer to the procedures, standard operating procedures (SOP), training, or policies implemented in the industry. This means that if the system, procedures, or SOP is flawed, this can lead to failure. Our investigation does not end here. The Principal Investigator and the Evidence-Gathering team should explore and dig deeper into why the system was flawed until the latent cause of the problem is finally exposed. Latencies are where change needs to occur within us.

Important Elements of a Root Cause Failure Analysis Investigation Includes

• Identification of the real problem to be analyzed.
• Identification of the cause and effect relationship that combines the cause and effect outcome of the problem.
• Disciplined preservation of evidence to verify the hypothesis and failure modes.
• Verifying the evidence to determine those that connect the problem and create a sequence of events.
• Identification of all physical, human, system, and latent causes associated with the problem and undesirable outcome.
• Developing corrective actions and countermeasures to prevent the same and similar causes in the future.
• Effective communication and translating the lessons learned to others in the organization.

7.2: Ending Our Probe on the Latent Cause of the Problem

One of the biggest confusions in performing a thorough Root Cause Failure Analysis is understanding how deep we should pursue with the analysis or simply stated, where do we stop the investigation in performing a Root Cause Failure Analysis? Going too deep will lead us to the Christian bible, Timothy 6:10, "For the love of money is a root of all evil." Going too shallow will allow the problem to resurface again and again. Although many analytical tools are currently being used by industries in the market today, are they really meant to uncover the root cause or simply what we term as the Physical and Human Cause of the problem only? One industry found out that John Doe was responsible for closing the valve that completely wrecked one of the turbines. John Doe was suspended from work for one month without pay. The lawyers, insurance people, and managers were happy that the culprit was finally condemned. A year later, the same problem occurred and this time by a man named Johnny Thorr, and he quote, "I just swept dirt around here, but the supervisor instructed me to close the valve, and I didn't know which valve to close, so I closed them all. The question being raised here is John Doe and Johnny Thorr, the Root Cause of the problem. Are we certain if John Doe is punished, the problem will be gone for good? Therefore, before we can begin any further with our analysis, we need to ask why we need to perform a Root Cause Failure Analysis investigation.

- Why do maintenance squirrel parts? Emergency buying

Why

- Why do we have a lot of emergency buying? Stock outs

Why

- Why is there a lot of stock outs? Physical and system inventory don't match.

Why

- Why does the Physical and system inventory don't match? Maintenance can access the storeroom during peek hours.

Why?

- Why does Maintenance have access to the storeroom during peek hours? The storekeeper is from 8 to 5 pm, while operations is 24 hours.

Why?

- Why is the storekeeper just from 8 to 5 pm if operations is 24 hours?   Decision makers think they are saving money by just hiring one shift for the storeroom

Why?

- Why does decision makers think that they are saving money by just hiring one shift for the storeroom?

Why

- From the Bible on 1 Timothy 6:10, For the love of money is the root of all Evil.

Answer

Figure 7.1: When Do We End Our Probe on RCFA?

Figure 7.2: Valve that John Doe and Johnny Thorr Closed

For example in figure 7.1, one of the most common problems experienced in industries is maintenance squirreling spare parts on their own. Squirrelling is the act where maintenance keeps the spare parts themselves. Why does maintenance do this? It is because there are a lot of stock-outs, which means that the part reflects in the system, but physically it is out of stock. The reason behind this is that both the system and physical inventory do not match. The next question is why the physical and system inventory do not match? It is because the maintenance can access the storeroom during peak hours since the storekeeper is from 8 to 5 pm and operations is 24 hours. Why is the storeroom only from 8 to 5 pm, while operations are 24 hours? The reason is that the management thinks that they are saving money if the storeroom will only be manned for 8 hours. If we continue with this case, then it will end up again on some cost-cutting schemes where money is the root of all evil, but what management and decision-makers do not know is that they are losing money. If you own a 7-11 Convenient Store, try to be open for 24 hours and just provide an 8 to 5 pm crew and you will understand what I'm saying.

Do we simply do root cause investigation to comply with customer's requirements, such as manufacturing plants? Do we perform a Root Cause Failure Analysis because our management wants to know what happened, or do we perform a Root Cause Analysis because we want to know what simply caused the problem? Industries must perform root cause for the right reasons, and there is no better reason than learning from the failure itself. If we want to learn from the failure, we need to change how we do things around the plant. We need to learn from the things that go wrong in our industry.

Again, do we learn from the failure when we discipline someone, or do we just expand the gap between the people and us? **John 8:2—11** and the scribes and Pharisees brought unto Him a woman taken in adultery; and when they had set her in the midst, they said unto Him, "Master, this woman was taken in adultery, in the very act. Now Moses in his law commanded us that such should be stoned: but what sayest Thou?" This they said, tempting Him that they might have to accuse Him. But Jesus stooped down, and with His finger, wrote on the ground as though He heard them not. So when they continued asking Him, He lifted up Himself and said unto them, "He that is without sin among you, let him first cast a stone at her." Before we can understand Latency, we need to ask ourselves, what is it about the way we are that contributes to our problems? What is it about the way I am that contributes to our problems? This means that before we condemn someone are we, or are we not also part of the problem?

If we are part of the problem, then we must be responsible for being part of the solution. Engineers, technical people, maintenance can easily arrive at the physical cause of the problem. Suppose some mechanical component such as bearing fails, these people can dig up evidence that will eventually lead to its physical cause, but digging deeper into the latent cause; I strongly recommend that the principal investigator and evidence gathering team must be an independent and unbiased person unless you can carry out the probe yourself because this matter is very sensitive. If you really want to get to the real root cause of the problem, you'll end up probing a person's life, their parent's personal lives, their grandparent's lives, the origin of the original sin, and all the way to "**Adam and Eve**." The key to any root cause investigation is knowing when to stop and when the latencies of the problem have been discovered and addressed, then this is when we end our probe on the latencies. There is always a deeper reason why the failure occurred, and industries should explore the depths of the latencies to find the truth and finally learn from the failure itself.

7.3: Are Latencies Good or Bad?

When we reached our probe and end up with the latencies, one question that perks upon my mind is if latencies are good or bad. We do not need to address latency as something bad or something good since latency is both good and bad. Bad in the sense that these latency issues had provided our problems in our industry in the past, such as equipment failures, near misses, accidents, downtime, safety issues, environmental issues, operational issues, and so on. It took a great amount of time to finally understand and learn from them. I believe that this is not something good.

Latencies can only be considered good if we learn from them; otherwise, we will revert back to the same things we used to do until the problem come. As Thomas Edison said, let us not call it a mistake; call it an education. Now we know of 10,000 things on how not to make a bulb light.

Figure 7.3: What Do You See?

The difficult part about latencies is admission, or admitting our fault. Nobody wants to do that, since we have our pride and ego. Our human nature tells us that if we do that, then we will be weak in the eyes of others. I remember my first work, when I was an oiler onboard MV Sea Raider in 1987, during one of my duties, the engineer on duty told me to close the valve for the air, but what I closed was the valve of the cooling water and the diesel engine almost overheated. The Chief Engineer noticed that there is no cooling water exhaust and immediately went down to the engine room and shouted who was sabotaging his ship. One shift later, I knocked on his room and admit that I was the one who did it and said that I was not familiar with all the valves or which is which.

Bad may not be bad, it may be good, and good may not be good; it may be bad, depending on how we look at the situation. Just think about figure 7.3; the question is, do you see a beautiful young woman or an old woman? It's our perceptions that make us see a situation as good or bad. Every adversity, failure, handicap, physical ailment, adversities, unpleasant circumstances, and experience carries with it the seed of an equivalent benefit, often in some hidden form.

A story was told that an old man once lost a horse. His neighbors all came to console him, but the old man said it may turn out to be a good thing. A few months later, the horse came back by itself and brought another horse with it. This time, the old man's neighbors came to congratulate him, but the old man said this may turn out to be bad. One day, the old man's son went out riding on the new horse. The horse ran too fast, and he fell down and broke his leg. He spent 6 months in the hospital recuperating which is a bad thing. He met this beautiful nurse who helped and guided him in his struggle to walk again. They fell in love and got married, which was a good thing.

After a few years, the wife has an affair, which the old man's son found out and killed her wife. The old's man's son spent the remainder of this time in jail, which was a bad thing. So are latencies really good or bad? Perhaps latencies may be bad initially, but it can turn out to be good if we can learn from it to become better. In my experience, and in most cases, latencies will involve high people from your organization as they are the ones who make decisions most of the time. And if their decision is wrong or flawed, it seems that nothing ever happened, since no one will admit to it. Perhaps this is one of the reasons why most industries never learn from the failure since their belief is admitting a fault is a sign of weakness, but if we look at it positively, it is actually a sign of strength.

Perhaps if we called the operations manager and asked him to observe that piece of equipment and ask him if the equipment running is good or bad, the operations manager might conclude that it is a good thing since they are producing products. If I look at the equipment, I might not be in agreement with him. Perhaps my thinking is that equipment is in a terrible situation even if it is running since it has some missing bolts and is too dirty.

7.4: Classic Case of Latent Cause: The Challenger Disaster

In the previous chapter, we discussed a classic case of human error that led to the sinking of the RMS Titanic. In this chapter, let us discuss another classic case on Latent Cause about why the Challenger Disaster Exploded.

Morton Thiokol, the OEM of NASA, provided the left and right Solid Rocket Booster to NASA Space Shuttle Program. Morton Thiokol was located in Utah, USA. The SRB is used to provide the explosive force to lift the space shuttle above the ground, during the launch. Kennedy Space Center is where the Space Shuttle Challenger will be launch. They were responsible for all launches and were located in Florida, USA. The distance between Thiokol in Utah, and

Kennedy Space Center in Florida is more than 2000 miles or roughly around 34 hours driving. The Solid Rocket Booster that Thiokol manufactured was delivered via rail in segments, as shown in figure 7.5. Each of these segments has two O-Rings. This means that Thiokol will deliver the left and right Solid Rocket Booster (SRB) segments and assembled them at Kennedy Space Center in Florida, where the launch takes place. Every time a shuttle is launched, the shuttle will be initially assembled with the main fuel tank where the shuttle is attached together with two Solid Rocket Boosters (SRB) on the left and right sides of the main shuttle. After a certain distance is achieved, the shuttle will separate from the left and right Solid Rocket Boosters and from the main fuel tank and proceed with its destined travel into space, while both the left and right SRB's together with the main fuel tank, will return back to earth via parachute and retrieved in the sea. This was the procedure during that time by NASA with Morton Thiokol.

After they have retrieved the Solid Rocket Boosters from the sea, the Thiokol engineers will inspect them for any traces of problems and wear in which the SRBs will be reused once more in their next voyage. This was the standard arrangement by NASA with Morton Thiokol. Roger Boisjoly was the main engineer responsible for the inspection of the Solid Rocket Booster during that time. Once the space shuttle is ready for launch, it will travel a distance of around 3.5 miles to the launch pad using a crawler. It will take around 6 to 8 hours for the crawler to reach the launch pad, refer to figure 7.4.

Figure 7.4: Space Shuttle Travelling to the LaunchPad

The solid rocket boosters for the Space Shuttle are manufactured in Utah and transported by rail. The outer shell of the rocket is made in segments, and each joint between the segments is sealed by two O-Rings. The solid rocket boosters are recovered by a ship when they land with their parachute on the sea and inspected for wear, which will again be reused on the next shuttle flight mission. The job of Roger Biosjoly was to inspect both the Solid Rocket Boosters if they can still be used or needs to be repaired, or replaced.

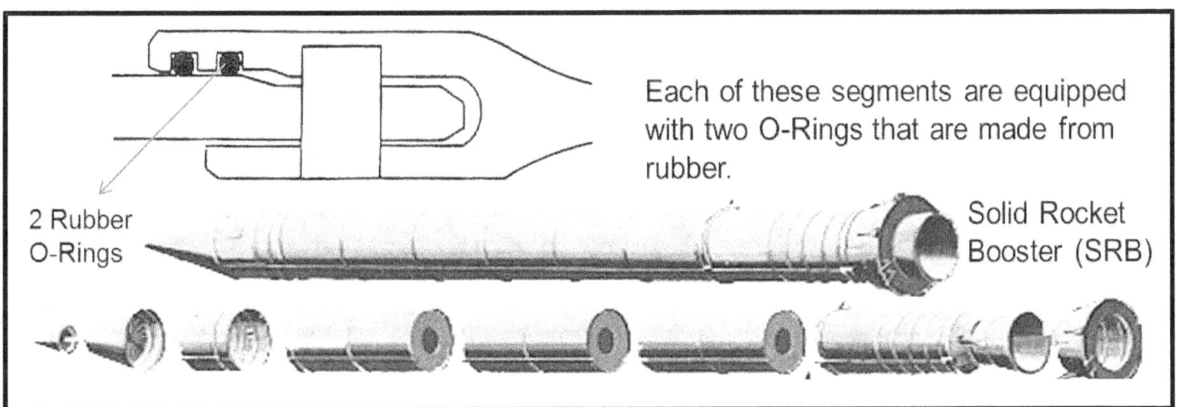

Figure 7.5: Morton Thiokol's Solid Rocket Booster (SRB)

On the other side of the story, this Challenger Mission was unique and has drawn much attention from the public. This was the first time that NASA will allow an ordinary citizen to join the astronauts in space. Only professional astronauts had flown the shuttle, although before the Challenger Mission, except for one exemption were Senator Jake Garn, part of the committee that funds NASA and Sultan Soliman, a Saudi Prince, has joined the shuttle mission previously, but they draw little attention from the mass public. NASA wanted something new that would bring them the needed attention. NASA knows that putting an ordinary citizen who will join the astronauts in space will draw much larger attention.

Figure 7.6: SRBs to Be Retrieved and Inspected

Figure 7.7: Christa Mc Auliffe Winner of the Teacher in Space Program by NASA

NASA decided to launch the Teacher in Space Program after Ronald Reagan announced it on the radio; Reagan said, Today, I am directing NASA to begin a search in all of our elementary and secondary schools and to choose as the first citizen passenger in the history of our space program, one of America's finest, a teacher. After President Ronald Reagan's announcement, around 11,500 applicants for the Teacher in Space Program submitted their applications, and one of them was Christa McAullife. These applicants were screened and Christa makes it to the next round of 114 teacher applicants until the final 10, where Christa McAullife finally wins the Teacher in Space Program to join the Challenger Mission. It was Vice President George Bush Senior who announced the winner. Christa now starts training together with the Challenger crew.

On the other side, during one of the inspections by Roger Boisjoly, he discovers a massive amount of erosion on the O-Ring on previous Flight 51C. This means that the primary seal or O-Ring did not work, and hot gas escaped; luckily, the second rubber O-Ring finally functioned. Although both NASA and Morton Thiokol had data from earlier flights showing that solid rocket gases had burnt and passed the first O-Ring seal on a previous cold-weather launch, they were not aware of the real threat of the problem. Moreover, they were unaware of what it can do with the space challenger and the crew of the Challenger Mission because they always escaped with the problem. Roger Boisjoly called his manager Robert Lund and requested to assign a group to investigate the problem, but in the end, he was dismayed over his management's lack of actions regarding the O-Ring problem. His manager even told him to keep the problem to himself as this might leak to the press. In an interview with him, he said that one month, one lousy month, we realized that we had no power, no authority, no resources, and no management support. How in the heck can you have anything done if you do not have any of those items that I have mentioned? You can't, and the frustration just absolutely goes up minute by minute, day by day.

The problem with the O-Ring had not been resolved, and a schedule was already provided for the new mission on the Shuttle Space Challenger in which Christa Mc Aullife will be joining. The Shuttle Challenger will be launched on January 26, 1986. However, during that day, thunderstorms hit the vicinity of Florida. NASA decided to postpone the launch the following day. On January 27, 1986 (Monday), the external handle used to lock the access hatch was jammed. At around 12:36 pm, the handle was fixed using a hacksaw. The crew was ready to go, but strong winds stop them from launching. NASA decided to scrub the launch. The launch was finally scheduled for the next day, January 28, 1986.

Remember that this is a publicized event because of the teacher Christa Mc Aullife since she was the main attraction of the Challenger Shuttle Mission. The press and media were blasting NASA because of the delays and inability to cope with the launch schedule. President Reagan was also scheduled to speak with the teacher once they are in space. The schedule was again set on the following day, Tuesday the 28th of January, 1986. However, a new problem emerged on the evening of January 27, one day before the launch. The weather forecast predicts extremely cold weather the following day. The following day, the temperature was 10 degrees below freezing; a shuttle had never launched before on such a low temperature. Hence, a decision to fly the following day has to be made. Figure 7.10 shows the temperature during the previous launches of the shuttle mission, with January 28, 1986, being the coldest so far.

On January 27, 1986, the weather forecast indicated a temperature below freezing of around -6.67 ° Centigrade the following launch day. The Morton Thiokol engineers spearheaded by Roger Bosjioly (figure 7.8 Roger Boisjoly inspecting the rubber seal) raised their concern to their managers about launching at a cold, freezing temperature. That evening, which is one day before the actual launch, a conference call (Ms Teams and Zoom were not yet available) between NASA, Morton Thiokol, and Marshall Space Flight Center took place to make a very important decision on whether to launch the Challenger into space or not. I hope you don't mind if I expand a little bit so the reader can follow the story.

Figure 7.8: Erosion on several segments of the SRB

Marshall Space Flight Center was responsible for all rocket and engine systems of the Challenger shuttle, and their plant was located in Alabama. During the conference call were George Hardy, Deputy Director of Science and Engineering; Judson Lovingood, Deputy Manager of Shuttle Project; Ben Powers Engineering Structure, and Propulsion. **Morton Thiokol** is the manufacturer and OEM of the Solid Rocket Boosters. Their plant was based in Utah. During the conference call, present were Jerry Mason, Senior VP of Wasatch Operations; Carl Wiggins, Vice President and General Manager of Space Division; Joe Kilminster VP Space Booster Program; Robert Lund VP of Engineering; Roger Boisjoly Engineer Structures Section and Arnold Thompson, Supervisor of Rocket Motor. **NASA (Kennedy Space Center)** was responsible for all space shuttle launches and was located in Florida. During the meeting, Allan McDonald Space Booster Project from Morton Thiokol, but was currently in NASA during that time, Lawrence Mulloy, Manager of SRB Project, Stanley Reinartz Manager Shuttle Project,

Jack Buchanan Manager, and Cecil Houston Resident Manager. The meeting was called to answer one important question which is, do we launch the Space Shuttle Challenger on a cold freezing temperature tomorrow or not?

Morton Thiokol in Utah manufactured the Solid Rocket Booster. It was transported by rail to NASA in Florida with a distance of more than 2000 miles. The SRB (Solid Rocket Boosters) was done in segments and assembled at NASA in Florida. The purpose of the SRB is to provide the explosive lift of the shuttle from the ground. The Challenger shuttle was attached to the main fuel tank during lift-off. Each segment of the SRB had two O-rings made of rubber that seal to prevent the flammable fuel from leaking. Due to the tremendous amount of vibration during liftoff, there will be a bit of play from each left and right solid rocket booster segment. The SRB's outer shell is made in segments, and two rubber O-rings sealed each joint between the segments. After both the main tank and solid rocket boosters separate from the space shuttle, the shuttle will proceed with their mission, while the fuel tank and SRB will fall back to earth and land on the sea through a parachute. A ship will recover the solid rocket boosters (SRB) as they land with their parachutes on the sea and will be returned back to Morton Thiokol for inspection for any traces of wear, especially erosion so that it can be re-used again during the next shuttle flight mission. That was the protocol from NASA and Morton Thiokol. NASA and Morton Thiokol had data from previous earlier flights showing that hot gases had burnt and escaped the first O-Ring seal on a previous cold-weather launch; luckily, the second O-ring caught the leak. During the previous flight, Morton Thiokol engineer Roger Boisjoly inspects the solid rocket booster and discovers a massive amount of erosion on the O-ring on the previous Flight 51C. The primary seal did not work, and hot gas escaped. Luckily, the shuttle completed its missions and returned back to NASA.

Figure 7.9: A Decision Whether to Launch the Shuttle Challenger or Not

On the night of January 27, 1986, Arnie Thompson, Boisjoly's colleague from Morton Thiokol, received information that the temperature the following day, which will be the day of the launch, will be very cold which is below freezing. They informed their Top Management team and recommended not to launch on this cold temperature in which their management was in agreement. Thiokol informed NASA of their concern not to launch the Space Shuttle Challenger due to the cold temperature. The coldest temperature was on STS-51C, in which the temperature was at 53 ° F (11.67 °C). Roger Boisjoly's inspection on the solid rocket booster on the previous flight 51C on January 25, 1985, indicated massive erosion on the Solid Rocket Booster, which means that the O-Ring malfunction and did not seal the hot gases. Figure 7.10 shows the temperature from the previous Shuttle Missions by NASA. On January 28, 1986, the temperature was at -6 ° C the following morning, which was a cold, freezing temperature.

Temperature During the Shuttle Launch

No.	Flight	Date	Temp.	No.	Flight	Date	Temp.
1	STS - 1	04/12/81	66 deg F	13	STS – 41G	10/05/84	67 deg F
2	STS - 2	11/12/81	70 deg F	14	STS – 51A	11/08/84	67 deg F
3	STS - 3	03/22/82	80 deg F	15	STS – 51C	01/24/85	53 deg F
4	STS - 4	06/27/82	80 deg F	16	STS – 51D	04/12/85	67 deg F
5	STS - 5	11/11/82	68 deg F	17	STS – 51B	04/29/85	75 deg F
6	STS - 6	04/04/83	67 deg F	18	STS – 51G	06/17/85	70 deg F
7	STS - 7	06/18/83	72 deg F	19	STS – 51F	07/29/85	81 deg F
8	STS - 8	08/30/83	73 deg F	20	STS – 51I	08/27/85	76 deg F
9	STS - 9	12/28/83	70 deg F	21	STS – 51j	10/03/85	79 deg F
10	STS – 41B	02/03/84	57 deg F	22	STS – 61A	10/30/85	75 deg F
11	STS – 41C	04/06/84	63 deg F	23	STS – 61B	11/26/85	76 deg F
12	STS – 41D	08/30/84	70 deg F	24	STS – 61C	01/12/86	58 deg F

Coldest Temperature ⟶ 25   STS – 51L   01/28/86   31 deg F

A decision had to be made on whether to launch or not

Figure 7.10: Temperature Comparison for All the Shuttle Launches

The conference call was made to answer one important question: do we launch the Challenger at a temperature below freezing? Note that Morton Thiokol, the manufacturer of the SRB at that time, recommended NASA to abort the mission. Morton Thiokol can only recommend, but the final decision will still have to be from NASA. Although this had always been the case, NASA had no history where the OEM was overruled. This means that NASA always followed their OEM recommendations, just like in most industries. As the evening on January 27, 1986, slowly passed by, the lives of the seven astronauts hang on one single question. How will the O-rings react to a cold, freezing temperature launch? This question was answered by Roger Boisjoly, one of the main engineers at Morton Thiokol, and he said; *at low temperatures, the seals are not so elastic. It won't seal properly. It will be like putting a brick versus a sponge. The grease, too, will be ineffective. It will be thicker and not a slick. We will be having a higher O-Ring actuation time. If that happens, then we will approach a threshold of secondary seal pressurization. This means that we have a situation where neither the primary seal nor the secondary seal will function.*

Figure 7.11: Icicles Forming at the Building and on the Shuttle

Roger Boisjoly said that if we launch tomorrow at a freezing cold temperature, the O-Ring will not function since rubber tends to lose its resiliency during cold. This means that the rubber O-Ring will not be so elastic. When the rubber seal does not function, then hot gases from the Solid Rocket Booster will leak. If we analyze the risk involved in this situation, it asks what harm it can do if the failure happens. This means that the failure had not yet happened; Morton Thiokol," NASA, and Marshall Space Flight were still in the conference call meeting and discussing the outcome. Joe Kilminser, the Vice President of the Space Booster Program, supported Roger Boisjoly's and stated that we recommend to NASA that we do not fly below our database of 53 °F (11.67 ° C), which was the coldest temperature the shuttle had ever flown where Boisjely found the erosion problem.

NASA then asked George Hardy from Marshall Space Flight Center, and Hardy said, Thiokol, I am appalled, but I won't go against the contractor. In that case, I cannot recommend a launch. After a few minutes, Lawrence Mulloy from NASA spoke up and said, for God's sake Thiokol, when do you want me to launch next April? You guys are generating new launch criteria here. Joe (he was referring to Joe Kilminster), we have been flying for 4 years with a known condition on these joints considered and accepted by Thiokol accepted by me and all levels of NASA management. Think about this; think about your data. Mulloy's message was very clear that he is pressuring Thiokol to agree with him to launch. Although NASA will be the one who will decide to launch, all Morton Thiokol can do was to recommend. But never in the history of NASA that they disagree with their OEM and contractors. Jerry Mason, Senior Manager from the Thiokol, asked Joe Kilminster to request NASA for a 5-minute break, and NASA agreed.

Figure 7.12: People Involved in the Challenger Disaster

The Morton Thiokol Managers knew that their billion-dollar contract with NASA is still under negotiations since their previous contract was already coming to an end, and Thiokol knows that other contractors can supply the Solid Rocket Booster besides them. There was even one contractor that can manufacture the SRB in full and not in segments. They know that if they insist on their decision, their contract will be affected. During the conference break, the Morton Thiokol managers voted to decide whether they will recommend a launch or not with just a few hours left. During the Thiokol, a separate meeting by the Top Management people, the engineers were neglected and ignored. Arnie Thompson intervened and explained to the management that the shuttle should not be launch. Jerry Mason, Morton Thiokol's Vice President, just looked at Arnie Thompson, and Thompson just returned to his seat and retained his silence. Roger Boisjoly also intervened and showed photos of evidence of erosion which again Jerry Mason and the managers ignored him. It was clear at this very moment that the managers of Thiokol were changing their recommendation to go for launch. After the Thiokol managers' private meeting, the conference continued, and Joe Kilminster told Lawrence Mulloy of NASA that they agreed to launch the Challenger. He said that the data regarding the blowby is inconclusive. Finally, Lawrence Mulloy spoke and asked all those present in the conference meeting if there were any concerns or disagreements regarding launching the Space Shuttle Challenger. There was absolute silence; nobody dares to speak up, not even Roger Boisjoly. It means that the silence means the Shuttle Challenger will be a go for launch despite the cold temperature.

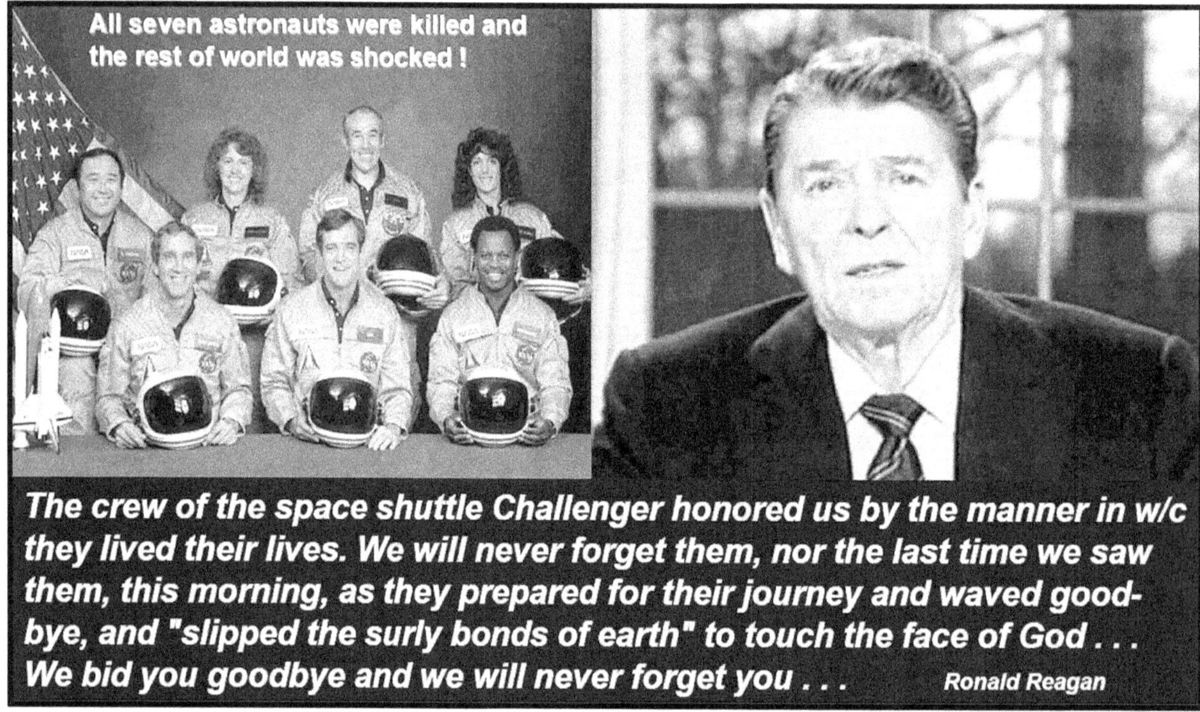

Figure 7.13: President Reagan Pays Tribute to the Challenger Crew

On January 27, the night before the launch, Christa Mc Aullife said her last goodbye to her husband. Challenger now goes for launch, and the crew prepares to enter the Space Shuttle Challenger. Dick Scobee, the Challenger Commander, even informed his wife that he would not be flying because of the cold weather, but he was surprised when he heard the Challenger will be launched despite the cold weather that morning. In the launchpad, icicles as long as 18 inches are now forming on the shuttle. During the day of the launch, January 28, 1986, the crew finally enters the shuttle, and the final countdown begins. They delayed the launch for a couple of hours more to let the temperature rise. Finally, the Shuttle Challenger launched from the pad, and after 74 seconds, it exploded. The shuttle was completely destroyed, and the world was shocked. The capsule where the crew was located was found several days at the bottom of the sea. Nobody survived, and the world was shocked. The Morton Thiokol engineers Roger Biosjoly and Arnie Thompson were right, but the management simply ignored them and agreed to launch. Jerry Mason even said to Robert Lund, who was the boss of both engineers that you need to remove your engineering hat and wear your management hat. This is a metaphor that simply means to think of our business first.

On January 28, 1986, the Challenger was launched despite the Thiokol engineers' recommendations to postpone the launch because of the O-Ring problems and the cold temperature that prevailed during that day. Seventy-four seconds after its lift-off, everything was gone. Two days after the disaster, President Ronald Reagan addressed a grieving nation and said that the crew of the space shuttle Challenger honored us by how they lived their lives. We will never forget them, nor the last time we saw them, this morning, as they prepared for their journey and waved good-bye and "slipped the surely bonds of earth" to touch the face of God. We bid you goodbye, and we will never forget you. *Ronald Reagan*

7.4.1: Physical Cause of the Challenger Disaster

A few weeks after the Challenger Explosion, President Ronald Reagan formed the Rogers Commission after its chairman William Rogers. Their main objective is to investigate the root cause of why the challenger exploded that killed the 7 astronauts. The members of the Rogers Commission that investigated the Challenger Disaster include:

- William P. Rogers, chairman and former United States Secretary of State (under Richard Nixon) and United States Attorney General (under Dwight Eisenhower)
- Neil A. Armstrong (Vice-Chairman), a retired astronaut. He was part of the Apollo 11 mission.
- David Campion Acheson, diplomat, and son of former Secretary of State Dean Acheson
- Eugene E. Covert, aeronautics expert and former Chief Scientist of the U.S. Air Force
- Richard P. Feynman, theoretical physicist and winner of the 1965 Nobel Prize in Physics
- Robert B. Hotz, Editor, Aviation Week and Space Technology
- Donald J. Kutyna, Air Force general with experience in ICBMs and Shuttle management
- Sally K. Ride, American engineer, astrophysicist, and the first female American astronaut in space, flew on Challenger as part of missions STS-7 and STS-41-G
- Robert W. Rummel, Trans World Airlines executive and an aviation consultant to NASA
- Joseph F. Sutter, engineer for Boeing and part of the team that developed the Boeing 747 aircraft
- Arthur B. C. Walker, Jr, solar physicist, and Stanford University professor
- Albert D. Wheelon, physicist and developer of Central Intelligence Agency's aerial surveillance program
- Charles E. Yeager, retired Air Force general and the first person to break the sound barrier in level flight
- Alton G. Keel, Jr., executive director of the commission

[8]The events and communication that followed during the lift-off were sequenced below:

---

[8] Rogers Commission Report, ***Report to the President by the Presidential Commission on the Space Shuttle Challenger Accident***, (Washington DC, 1986)

- 6.6 seconds: Space Shuttle engines ignition

+ 0 seconds: Solid Rocket Booster ignition

+ 7 seconds: "Roll program." (Challenger), Roger, roll Challenger. (Houston)

+ 24 seconds: Main engines throttled down to 94%

+ 42 seconds: Main engines throttled down to 65%

+ 59 seconds: Main engines throttled up to 104%

+ 65 seconds: "Challenger, go at throttle up. (Houston) "Roger. Go at throttle up." (Challenger)

+ 73 seconds: Loss of signal from Challenger

In figure 7.14, black smoke was evident on the right SRB during the lift-off due to the amount of vibration on the challenger and the cold temperature during that day. The SRBs are connected via segments, and each of these segments contains two rubber O-Rings. The main function of the O-Rings is to seal these joints so that hot gases can be sealed and would not escape the joints. During the lift-off, there would be a tremendous amount of vibration from the shuttle, this means that there will be a bit of play from the segments of both the left and right solid rocket boosters. In the picture below, the black smoke means blow-by or a leak. This simply means that the O-Ring did not function on that particular segment to create a seal.

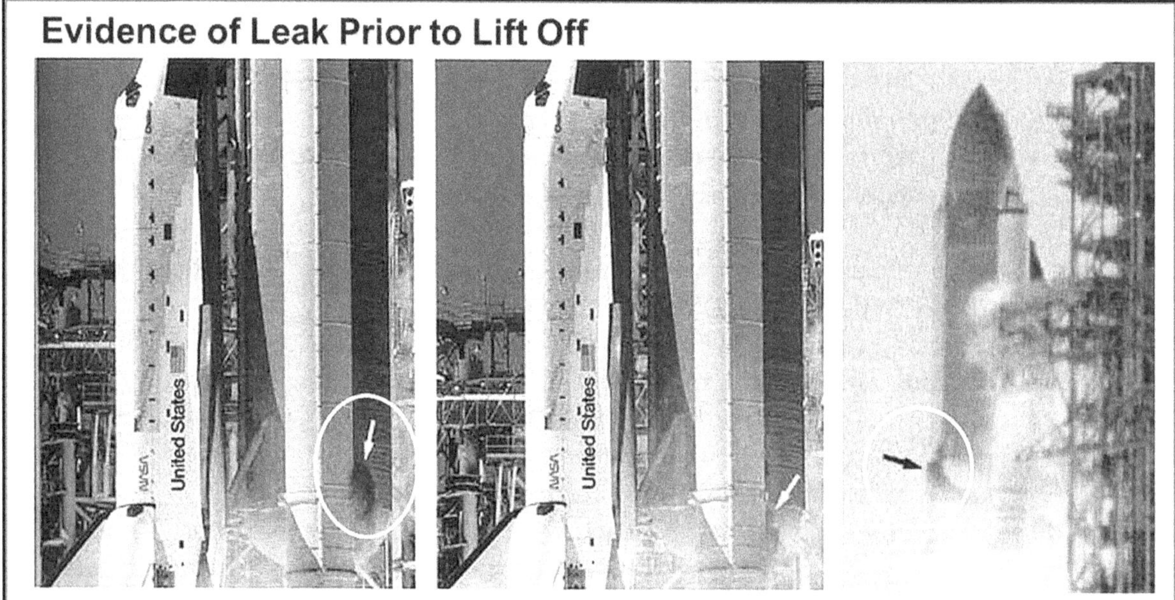

Figure 7.14: Black Smoke Before the Lift-Off

One of the key people in the Rogers Commission was Richard Feynman, a physicist, and a Nobel Prize Winner. He requested an actual sample of the rubber O-Ring and conducted his experiment during one of the hearings at the Rogers Commission to finally get to the bottom of the problem. Feynman used a simple experiment using a clamp and the sample of the O-Ring used in the Solid Rocket Boosters. During his interview with Lawrence Mulloy, the Manager of the SRB Project from NASA, Feynman ordered a glass of water with ice. He got the rubber O-Ring and applied force by locking it up in the clamp. After taking a sip of his ice-cold water, he inserted the clamp on the glass of ice water. During the interview, Feynman said that the O-Ring should be a rubber material since it needs to expand back to create a seal since there will be a play due to the vibration and pressure induced by the SRBs. He said it had to be rubber since the seal should have the property to expand and contract, which Lawrence Mulloy confirmed. Feynman said that if the rubber would not be resilient, then that would cause a leak. Feynman got the O-Ring from his ice-cold water and explained that he discovered that when you applied pressure on this rubber seal and place it on ice-cold water with a temperature of 32 ° F (0 ° C) and remove the rubber O-Ring from the clamp, it did not stretch back and it stays on the same dimension. This means that there was no resilience or expansion on the rubber for a few seconds at 32 ° F. Feynman said that he believed that it had some significance to our problem. Feynman's simple experiment of the O-Ring and the clamp made it simple for everyone to understand the problem, including the non-technical people present during the Rogers Commission hearing.

Although the NASA officials claim that the chances of failure from the shuttle were about 1 in 100,000, Feynman found that this number was actually closer to 1 in 100. He also learned that rubber used to seal the solid rocket booster joints using O-Rings failed to expand when the temperature was at or below 32 ° F (0 ° C). The temperature at the time of the Challenger liftoff was around 32 ° F. [9]The consensus of the Commission and participating investigative agencies is that the loss of the Space Shuttle Challenger was caused by a failure in the joint between the two lower segments of the right Solid Rocket Booster. The specific failure was the destruction of the seals that are intended to prevent hot gases from leaking through the joint during the propellant burn of the rocket motor. The evidence assembled by the Commission indicates that no other element of the Space Shuttle system contributed to this failure.

Figure 7.15: Feynman's Explanation on the Resiliency Loss on the O-Ring

Although Roger Boisjoly already has pieces of evidence of erosion problems on the joint during their last shuttle flight on STS-51c on January 24, 1985, in which the temperature was at 53 ° F or 11.67 ° C. This was so far the coldest temperature recorded on all the shuttle flights during that time. The shuttle remains intact, and the crew returned back safely. Roger Boisjoly

---

[9] Rogers Commission Report, ***Report to the President by the Presidential Commission on the Space Shuttle Challenger Accident***, (Washington DC, 1986)

knows that the rubber O-Ring did not work, NASA was lucky and escaped the failure. The erosion problem had to be due to the coldest temperature they experienced on all their 24 shuttle flights (refer to figure 7.10). This means that the rubber O-Ring did not function and lose its resiliency. However, on January 28, 1986, the temperature was way lower at 36 ° F or 2.2 ° C. No shuttle has ever flown on that temperature. The problem was that NASA was reluctant to abort again unless Thiokol can have solid data that can prove otherwise. The leak that occurred in the Shuttle Challenger was directed towards the main fuel tank.

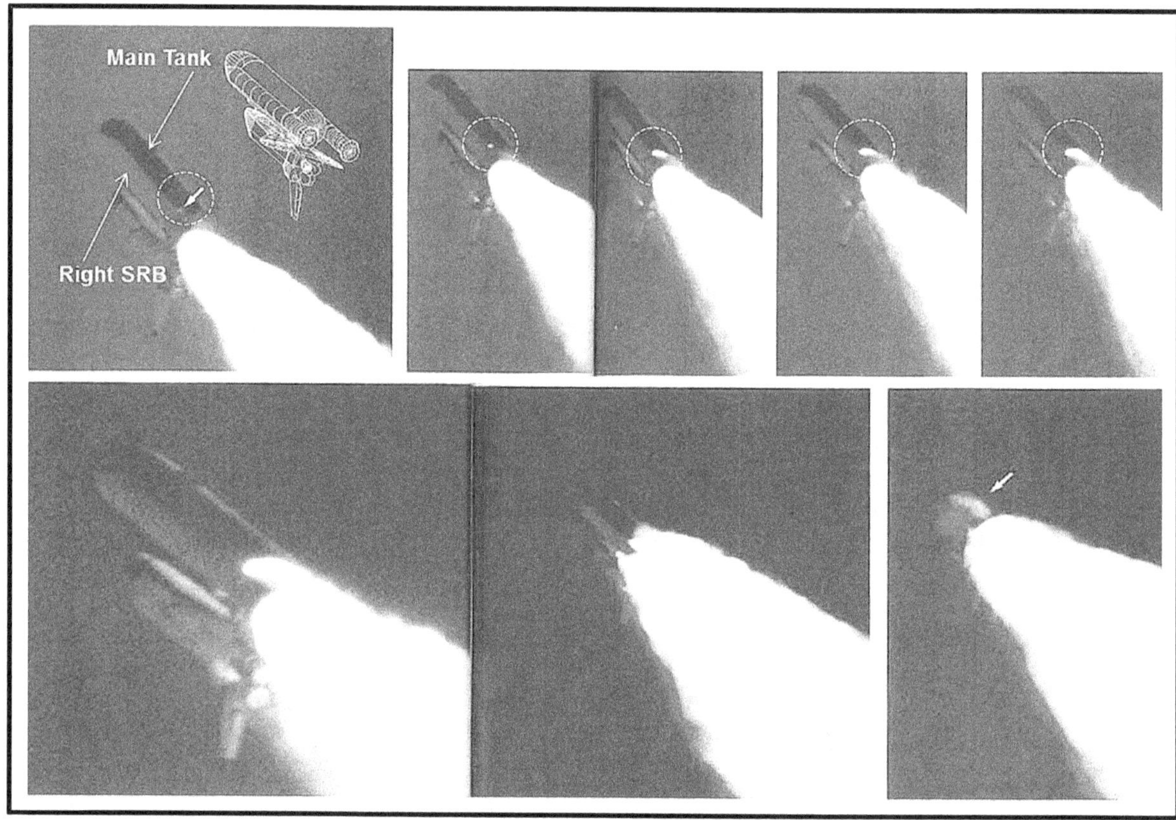

Figure 7.16: Leak from the Shuttle Challenger Directly in Front of the Main Tank

Suppose the leak perhaps occurred on the opposite side of the main fuel tank; in that case, the only problem the crew will have is they will run out of fuel much quicker in which Dick Scobee, the commander, can detached from the main tank and return back to Kennedy Space Center or continue their voyage in space. The problem was nobody knows where the leak will occur. Roger Boisjoly knows that a leak is imminent to happen and warns his managers. The only question that Thiokol's engineers cannot answer is where the leak will occur since the leak can happen anywhere on the segments. NASA was unconvinced of the potential risks as there were also cases of leaks and blow-by during their previous flights that did not harm the shuttle crew. But unluckily this time, the leak on the Shuttle Challenger was facing the main fuel tank directly, as seen in figure 7.16. The cameras on the ground provide strong evidence of blowby on the right Solid Rocket Booster and show a plume of fire. At +72.9 seconds, the instruments malfunctioned. At 73.1 seconds, the bottom of the hydrogen tank falls of the main tank. At 73.5 seconds, the engines of the Challenger shut down, and people on the ground watching realized that something is wrong with the Challenger. Finally, an announcement on NASA stated that there was a malfunction on the shuttle and the bits of pieces of the challenger rumbles into the sea, and the world was shocked.

Therefore, the physical cause of the Shuttle Challenger explosion is that the rubber O-Ring used to seal the joints did not function and lost its property of expansion and resiliency due to the cold temperature during lift-off that allows the hot gases to escape and burn the main flammable fuel tank. Both the primary and the secondary O-Ring did not function and lost their property of resiliency.

7.4.2: Human Cause of the Challenger Disaster

Although during their teleconference with NASA and Marshall Space Flight Center, both the Morton Thiokol engineers and the Management team present during the conference call were unanimous and recommended to NASA to abort the launch on January 28, 1986, with a cold freezing temperature. It was evident that a management decision made by NASA and their contractors was the human error that caused the loss of the Challenger Disaster and the 7 crew including Crista Mc Aullife, the winner of the Teacher in Space Program. Although both Marshall and Morton Thiokol can recommend aborting the launch, it was still NASA that will be the main body to decide for a go or no go for launch. The question was, what caused the Morton Thiokol management to change their decision from not wanting to launch, and agreeing with NASA to finally launch the Space Challenger the following morning on January 28, 1986?

If we follow the story on the Challenger Disaster in section 7.4, Lawrence Mulloy was pressuring Morton Thiokol to agree with him to launch the Shuttle Challenger. Remember that both the Thiokol engineers and management were unanimous about not launching on this cold temperature the following day, on January 28, 1986. Joe Kilminster recommended to NASA during their conference call not to launch below 53 ° F (11.67 ° C) since this was the coldest temperature that they have ever flown the shuttle, but NASA was not convinced unless they can show real evidence of data which the Thiokol cannot provide. So why was it that Thiokol engineers and management do not have any data to support their claims even if the engineers know the problems? Two of the senior engineers from Morton Thiokol were Arnold Thompson and Roger Boisjoly. These people report to Robert Lund, who was the Vice President of Engineering during those times. Both Thompson and Boisjoly have repeatedly informed their boss to investigate the seal problem involving the O-Ring, but their request falls on deaf ears from their boss. Hence there was no concrete data on Rogers Boisjoly's claims regarding the blowby. Although Boisjoly's claims were true, he just does not have a conclusive report to support his claims because he was never given that chance to form a team to investigate the problem with the O-Ring. Another problem, in this case, is that even if Roger's Boisjoly claims were true, it is just not possible to predict where the blow-by or leak will happen on both the left and right solid rocket boosters. If the leak happens opposite the main fuel tank, it will not explode and the only consequence, in this case, is that the Challenger will run out of fuel and may detach and return back to NASA or probably continue their voyage in space.

But the main question here is, why did the Morton Thiokol management team change their minds and agree to launch. They can still refuse, and NASA will decide on the fate of the Challenger. Before we dwell on this question and with all fairness, just try to be in the shoes of the Thiokol Management. During my class on RCFA, I usually explain the Challenger Story, and at the end, we have a team workshop in which one of the questions raised is if you are one of the Morton Thiokol's management present during the conference, would you also agree with them not to launch the Shuttle Challenger? Morton Thiokol had been the contractor and OEM of NASA for their Solid Rocket Boosters for 4 consecutive years, and their contract with NASA is nearing to an end. This means that they need to renew their contract with NASA soon. The management team of Thiokol knows this. During the conference call, when Lawrence Mulloy was pressuring Thiokol to agree with him to launch, Jerry Mason asked Joe Kilminster to request a break to discuss the matter privately. During the break, which took around 30 minutes, the Morton Thiokol management voted on whether to launch or not. Jerry Mason, the senior vice-president, told his management team that they need to vote during that time. He even told Robert Lund, the boss of Boisjoly and Thompson, to remove his engineering hat and put on his management hat. Jerry Mason was telling Robert Lund that we need to be in business with NASA; remember that we are not only the manufacturer of SRBs on the planet, and we might not be able to renew our billion-dollar contract with NASA if we refuse to launch. After they voted and the conference meeting was resumed, Joe Kilmister advised NASA that the data on blowby was inconclusive, and they agreed to launch, which greatly disappointed the engineers, especially Boisjoly. After hearing the decision of the Thiokol team, Lawrence Mulloy said if there were any disagreements, no one replied, not even Boisjoly, there was complete silence. If Roger Boisjoly voiced his concern during that time, perhaps the events would be different, which I cannot tell.

The human cause or human error committed was that the Morton Thiokol management changed their minds and decided to launch for fear of losing their multi-million dollar contract with NASA. The management team of the Morton Thiokol was thinking about losing their contract with NASA if they refuse and stick with their decision not to launch. Another human cause I see in this incident is the refusal of Engineering Vice President Robert Lund to provide a root cause investigation into the previous flights where Roger Boisjoly found erosion on the previous flight 51C. Boisjoly's request was not taken seriously by his management. In an interview with Roger Boisjoly, on one occasion, his management even told him to keep the problem to himself as this matter is too sensitive and might be leaked out, especially to the media press. He was shocked to hear this from his management.

7.4.3: Latent Cause of the Challenger Disaster

To answer the latent cause is to ask what it is about the way we are as an organization that contributed to the problem and what it is about the way I am that contributed to the problem. We need to look ourselves in the mirror and accept the fact that we also contributed to the problem. When we talked about the latency of the problem, this is where change needs to take place.

Richard Feynman provided a separate report from the Rogers Commission and insisted that his report should be included. It was later on decided to include Feynman's Report in Appendix F on the official report submitted by the Rogers Commission to President Reagan. Feynman not only discussed the problem of the Solid Rocket Booster but also covered the previous failures the shuttle experienced from avionics, certification for these engines, Liquid Fuel Engine, and others. He said that Engineers at Rocketdyne estimate the total probability as 1:10,000. Engineers at Marshal estimate it as 1:300, while NASA management, to whom these engineers report, claims that the probability of failure in a flight is 1:100,000. An independent engineer consulting for NASA claims that the probability of failure is between 1 to 2 per 100, which Richard Feynman believes to be a reasonable estimate.

In the conclusion of Richard Feynman's report regarding the Challenger Disaster, he wrote that if a reasonable launch schedule is to be maintained, engineering often cannot be done fast enough to keep up with the expectations of originally conservative certification criteria designed to guarantee a very safe vehicle. In these situations, subtly and often with apparently logical arguments, the criteria are altered so flights may still be certified in time. They, therefore, fly in a relatively unsafe condition, with a chance of failure of the order of a percent which is difficult to be more accurate.

On the other hand, official management claims to believe the probability of failure is a thousand times less. One reason for this may be to assure the government of NASA's perfection and success to ensure the supply of funds. The other may be that they sincerely believed it to be true, demonstrating an almost incredible lack of communication between themselves and their working engineers.

In any event, this had very unfortunate consequences, the most serious of which is to encourage ordinary citizens to fly in such a dangerous machine as if it had attained the safety of an ordinary airliner. The astronauts, like test pilots, should know their risks, and we honor them for their courage. Who can doubt that McAuliffe was equally a person of great courage, who was closer to an awareness of the true risk than NASA management would have us believe?

Figure 7.17: The Challenger Crew

In an interview, Richard Feynman said that, let us make recommendations to ensure that NASA officials deal in a world of reality in understanding technological weaknesses and imperfections well enough to actively try to eliminate them. They must live in reality in comparing the costs and utility of the Shuttle to other methods of entering space. They must be realistic in making contracts, estimating costs, and the difficulty of the projects. Only realistic flight schedules should be proposed, schedules that have a reasonable chance of being met. If, in this way, the government would not support them, then so be it. NASA owes it to the citizens from whom it asks support to be frank, honest, and informative so that these citizens can make the wisest decisions using their limited resources. For a successful technology, reality must take precedence over public relations, for nature cannot be fooled.

Morton Thiokol Latent Cause: The latent cause in the Space Shuttle Challenger Disaster was that the Morton Thiokol Management prioritize their business by compromising the safety of the Challenger and the crew. Safety was superseded by their business ambitions to remain the contractors of NASA for the space shuttle mission. This was a tough call for the management of Morton Thiokol, and the human error was committed due to the wrong decision. Today, this problem still exists within industries as competition becomes a global problem for industries, sacrificing everything to remain in business and operations.

NASA Management Latent Cause: The latent cause in the Space Shuttle Challenger Disaster is that NASA exaggerated the probability of their failure of 1:100,000 and let the public think it is safe to fly by placing an ordinary citizen into space so that the government can fund the needs of NASA.

Before we condemn someone, are we sure that if we are in the shoes of these people who committed the mistake, will we do otherwise or commit the same mistake? What was in the person's mind that eventually leads him to commit that mistake? The flight on January 28, 1986, ended the life of the six astronauts and one civilian. What amount of pressure had management been under that led them into this tragedy since it is being covered worldwide? Are we sure that we had done it differently or otherwise the same if we are in their shoes at that particular time when they made that decision? Think about it. People make systems, management make decisions, and we are all humans. Latency understands why people did what they did, which eventually leads to this problem. We can only learn from the things that go wrong if we accept the truth that we are also part of the problem itself. RCA philosophy tells us that it has no room for pinpointing mistakes and blaming others.

In an interview, the wife of Dick Scobee said that Dick was concerned about placing ordinary citizens into space since astronauts still need training. No one at NASA was held accountable for the disaster. June Scobee, the wife of Dick Scobee, the commander of the Challenger said in an interview that Christa Mc Aullife asked her if the Space Shuttle Challenger was safe to travel and she said that if it was safe for my husband, then it would be safe for you as well. How she regrets and resents herself after that tragedy. At the end of 1986, Larry Mulloy of NASA and George Hardy of Marshall retired. Robert Lund was promoted to a more senior position at Morton Thiokol. Robert Boisjoly's testimony created some friction between his co-workers and his management. He resigned from Thiokol. NASA was shut down for a few years, and a modification was done on the shuttle orbiter, which provided an escape system for the crew during an emergency. For the O-Rings, a heated layer was placed to counter the effects of cold, and for Crista Mc Aullife, a moon crater and asteroid were named in her honor. Richard Feynman died of cancer two years after the Challenger Disaster that he was battling for several years. He had once again done a great honor for the Challenger Crew and his country.

In the case of the Challenger explosions, we all know the aftermath of what happened, which destroyed the space shuttle and killed all the 6 astronauts and the teacher Crista Mc Aullife. If we go back in time and asked what if things were different? What do you think will happen? What if the Space Shuttle Challenger did not launch? What if the top management of Morton Thiokol insists on their decision not to launch the Challenger? What will be the outcome of the situation? What happens is that NASA will wait for a normal day to come where the temperature will not be so cold. It may be next week, next month, or after a few months. NASA may terminate their contract with Morton Thiokol on the Solid Rocket Boosters and look for other vendors. If this happens, Thiokol's business will weaken and might be forced to lay off several employees or perhaps completely shut down their operations. This might be what is on the minds of the management team of Morton Thiokol that force them to agree with NASA to launch the Challenger.

7.5: Situation, Filter, and Outcome

According to Cambridge dictionary, a situation is a set of things happening and the conditions that exist at a particular time and place. Merriam Webster dictionary states that situation is how something is placed concerning its surroundings. The outcome is something that follows as a result or consequence of the situation. It is a result or effect of an action, situation, or event. When a situation presents itself, our brain will filter it out until we can arrive at the outcome of what we are going to do. There are three ways the brain can filter out a particular situation.

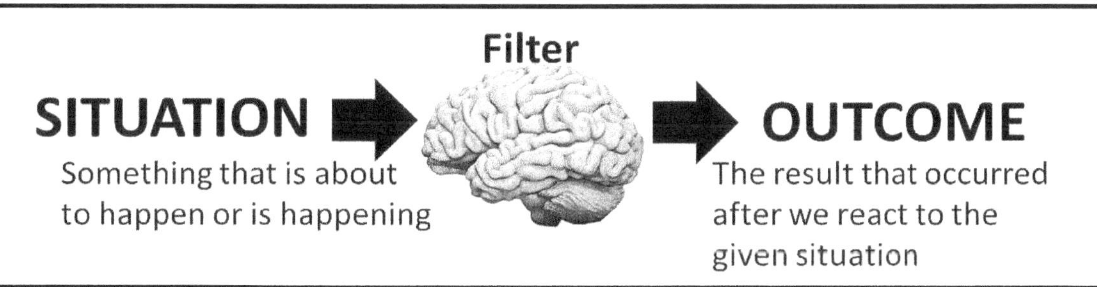

Figure 7.18: The Situation and Outcome

Brain is on Auto-Pilot: As human beings, when a particular situation occurs, we react. The actions we take are said to be the outcome. Some situations will present themselves in which the reaction of the brain will be fully automated. This means that our brain is in auto-pilot mode when the situation actually happens. In fact, we do not have to think to make a decision. For example, when we drive and the traffic light turns red, we automatically step the brakes to stop the car without thinking. If we see someone crossing the street while driving, we slow down and press the brakes. When it's the birthday of your wife or husband, we purchase something to celebrate the event. When we wake up in the morning, we brush our teeth. If you are a lead guitarist in a band and playing solo, you don't need to think about the next note to pluck as it will come naturally. Our brain is already programmed that if this situation or event happens, this is what we do. We go to a gas filling station since our indicator on the fuel tells our brain that if we continue to travel, the car will run out of gas. In this situation, our brain interprets the situation, and our reaction is already automated or on auto-pilot. In short, when a situation presents itself, we automatically know what to do even without thinking. The reason the brain is on auto-pilot is because it knows that this is the right course of action to do.

Brain is Programmed to Comply: In situations like this, our brain will filter and interpret the situation and then provide us with a course of action. It means that this is what we have been told and taught to do. The actions we took may be right or wrong. We continue to do these things we used to do because our brains have been programmed to do it. Our SOP, procedures, policies, rules, regulations, and protocols indicate that this is what we need to do. In situations like this, our brain will follow a specific course of action based on what is written or taught to us, and we need to comply. It does not necessarily mean that what we are doing is the correct course of action. For example, a Preventive Maintenance task will indicate replacing a particular spare part, even if we know that the spare part is not yet in a failed condition. We change the oil in our car every 5000 km because it is what the car's operating manual and OEM told us to do. In this case, the driver does not know if the additives are already depleted but will continue to follow what is written on the driver's manual. When equipment fails, a job order will be issued in which it will require the maintenance to repair the failure, even if performing root cause is the better course of action to undertake.

Brain Needs to Make a Decision, but the Decision was Flawed: We also have cases where a particular situation is given. A course of action needs to be done. There are currently no rules, regulations, or policies to follow; and our brain interprets the given condition and performs a particular action. The actions we took may be right or wrong. In situations like this, the brain finds it difficult to distinguish what is right and what is wrong because either way, there will always be consequences of their actions. The brain will think both ways, determine the risks and consequences for both possibilities, and finally decide. This usually occurs with top-level people in the organization. This **covid-19** pandemic affected the company's revenue by 70%, the management decided to retrench one-third of its employees. 10 candidates applied for a particular position, and it was trimmed down to the final 3. The management team hired what they thought was the best candidate and resigned after a year. Today, only around 52 US companies have made it to the Fortune 500 companies since 1955. This means that only 10.4% of these companies have remained until today after 64 years since 2019. More than 89% of these companies have gone bankrupt, merged, or acquired by other industries or have fallen out from the original Top 500 Fortune Companies in the United States. Many of these companies were big shots during their time in 1955. Still, today, they are now unrecognizable, forgotten, or merely gone.   What do you think happened with these companies?

7.6: Understanding the Latent Cause of the Problem

As we discussed previously, all physical failures are triggered by humans. But humans are negatively influenced by latent forces. Therefore, the goal is to expose and identify these latent causes. We also explained that in uncovering the latent cause, we need to answer the following questions;

• What is it about the way we are as an organization that contributed to the problem?
• What is it about the way I am that contributed to the problem?

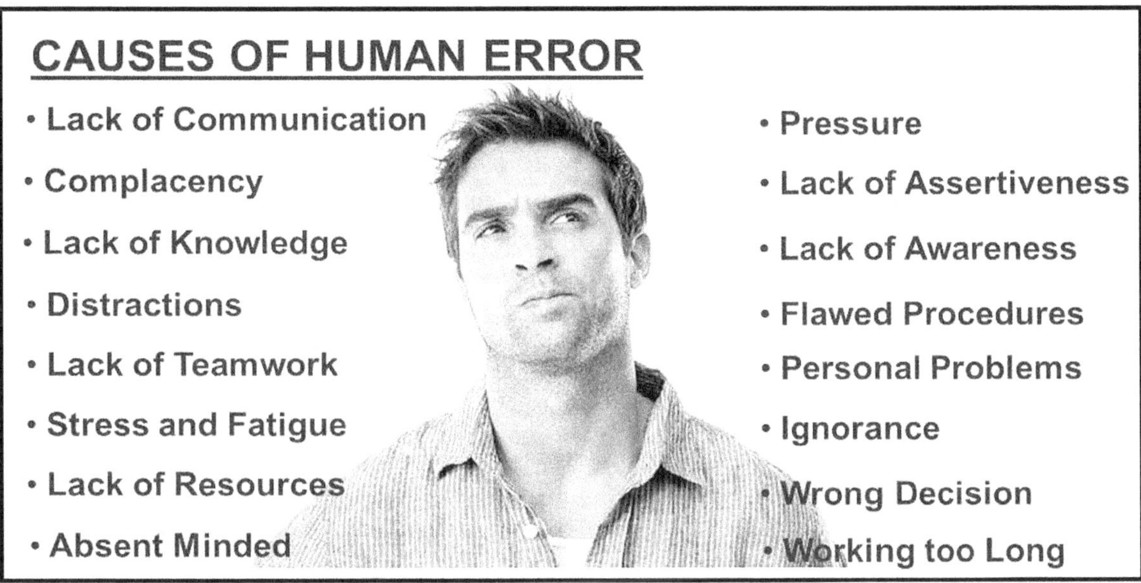

Figure 7.19: Causes of Human Error

We have explained that the word latency means hidden. This means that it exists, but nobody wants to expose or talk about them. Perhaps we are not even in authority to do something about it. There are two kinds of latencies: our personal latencies and the other is about organizational latencies. In my own interpretation, I believe latencies are the main reason behind the human error. This is a deeper reason why the human error was committed. It answers, why did you decide to do that? What was on your mind? There is a difference between a human error and a latent cause, but these two are connected. When a person commits an error, there are several reasons why it happened, and why he did it as in figure 7.19. The difference between a human cause and a latent cause is that the human cause will be limited to the person who committed the mistake, while a latent cause will involve other people who have contributed to the error or mistake. A human error can be a single person, but a latent cause will involve several people in the organization.

Human error happens because our judgment was clouded. As discussed in section 7.6, the brain can be on auto-pilot, programmed, or needs to decide, but we are not focused; or perhaps we are thinking of something else. Although in the majority of cases, the human error was unintentional. In my home, I have a portable hot and cold water dispenser. It is color-coded where the hot water is red, and the cold water is blue. I had lost count on how many times I placed the cup on the blue faucet, which is the cold water when I wanted to have a cup of coffee. I need to throw away the cold coffee, wash my cup again, and focus on the red faucet for the hot water on my next attempt.

On the other side, we have the latent cause, and it is connected to the human error. The main difference is that in the human cause, we limit ourselves to the person who made the error. They are the culprit, the fall guy. They were the ones who fall into the trap. On the other hand, we have latencies; they are invincible. These people made the trap. It has something to do with the deeper meaning of why the human error was made. These latencies are the deeper reasons for human error and are induced not just by one but by several people. Let us provide an example; in this case, we have a robber caught by the police. He was sent to a court hearing, and the judge sentenced him to 5 years in prison. If we probe the life of the robber, he started stealing when he was 7 years old. What caused him to steal? His father was an alcoholic who always beat him, while his mother was a drug addict who left him to starve. Some gangs befriended the boy, which taught him how to steal. During his lifetime, until he reached manhood, he was exposed to many bad influence people in his entire lifetime. The fact is, he was exposed to a lot of people that made him what he was.

The same thing happens in industries; several people influence the culprit to commit an error. The problem is we go after the culprit that committed the error, while the people who influence the culprit remain unknown and wash their hands as if nothing ever happened and business as usual. Remember that when someone commits a mistake or error, there are others also involved. This means that probing the latencies means that we are just making the invisible or hidden to be visible. Engineers with technical expertise can derive the physical cause of why the failure happened. Management people can react and condemn the person who committed the error, and in most investigations, it stopped here. The problem is that there is still one more cause much deeper than the human cause, and the people involved in this are good at cloaking themselves, making themselves invincible, just like the movie Predator. The goal of any root cause failure analysis is to expose and make them visible. Let us discuss about the latent cause cases in this book.

Morton Thiokol's management team was forced to make a wrong decision and recommend NASA to launch the space shuttle that killed the astronauts and Crista Mc Aullife. What made Thiokol's management to change their decision and agree with NASA to launch? It is because they were pressured by Lawrence Mulloy of NASA. Thiokol's management knew that there might be repercussions on renewing their contract with NASA, which was nearing its end. NASA might go for other OEM contractors on the Solid Rocket Boosters. Were there other reasons the Thiokol management change their mind to launch the Challenger? During the conference, Lawrence Mulloy of NASA told Thiokol that we have been flying with these known conditions for several years. This means that the shuttle had never exploded in the past. Although they knew there were risks, the shuttle always came back to them in one piece.

On the other side, why did Lawrence Mulloy pressured Thiokol management? This Challenger Mission draws the attention of many people not only in the US but worldwide since this will be the first time that NASA will include an ordinary citizen in space. There was heavy coverage from the media. Remember that the original schedule for the launch was January 26, 1986. It was postponed to January 27, the following day. The media were blasting NASA for the delay. There was also a plan that President Reagan will be talking to Crista Mc Aullife in space; hence, President Reagan supposed communication with the teacher, Crista Mc Aullife, was delayed twice, and in baseball, NASA knows that they were strike 2. Remember that it was the US government that is funding NASA. So, in this case, we have the media people blasting and saying not-so-good things about NASA because of the delays. We also have the President of the United States who cannot communicate with Crista Mc Aullife due to the flight delays, and perhaps the millions of people in the US anticipating the launching who paid their taxes that funds the NASA program.

7.6.1: Latent Cause of the RMS Titanic Sinking

We have discussed the human errors that led to the sinking of the RMS Titanic in the previous chapter of this book. If we probe about the depths of the latencies, we asked why they made those decisions during those times. If we go back to history, RMS Titanic belongs to White Star Liner, who owned two luxurious ships named RMS Olympic and RMS Titanic during those times. In September 1911, during its fifth voyage, the Olympic collided with the HMS Hawke. There was fierce competition in the shipping industry during those times. JP Morgan owns the white starliner, and Bruce Ismay was the Managing Director during those times. With the loss of one ship that must undergo repairs, their revenue mainly depends on the RMS Titanic.

Although we all know that the RMS Titanic hit an iceberg, the ship was traveling too fast on icy waters. Why was the ship traveling too fast in that situation? It is because Bruce Ismay, the managing director who was also a passenger during the RMS Titanic maiden voyage, was pressuring Captain Smith to light up all the boilers for the ship to run faster? Why did Bruce Ismay pressuring Captain Smith to run the ship faster at icy waters? It is because Bruce Ismay wanted to create an impression that RMS Titanic was not only the biggest and most luxurious but also the fastest. Why did Bruce Ismay want to create this impression? Because Ismay and JP Morgan knew that there is fierce competition in the shipping industry during those times. They wanted the world to know that RMS Titanic was not only the biggest and most luxurious ship, but he wants it also to be fast. Traveling from Southampton, England, to New York would take 5 days; if RMS Titanic can make it in four days, Ismay knows that more passengers would take the RMS Titanic. Although RMS Olympic was already repaired when RMS Titanic's maiden voyage, White Star Liner knew they needed to cope with the losses. RMS Britannic, the 3rd ship of White Star Liner, did not yet exist during those times. Like the RMS Titanic, it also sank on its 6th voyage on November 21, 1916.

Hence, the latencies on the RMS Titanic have something to do about being number 1 in the shipping industry so that White Star Liner can become more profitable and earn revenue at all means even at the expense of sacrificing the life of the passengers and crew.

7.6.2: Latent Cause of the Bhopal Tragedy

On December 2, 1984, 40 tons of highly poisonous methyl isocyanate gas leaked out of the pesticide factory of Union Carbide in Bhopal. A cloud of methyl isocyanate gas escaped from the Union Carbide plant in Bhopal. More than 6,400 died, many from the effects several years after the disaster, and 30,000-40,000 were seriously injured in the world's worst chemical disaster. More than 500,000 men, women, and children have been exposed to the poison clouds, and at least six thousand people died within the first week of the disaster. The current death toll is well over 16,000 and rising. Pregnant women were miscarried, and babies were stillborn. The poison affected some people's lungs and nervous systems. So the question is what caused the methyl isocyanate gas to leak, leading to the deaths of thousands of people on December 2 to 3, 1984, in Bhopal, India.

The physical cause of the failure is that MIC (Methyl Isocyanate) gas mixed with water, creating a very violent reaction which caused the MIC to heat up, boil, turned into gas, and escape the Union Carbide factory. MIC gas is dense and is heavier than air; which mostly stays in the ground. There were too many mistakes and errors to mention if we speak about the human causes of the failure. There were already a massive amount of errors committed before the disaster, which includes the following:

Figure 7.20: Union Carbide Defenses

• The maintenance manager was laid off from his work months before the tragedy.
• Instrumentations were not calibrated due to a lack of resources and budget.
• The scrubber was not working during the time of the tragedy for several weeks. Although, the scrubber can only contain a fraction of the leaked MIC gas that escaped.
• Gauges and instrumentation were uncalibrated and provided false reading. There was no one to check since the maintenance manager was laid off.
• The corroded pipe was not replaced on the flare tower. Should the flare tower was operational, it might have averted the tragedy.
• Emergency water hose spray insufficient pressure to reach the vapor exiting the exhaust stack.

• Nitrogen gas was removed from the MIC tank. It serves as cooling for the MIC. If the Nitrogen gas was not removed, perhaps we can either prolong the leak, or the leak would never happen as it cools the MIC. It was removed due to a cost-cutting scheme by management.
• A declogging operation took place before the tragedy. A slip-line should be placed before their declogging operation. However, it was not done.

The plant is located in Bhopal where many people were residing near the plant. Most of them were slums. There was no orientation done to the people living near the plant just in case some emergency cases, happened. One of the employees from the EHS was proposing to provide an orientation to the community but was rejected by his superiors. The person resigned from Bhopal. If people just knew what to do, perhaps even if the MIC gas escapes then there will be fewer casualties. Since MIC gas is denser or heavier than air, then people can inhale it. If you are in your home, you just need to close down the windows and door. Place a wet towel on the bottom of the window or door, the MIC gas that escaped will be absorbed by the towel and will not be inhaled by the people inside the house. This was actually what one of the survivors did when she called the person who knew what to do in case of the MIC gas escaped and survived the tragedy.

Figure 7.21: The Aftermath of the Bhopal Tragedy at Union Carbide

The CEO of Union Carbide, Mr. Anderson insists that the incident was caused by sabotage; however, with the pieces of evidence presented, this was a clear case of top management negligence. MIC (Methyl Isocyanate) was a lethal and toxic liquid that, when turned to gas and inhaled, can cause many problems to a human being. Small leaks were already experienced before the incident, yet nothing was done, even if one employee had already died. After the

Bhopal Tragedy and litigation in court in 1989, the Supreme Court of India approved a settlement of the civil claims against Union Carbide for $470 million to pay damages to the people affected by the tragedy. This is approximately around 400 US dollars for every individual who died and been affected by the disaster.

Probing on the Latent Cause: It is very evident based on the events that unfolded that the latencies in the Bhopal tragedy and disaster have something to do about Top Management cutting costs on corners, especially for the maintenance department. The reason for the cost reduction was that business profit for Union Carbide was on a decline. They fired the maintenance manager months before the tragedy; the scrubber and the broken pipe on the flare tower were never restored. The nitrogen gas on the MIC tank used to cool the MIC liquid was discontinued, again due to cost-cutting schemes by Top Management.

Report also indicates that Union Carbide denied reports of safety lapses and started cutting costs, including reducing manpower from the maintenance department as the company was losing money. In 1984, the Bhopal reported a loss of $4,000,000.00 million while Union Carbide's profit fell from $800,000,000.00 three years earlier to just $ 79,000,000.00. The report indicates that Union Carbide cut training for MIC plant operators from the original six months to only 15 days. On November 26, the company eliminated the position of maintenance supervisor on the second and third shifts. They shut down the MIC refrigeration plant to save $ 70.00 a day.

Latencies always involved more than one person that needs to be exposed. These people involved may sometimes be shocked to learn of their involvement in the problem. They may be unaware that they are involved in the problem. The reason for conducting a thorough Root Cause Failure Analysis investigation is to find the truth and learn from the things that go wrong in our industries, whether this is a big scale, medium scale, or small scale event. By learning from the failure, we can sharpen our decisions on what to do and what not to do. It helps us understand what is right and wrong for our industries.

7.7: Decoding the Song Man in the Mirror by Michael Jackson

I usually play this song during my RCFA class when we discuss the latencies. This is mostly the end song at Michael Jackson's concert. Michael Jackson usually composed his songs but not this one. This song was composed by one of her female backup vocalists, Seidah Garrett, with the help of Glen Ballard. This is a very sentimental song that contains a lot of metaphors and hidden messages. Let me decode this song in the best way I can and how it reflects what is really happening in industries.

- [Verse 1]
  Hmmm, I'm gonna make a change, for once in my life
  It's gonna feel real good,
  Gonna make a difference
  Gonna make it right

• [Verse 2]
As I, turn up the collar on, my favorite winter coat
This wind is blowing my mind.
I see the kids in the streets with not enough to eat
Who am I to be blind pretending not to see their needs?

• [Verse 3]
A summer disregard, a broken bottle top, and a one-man soul
They follow each other on the wind ya' know, cause they got nowhere to go
That's why I want you to know

• [Refrain]
I'm starting with the man in the mirror
I'm asking him to change his ways
And no message could have been any clearer
If you want to make the world a better place
Take a look at yourself, and then make a change

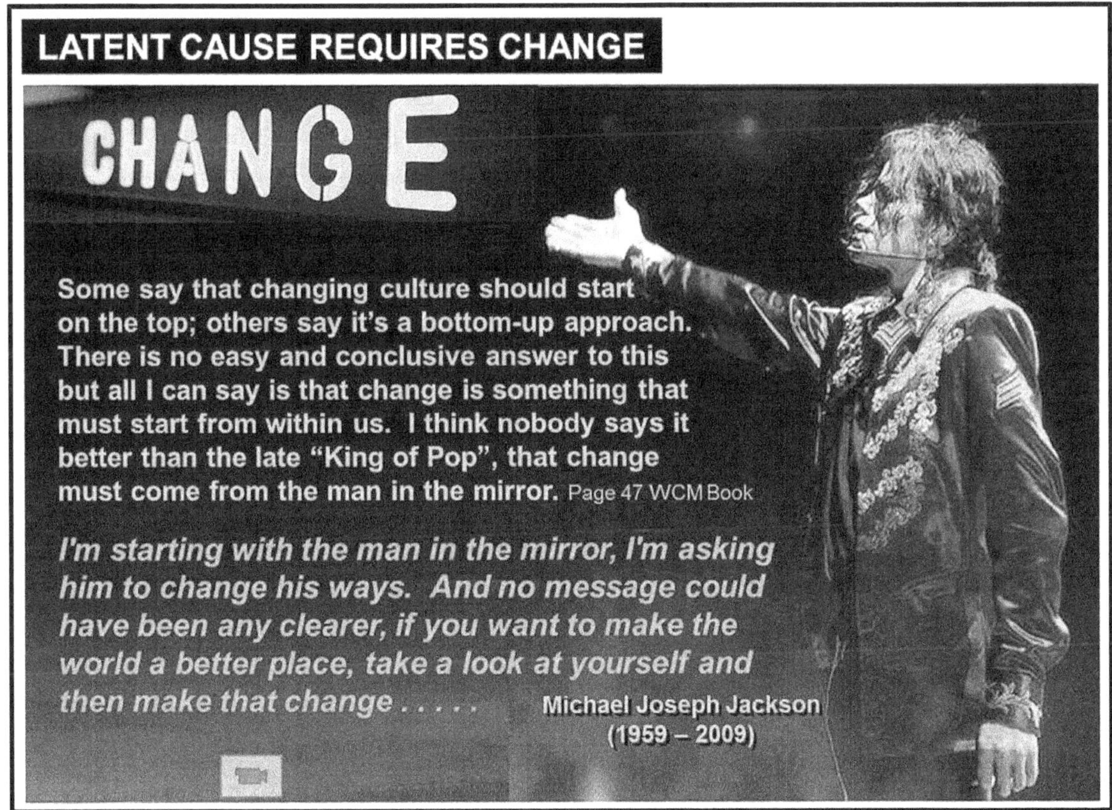

Figure 7.22: Latencies is About Looking in the Mirror

Verse 1: We All Want to Change Something: We are all eager to change something. The moment we were hired in the industry, we knew that among the rest who applied for the job, we have been selected since we got some experience in doing it. We knew from the start that we will contribute something good that's why they hired you. In your mind, you were so excited about the upcoming challenges that await you. You know that your ideas would make some difference and it feels good.

Verse 2: Now You See the Problem: *As the start of the verse says; as I turn up the collar on, my favorite winter coat, this wind is blowing my mind.* You dress up, put on your favorite winter coat and report for work. As this was your first day, you observe what's going on around and what's on your mind was, Oh My God, this was not what I'm expecting. As you stay a bit longer, it's no different from the first day you reported for work. And the song goes*; I see the kids in the streets with not enough to eat, who am I to be blind pretending not to see their needs.* Although it is clear that Michael was referring to the many people from around the globe who are struggling as they do not have enough to eat. We see the problem, however, people are deaf and blind to do something about the problem. The same goes with industries, where management is deaf and blind to see the needs of their kids. In these cases, these kids represent the people working in industries, especially the maintenance people. Maintenance needs education, knowledge, training, tools, instruments to build their skills so that they can do their jobs correctly and right the first time around. But due to management cost-cutting schemes, all we got is a bunch of WD40 and Duct Tape. So the question is how on the planet can you do your job? As Roger Boisjoly said, I was requesting my management for a team to investigate the erosion problem on the Solid Rocket Booster and all I got is just promises that never came. This means that if there is no support nor commitment from the management, then how can you improve things around here? You just can't. We all know the needs of the maintenance function, however, they have been deprived of their needs and who am I to be blind not to see their needs. But as the saying goes if you want to stay, then just pretend. This means that you see nothing, you hear nothing and you speak nothing about it.

[Verse 3]: It's All About Our Culture: *A summer disregard, a broken bottle top, and a one-man soul. They follow each other on the wind ya' know, cause they got nowhere to go. That's why I want you to know*. Since nothing had been provided to their people as in verse 2, hence, we are left with so many problems, failures, and breakdowns. Given this situation, we are left with nowhere to go but to repair the failure since this is what everyone else is doing, we just follow the pattern of the wind ya know that every time a failure struck us, our only option is to repair the failure and this is our culture since the very beginning of time. In fact, no one here knows how to analyze failures since we were never taught about how to perform a root cause. A broken bottle top. And a one man's soul means that if a problem struck it means every man for himself and you are on your own. That's the culture here, and you need to accept that.

[Refrain]: What We Need is for Us to Change*: I'm starting with the man in the mirror. I'm asking him to change his ways. And no message could have been any clearer if you want to make the world a better place, take a look at yourself, and then make a change*. Everyone is expecting some changes for things to get better. We want that person to change. We want our boss to change. We want our Top-Management and C-Level people to change. The message here is before we expect others to change, let it begin with me. If you want the world to be a better place, if you want your family's relationship to be better, if you want your industry to become better, or if you want the whole maintenance function to be better, don't wait for someone to start but rather start it with yourself,. If everyone is repairing and troubleshooting breakdowns and failures, then start by analyzing them. If management won't provide the budget for training, find your way. Training is not the only source of knowledge. Spend time to educate yourself, there are so many articles, newsletters, books, or videos online that you can watch or read. Don't wait for your management to provide you the budget for training. There are far so many ways to gain knowledge. Once you have the knowledge, apply it and teach others for a start, share what you have learned with your people.

Latencies are hidden, veil, or concealed. This is far beyond the human and even the system cause. People commit mistakes and errors. Humans commit and will continue to commit errors since this is part of our human nature. Latency is about humility. We need to humble ourselves first before we can accept the fact we too are also part of the problem. It's all about leaving our pride, ego, politics, bureaucracy, arrogance behind before we can start looking ourselves in the mirror and ask ourselves what did I did wrong or how did I contribute to the problem.

Latencies are all about looking at the Man in the Mirror. In industries, as the saying goes, if you want to stay long here, just follow whatever your boss tells you to do. Apply Newton's Law, what comes in must come out. If you hear something unpleasant, let it just pass the other ear. If not and it mixed with your brain and heart, then you go nuts. If the boss wants his people to dance boogie, then everyone dances boogie, if not then you are not on my team. World Class industries were not born that way, they were previously reactive, but these industries struggle to achieve that feat.

Just like the lessons of life, it is when you're down that the urge to change comes. Humans have the desire to make a change. But if we think that our leaders will do the change for us, then that would not happen. That's why the song Man in Mirror is about focusing on oneself because we cannot control the world or even our organization, but we can control ourselves and our actions. This means that we focus our efforts on what we can control. Start small, and start with the Man in the Mirror. If one by one everybody is doing this then summing up all the changes will be one big change. As one of the lessons on root cause is that our failures are simply an accumulation of small problems that have been left unattended. Focus on the small problems and by doing this, we can now have a chance on addressing the big problems. As the late John F. Kennedy said in his speech. Ask not what your country can do for you, ask what you can do for your country.

7.8: Addressing the Organizational and Personal Latencies

Addressing the latencies will involve two things, first is understanding what changes does the organization needs to address the problem, and second are things that need to be changed within us. What I believe is that addressing these two latencies is not only intended to address the current problem we are dealing with but also on future problems that can happen as well. These are the lessons learned on the things that go wrong with our industry. Latencies are generic. An example of a latent cause is a lack of communication between management and their engineers. Dealing with these changes allows management people to communicate better and listen to their engineers. This means that we are not only dealing with this problem but with future problems as well. That is why latencies are not specific to that particular problem alone.

If industries will not address these latencies then there is the possibility of the problem resurfacing again.

The personal latencies of each stakeholder involved may differ from one person to another. It cannot be answered by the principal investigator or anyone else but only the stakeholder themselves can address these latencies. The role of the principal investigator is to facilitate the stakeholder meetings. Latencies are connected to the system and human cause, the difference is that in the human cause, this is about finding out the reason why the person committed the mistake or error. The people involved in the latent cause can be anyone in the organization that has contributed to the incident. The first step to address the latent cause is to identify these people, then asked them the reason why they did it that finally led to the problem. Most of these people involved have no idea or were unaware at the beginning about their involvement and contribution to the problem. They need to be exposed not to embarrass, or ridicule them, but to let them know how they become involved and what are they going to do about that. What was the reason they made that decision?

In conducting root cause, implementing these corrective actions generated will not solve the problem 100%. It also involves certain people to change and this is what latencies are all about. It's about looking at the Man in the Mirror.

Chapter **8**

Guidelines in Implementing an RCFA Investigation

> ***Suppose you really want to get to the real root cause of the problem; in that case, you will end up probing a person's life, his parents' personal lives, his grandparents' lives, the origin of the original sin, and all the way to the bible down to Adam and Eve. The key to any Root Failure Cause Analysis is to know where to stop, and we will only stop probing when we have reached the latent cause of the problem.***

8.1: Training on RCFA Investigation Process

The first and foremost activity that must be done is to hire a third-party consultant to train their people on how to conduct a Root Cause Failure Analysis investigation. Several qualified organizations can provide these services. It is important to ask the third-party provider the details and modules on what the training will compose to determine what they will teach your people. This book states that there is a big difference between conducting an analytical problem-solving tool and a Root Cause Failure Analysis investigation. What your people need to know is what it takes to be a principal investigator and how to gather the different types of evidence. This will be the primary requirement needed in conducting a root cause investigation. Only people trained in Root Cause Failure Analysis will qualify as a Principal Investigator and Evidence Gathering Team. Different RCFA/RCA consultants can have different training modules and details. Be sure to review the modules on what will be taken during the training process. My preferred lists of RCFA modules would include the following:

• Root Cause Methodology
• When to Conduct a Root Cause Failure Analysis Investigation
• Criteria for Selecting the Principal Investigator
• What to look for during the investigation.
• What are the different types of evidence?
• How to conduct the interview process
• Understanding the Physical Cause of Failures
• Mechanical Failures, Different Types of Wear
• Human Errors
• Understanding the depths of the Latencies

- Understanding the Physical Cause of Failures
- RCFA/RCA Case Studies
- RCFA Corrective Actions
- RCFA quizzes and workshops

8.2: Setting-up a Centralized RCFA Core Team or Council

There should be a centralized group of people that should consolidate all the RCA and RCFA cases performed by every single function in the organization in addressing every problem, whether it will be equipment or non-equipment-related problems in the plant. For industries implementing Total Productive Maintenance, these can be handled by the TPM staff office. For industries with a continuous improvement function, these people can handle these, or an independent RCFA Core team or Council should be established in the plant. The RCFA Core Team or Council members can be part-time people from different functions, plus an RCFA facilitator in their organization. They will serve as the lighthouse and provide the directions and guidelines in conducting a root cause investigation.

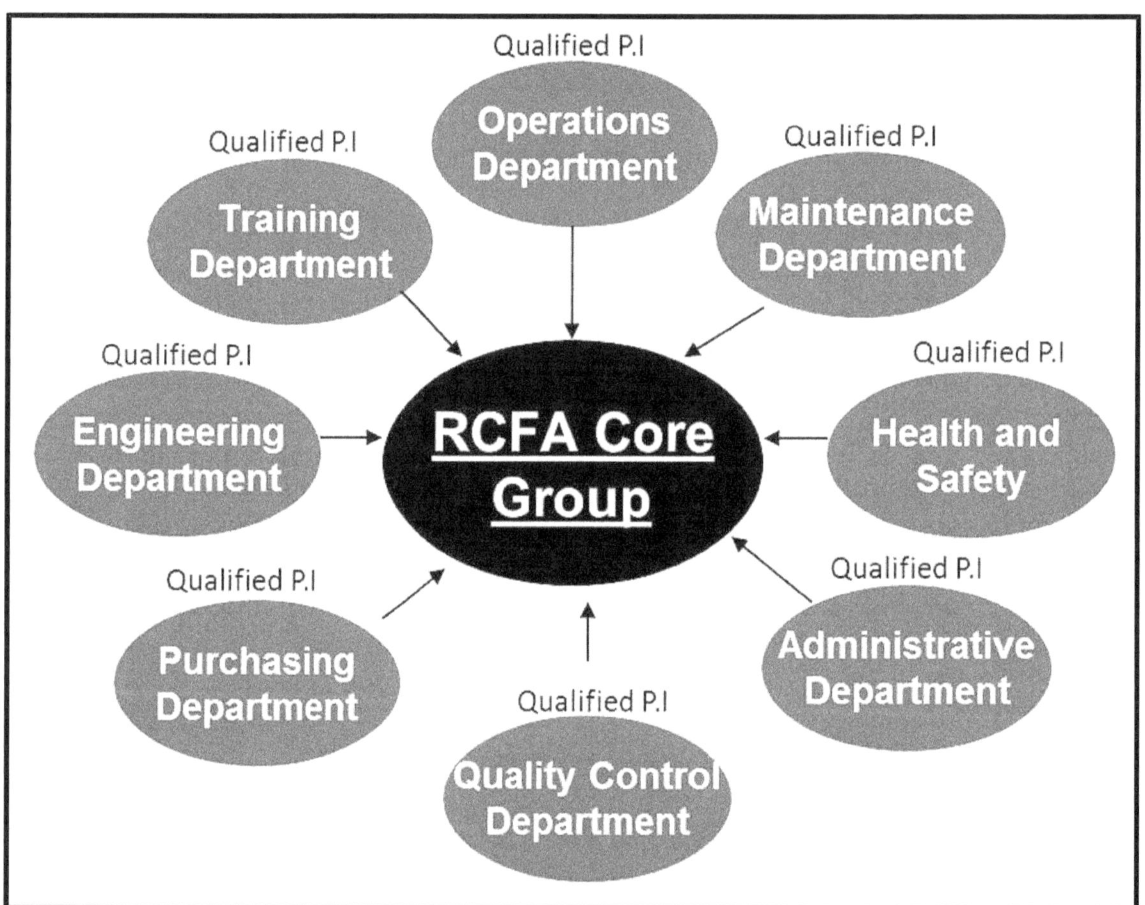

Figure 8.1: Creation of RCFA Core Group or Council

This can be applied to equipment failures, equipment failures, quality-related problems, customer complaints, safety investigation, accidents, near-miss, or any related administrative problems an industry encounters that warrant a root cause investigation. This will be a group of people with a passion and commitment to understanding and learning from the things that go wrong in their industry. These people should be trained in Root Cause Failure Analysis investigation by a qualified organization as a preliminary requirement. The RCFA or RCA Core Team or Council will be responsible for driving all RCA or RCFA efforts in their plant, whether the root cause investigation is a big or small scale event. These people should be committed to cultivating the seed for every employee in the organization to have a Root Cause mentality by challenging failures they experience and providing SMART solutions to their day-to-day problems to further avoid a repeat of the cause of the problem.

For a plant-wide RCFA implementation, industries can start with a centralized group. Let us call this the RCFA Core team or Council. They will be responsible for safekeeping all RCA or RCFA cases conducted within the plant. Conducting root cause investigation should not only be limited to technical but should also extend to administrative problems. This means that other organizational functions should likewise be trained on how to conduct root cause investigation. Here are the roles of the RCFA Core Team or Council:

- The Root Cause Core Team should arrange their people who will be trained in RCFA with a reputable training organization who will act as the Principal Investigator and Evidence Gathering Team. They need to agree with the basic guidelines on conducting an RCA or RCFA investigation. Not only should Health, Safety, Operations, Reliability, and Maintenance attend a root cause training, but those from the administrative part of the organization should also attend this training. Note for equipment-related problems, we will be using the term Root Cause Failure Analysis, while for non-equipment-related problems, we will be using the term Root Cause Analysis.
- They will be responsible for assigning the Principal Investigator to handle the case.
- The Root Cause Core team must understand that their main objective is to help adopt a culture of root cause mindset and mentality on problems, defects, failures, and breakdowns that warrants a root cause analysis in their entire organization.
- Agree to provide support and guidance to any organization's function initiating a probe or investigation within their scope of responsibilities.
- Provide the energy, motivation, enthusiasm, and convince other members and functions to carry out root cause investigation on related problems and issues.
- Serve as the organization's central hub and responsible for safekeeping all relevant RCA and RCFA cases conducted in their plant, either conducting an RCA, RCFA or any problem-related solving techniques.
- Provide a list of qualified Principal investigators capable to handle the specific problem within their plant. Whenever a problem occurred, the root cause council should assign the Principal Investigator to handle the investigation.
- They will be responsible for providing SMEs (Subject Matter Experts) or third-party consultants to conduct initial root cause training for their people.
- Provide roles and responsibilities for the RCFA Core team to be known to all employees in the plant.
- The Root Cause Core team will be responsible for monitoring all departments and functions within their plants on areas performing root cause investigations
- Set the initial thresholds for an RCA or RCFA case depending on the magnitude of the problem being investigated. (May differ from one function to another).
- Provide detailed and structured guidelines in conducting a root cause failure analysis investigation.

- Responsible for providing a one-day refresher course on RCA or RCFA for those who have already attended formal Root Cause or Root Cause Failure Analysis training.
- Provide criteria for determining if a failure is a small-scale, medium-scale, or large-scale event for both equipment and administrative problems that warrant a root cause investigation.
- The Root Cause Core Team should provide a time frame for completing the investigation. One of their roles is to check with the Principal Investigator regarding the progress of the case.

<u>8.3: Traits of a Good Principal Investigator and Evidence Gathering Team</u>

Everyone has their first time doing things, whether you are a detective handling your first case, a lawyer defending your first client, a resident surgeon doing your first operation, or a rootician handling your first investigation. Note that a rootician refers to the person handling the root cause investigation or probe. As we said that the preliminary requirement to qualify for being a Principal Investigator is that you need to be fully trained in the RCA or RCFA process. Although the company can have the decision to hire a professional RCA or RCFA investigator to handle the case if the company has the budget for that, however, it would be better if there are several people in the organization that can be trained on how to investigate problems and failures they experience on their equipment and assets. The Principal Investigator, in this case, will be an employee of an organization and has a certain responsibility depending on his job description. The person assigned to be the principal investigator may be involved in engineering, maintenance, production, safety, or other organizational functions, and performing a root cause investigation is not his full-time job but should handle the case full-time, or depending on the policies of the RCFA Core Team. He needs to advise his superiors to be replaced and find someone to take care of his responsibilities for the time being until the duration of the investigation, especially if this is a large-scale event. The person assigned may check his ongoing responsibilities from time to time. What is important is that the principal investigator should possess the following traits.

1. **Unbiased:** The Principal Investigator must be free from bias and should not immediately jump to conclusions. Any decision that has to be made must be based on the pieces of evidence uncovered. Typically if a major failure occurred, the industry would call their OEM, and the OEM will provide an expert or consultant. What is important in any investigation is to determine the hidden causes, which are the latent causes. We would refrain from hiring consultants to act as the principal investigator. The reason behind this is that if we discussed the events that took place, the consultant has already concluded what caused the problem, without even seeing the pieces of evidence. These consultants and experts often conclude their investigation on the physical and human causes only. We prefer that the principal investigator follow the lead based on the pieces of evidence unfolded. If the plant lacked the expertise to conduct the investigation, the best thing to do is gather the physical evidence and send it to a metallurgical laboratory to determine the physical cause of the problem. Once the physical cause is known, the investigator can move with the human, system, and latent cause of the problem.

2. **Good Communication Skills:** There will be cases where the person acting as the evidence-gathering team or the Principal Investigator will conduct the interview themselves.

What is important is that whoever is conducting the interview should have good communication skills. The person in charge of conducting the interview should understand the difference between an interview and an interrogation. He must have the skills and ability to listen to the people he interviews. People involved in gathering evidence must be skilled at listening, observing, and taking notes simultaneously. Recording of the interview will only be allowed with the consent of the person being interviewed. If the person interviewed would not want the conversation to be recorded, then the person in charge of conducting the interview should write important notes during the process.

3. **Be Able to Balance his Time:** As we said, the principal investigator may work full time, or part-time as he may have other responsibilities in the plant depending on the policies from the RCFA Core Team. Suppose his work is too important in the plant or the principal investigator has deadlines on his existing job; in that case, he may ask the Root Cause Core Team to be replaced. The principal investigator assigned must fully be focused on the completion of the investigation they are handling. The principal investigator should be committed to the time provided to complete and conclude the investigation process. Suppose he finds it difficult to carry on the investigation due to his existing workload; in that case, he should discuss this matter with the Root Cause Core Team to find a replacement. Regarding how many hours will the Principal Investigator will work on the RCFA case and his existing job will depend on the policies developed by the RCFA Core Team. If it requires the Principal Investigator full time on the investigation, then the people assigned must inform their management.

4. **Honest and Trustworthy:** The Principal Investigator and Evidence gathering team should be trustworthy and credible. We do not want any member of the RCFA or fact-finding team with a character or attitude problem, as this will pose a credibility and integrity issue to the people involved in the investigation process. We do not want people who will spearhead the Root Cause investigation who always complain about their work or spread rumors and gossip in the plant. Not everyone will qualify to be the lead Principal Investigator or Evidence Gathering team. This will be one of the responsibilities of the Root Cause Council or Core Team to list all qualified people to act as the Principal Investigator and Evidence Gathering team in the plant. It will be good if the Root Cause Core Team can provide a qualification list for both the Principal Investigator and Evidence Gathering Team. Note that both the principal investigator nor the evidence-gathering team cannot leak any part of their investigation if it is still ongoing, even if the management wants updates on the investigation. This must be clear to top management before conducting any root cause investigation.

5. **Capable of Thinking Outside the Box:** The principal investigator and evidence-gathering team is a fact-finding group. Their conclusion, judgment, and decisions are purely based on the facts and pieces of evidence gathered. Any decision made in prejudice can compromise the integrity of the investigation process. These people involved in the investigation should maintain independence and must be objective in their findings. The Principal Investigators must assess situations, gather evidence, formulate theories, analyze the facts, to reach conclusions. The decision arrived is based on pure facts and evidence and not on the opinions and experiences of other people. All physical causes result from human error, but it is not always the fault of the human who committed the error. There is always a deep underlying cause that influences the

person to commit the error. These fact-finding groups must be open to exploring all means to understand the reason behind the error that led to the physical cause of the failure.

6. **Willingness to Seek Outside Assistance when Needed.** A good principal investigator must be open and willing to seek outside assistance from another investigator or other people with experience and expertise when needed during the investigation process. He must have a boundary on understanding what he knows and what he does not know. Perhaps, more importantly, the principal investigator's decisiveness is important in identifying when outside intervention is needed. He must be decided whether the evidence collected is sufficient to move on to the next level or insufficient and knows what to do. The principal investigator must have a list of people he can talk to if he needs some additional information or whether he needs to verify something. For example, suppose the principal investigator is not familiar with the process; in that case, he can visit the production line and ask someone to explain the process slowly and in detail.

7. **Must be Neutral Taking No Sides During the Investigation Process**: There are always two sides to the coin. There will always be rumors and gossip, especially if someone has been pinpointed as the fall guy or culprit on why the failure happened. Although this is also similar to item 1, the principal investigator must refrain from favoring someone during the investigation process. The principal investigator must avoid relying on his own assumptions and biases during the investigation process. They must be objective, and conclusions must only be drawn if all the pieces of evidence have been unfolded. The thing to bear in mind for the investigator is that if the error is unintentional, it means that something triggers the person to commit the mistake. Be in their shoes and wait until there is conclusive evidence before deciding and moving on to the next step of the investigation process.

8. **Capable of Handling the Stakeholder Meeting:** This is an important trait of the Principal Investigator. Gathering the stakeholder in one place and discussing the problem and their involvement is very sensitive. Expect these people to be defiant and defensive during the initial meeting. There will be cases where some of these stakeholders involved are unaware of their involvement in the problem. This will be most especially true when covering the latencies of the problem. Some of the people involved in the meeting may be at the top management level. The Principal Investigator needs to understand how to handle this meeting. They must handle the meeting with the utmost care, professionalism, and humility as well. Remember that the stakeholders are human beings. Human nature states that nobody wants to admit their mistakes and errors. Provide some house rules at the very beginning of the Stakeholder Meeting such as when you enter this room, please leave your position, bias, ego, and those sort of stuff as we will be discussing a very important incident that needs to be addressed.

9. **Dedicated to the Completion of the Root Cause Investigation:** A small-scale incident may take a few days, a medium case may take a week or two, while a large-scale investigation may take months to complete depending on the number of evidence gathered during the investigation. The principal investigator and the evidence-gathering team assigned to the case must be passionate and committed to completing the investigation. Remember that the main goal of conducting the investigation is to find the truth, learn from the failure, and do something

about it to prevent recurring problems. When a root cause investigation case had been closed, we are doing something good for your industry. There is a feeling of contentment and satisfaction that the industry can benefit from.

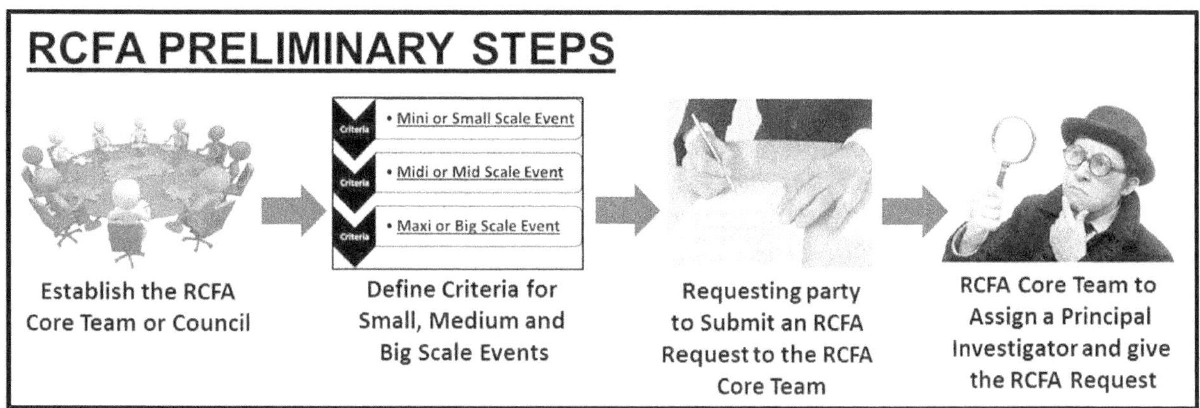

Figure 8.2: RCFA Preliminary Steps

8.4: Preliminary Steps Before Conducting a Root Cause Investigation

Before conducting the actual root cause investigation, the following should be well established in the plant or industry. These will be the preliminary requirements and steps needed to set up a root cause investigation. Once the people have been trained on the RCA/RCFA methodology, set up the following steps as part of the preparation process.

1) **Organize the RCFA Core Team:** As discussed in Section 8.1 of this Chapter, a Root Cause Core Team or Council should be organized. The roles and responsibilities of the core team or council are also covered in this section. This must be mandatory as there will be several investigations that will be conducted in the future. The department requesting a root cause investigation should submit an RCA or RCFA request to the Root Cause Core Team or Council. To create a boundary, all equipment-related problems will be eligible for Root Cause Failure Analysis, while non-equipment and administrative problems will require a Root Cause Analysis investigation. The department or function requesting a root cause investigation should submit an RCFA initial request form in figure 8.3 to the RCFA Core Team or Council. The Root Cause Council will be responsible for assigning the Principal Investigator to handle the investigation. Selecting and assigning the Principal Investigator should be done immediately, and as much as possible should just take a maximum of two hours or less. Remember that the longer the selection of the Principal Investigator is done, the bigger the chances of the people's evidence to evaporate as this is the most sensitive among the three types of evidences. It is highly recommended that the Principal Investigator assign must not come from the same function or department where the failure happened to avoid bias or any cover-up in the investigation process. It will be the discretion of the Principal Investigator (PI) to select his own people who will serve as the evidence-gathering team for both medium and large-scale events. Note that both the principal investigator and evidence-gathering team should be initially trained in the RCFA investigation process by a qualified provider as a preliminary requirement. Remember that 80% of the time spent on conducting root cause will be about gathering as much evidence as possible.

Another important thing to note in this case is that the department or function requesting an RCFA investigation should be made sure the equipment has not been repaired yet and the evidence has been preserved. Another option is that the equipment has already been repaired but the physical evidence has been preserved by the restoration team since they are also trained on how to collect the physical evidence. Hence, the principal investigator or someone assign will just collect the physical evidence on the area requesting the RCFA investigation. Note that if the area or department requesting the investigation has already repaired or restored the equipment and the evidence is not preserved, then we cannot perform a root cause investigation, but an analytical problem-solving tool can be adopted instead.

2) **Determine a Criteria for Small, Medium, or Large Scale Events:** The industry should create a criterion if the failure or problem is considered a small-scale, medium-scale, or big-scale event. These criteria may or may not be the same for equipment and non-equipment-related problems.

Small Scale-Events are performed on mini or minor events. The impact of the failure is small, but we still need to understand the cause of the failure and the need for the cause to be known. There are chronic or repeating failures in which the impact is small that can be subject to a root cause investigation. In most cases, chronic losses occurred because they have several causes. In conducting RCFA, treat one cause at a time. If a problem has two or more causes, separate the RCFA logic tree for each cause. Small-scale-event can be performed by a single person who will also be the one to gather the evidence. The principal investigator will also gather the three types of evidence physical evidence, people evidence, and paper evidence. If the situation needs, the Principal Investigator can assign just one person to assist him in gathering the pieces of evidence. This will not require a stakeholder meeting. The investigation should include the physical, human, system, and the latent cause of the problem. An example of small-scale criteria will be for failures whose impact cost is less than USD 2000.00.

Medium Scale Events is performed on midi or medium-scale events. They are led by a Principal Investigator with one or two persons who will be assigned as the evidence gathering team. This will require a stakeholder meeting. The investigation should include physical, human, system, and the latent cause of the problem. An example of medium-scale criteria will be failures whose impact cost will be more than $ 2000.00 but less than USD 10,000.00.

Large Scale Events is performed on failures with a large impact on the industry. There will be three persons or more who will be assigned as the evidence-gathering team, and this will require a stakeholder meeting. The investigation should include the physical, human, system, and the latent cause of the problem. An example of large-scale criteria will be for failures whose impact cost will be more than USD 10,000.00. For failures involving loss of lives, environmental consequences, or if the industry's lawyers and insurance people are involved, a third-party RCFA provider is highly recommended to act as the Principal Investigator. However, the industry can also assign their own investigation which can be compared later on to the investigation of a 3rd party if their conclusions matched up or not.

3) **Requesting Party to Fill-up the RCFA Request:** The requesting department or function should fill up the RCFA request form and submit it to the RCFA Core Team or Council. Refer to

figure 8.3. This should be immediately be done right after the failure occurred. Maintenance must be informed that the evidence should be freeze and preserved. This will be needed by the person in charge of the physical evidence. Again, suppose the failure had been repaired, and the evidence was not preserved; in that case, it is unlikely to perform a Root Cause Failure Analysis investigation. Both operations and maintenance people should be aware of this situation. Remember that the most important part of conducting a root cause investigation is preserving the evidence.

Request for RCA/RCFA Investigation

Control Number: ___________

Who: (This will be the department or function requesting an RCA/RCFA Investigation)

What: (State the Problem in its precise terms. Specify the asset to be investigated)

When: (State when did the failure occur? Specify the date and time)

Who: (Specify the people involved in the problem, no names just indicate the position)

Oddities: (Specify if there are any irregularities, deviations before the failure)

Person Requesting for the Investiation   Received by RCFA Core Team

_______________________   _______________________
(Indicate Position)   (Indicate Position)

Figure 8.3: Request Form for a Root Cause Analysis Investigation

4) Root Cause Core Team or Council to assign the main Principal Investigator (PI): The principal investigator assigned will head the investigation. As a general rule, the Principal Investigator assigned to handle the investigation should not come from the existing department or function requesting the investigation to avoid any form of biased, cover-ups, or any influence in the investigation process. If this is a big-scale event that will involve the company lawyers and insurance people, it is highly recommended that a qualified third party or independent body should conduct the root cause investigation, but again the industry can set up their own principal investigation at their discretion.

5) The Principal Investigator selects the Evidence Gathering Team: The Principal Investigator (PI) will have the freedom to choose their evidence-gathering team, especially if this is a large-scale event. It may or may not come from the department or function being investigated. For a small-scale event, the principal investigator can also be the one responsible for collecting the evidence. For a medium-scale event, the PI can assign one or two people to gather evidence, including the physical, paper, and people. For a large-scale event, the PI will assign three people to gather the evidence. One person is assigned to collect the physical evidence. One person shall be assigned to conduct the people's evidence and interview those, directly and indirectly, involved in the incident. And the third will be responsible for collecting all relevant paper evidence. The people's evidence must be conducted immediately as this is the easiest to evaporate. The person handling this evidence should ensure that his questions are consistent.

8.5: Detailed Steps in Conducting an RCFA Investigation

Employees from all functions should be trained on Root Cause Failure Analysis by a qualified independent group of third-party consultants. Management must set expectations as to why their people are trained on root cause. Start by educating your maintenance, safety, and technical people on root cause. Conduct management presentation and overview on the basics of RCFA and when it should be applied. Note that Root Cause Failure Analysis are done on current failures and not on failures experienced in the past. This means that a failure must happen first. Hence, people must understand the importance of freezing the evidence, taking photos, and preserving the part that failed.

Step 1: Determine whether the failure is a small, medium, or large-scale event: As discussed in Section 8.4, it is strongly recommended that the industry categorize which failures should be considered a small, medium, or large-scale event. The basis for consideration can include the cost of failure, impact, or consequences of failure in the plant. For example, failures that cost $ 2,000 or less will be considered small-scale events. Failures that cost more than $ 2000 but less than $ 10,000 will be considered medium-scale events, and failures that cost more than $ 10,000 will be considered large-scale events. There might be some differences in classifying the RCFA or RCA criteria for equipment-related and non-equipment-related problems that warrant a root cause investigation.

Step 2: Submit the RCFA Request: The area or department where the incident or failure occurred should submit an RCFA request to the RCFA Core team as soon as the failure occurred. As a general rule, the maintenance cannot repair the equipment if the failure will warrant an RCFA investigation. They can only proceed with the repair if the restoration team is knowledgeable on how to extract the physical evidence on the equipment.

Step 3: Set up the RCFA Team: The Root Cause Core Team will provide a principal investigator. It will be the discretion of the principal investigator to select their evidence-gathering team. For medium and large-scale events, this will require a principal investigator and an evidence-gathering team. The person assigned as the RCFA Principal Investigator and evidence gathering team should have been trained in the RCFA process. This should be a preliminary requirement. It is highly recommended that if the failure happened in your area, the Principal Investigator should not come from your area but from other departments or functions of the organization. It is strongly recommended that an outsider or third-party group lead or act as the principal investigator for large-scale events in which the lawyers and insurance people are involved. Another option is that the industry can set up their own investigation in parallel to the third party and both can compare their findings only after the investigation had been completed. The department or function requesting a root cause investigation must submit the RCFA request to the Root Cause Core Team since they will be the one to assign the principal investigator for the incident.

Step 4: Freeze the Evidence: RCFA can only be performed on fresh failures or when the parts that failed and debris had been preserved on the equipment or asset. This will be part of the physical evidence. Both operations and maintenance need to understand that if a failure warrants a root cause investigation, the failure or breakdown should not be repaired immediately since the pieces of evidence should be well preserved. The three evidence-gathering teams should be independent of one another. This means that they cannot compare their evidence with one another to preserve the investigation's integrity. One will be assigned to interview all people involved in the incident, both directly and indirectly. One evidence gathering person will be assigned to collect all physical evidence on the equipment, and the other evidence-gathering person will be assigned to collect all paper evidence which they think is important and relevant during the time of the incident, which may include recent Preventive Maintenance records, history records, recent Predictive Maintenance records, charts, operators logbook, and other relevant documents during the time of the failure.

Step 5: Proceed with the Evidence Gathering Event: Sufficient time will be allotted to the people in charge of gathering the three types of evidence. The principal investigator will provide a timeframe to gather evidence on the physical, people, and paper evidence. Based on their findings, all evidence collected should be summarized by the three evidence-gathering teams independently of each other by determining the physical, human, system, and latent cause. It is important for the three evidence-gathering teams assigned to the case not to compare their evidence to avoid being one-sided and biased. Each of the evidence gathering teams should gather the pieces of evidence assigned to them. After collecting all the evidence, each of the three evidence gathering teams (physical, paper, and people evidence) should pretend that their evidence is the only evidence, summarizing their findings in writing by stating if our evidence were the only available evidence, what will be the physical, human, system, and latent cause of the problem. The physical cause will be written in a paragraph describing the physics of the incident. It should be written in past tense and must be specific. The human cause can be bulleted lists stating who did what wrong that triggers the above physical causes. No names must be mentioned. If the operator was involved, just write operator and so on. It should be past tense and specific. The latent cause should be written in paragraph form and should indicate what does this evidence suggests about the way we are as an organization that might have led to this incident. The latent cause should be written in the present tense and must be generic.

Step 6: Principal Investigator and Evidence Gathering Team Meeting: Once the three evidence gathering team had completed summarizing their evidence, the principal investigator and the three evidence-gathering teams will initially meet for the first time. Each of these evidence-gathering persons will individually present their evidence to the Principal Investigator and other evidence-gathering people. Each person responsible for collecting all the pieces of evidence assigned to them must be completed before the meeting can take place. Any of the evidence-gathering persons can request some additional time from the Principal Investigator if needed. Each evidence-gathering person involved will be provided sufficient time to present their evidence to each other and to the principal investigator handling the case. The physical, human, system, and latent cause written by each evidence-gathering person should be presented to the principal investigator. Suppose any evidence-gathering person has completed gathering their evidence; in that case, they may assist the other evidence-gathering person. For example, the person in charge of gathering the physical evidence has completed gathering all evidence; this person can be assigned by the Principal Investigator to help and assist the person conducting the interview process or paper evidence to speed up the evidence collection.

Step 7: Principal Investigator and Evidence Gathering Team to Conduct the RCFA Logic Tree Block Diagram: The evidence gathering team and the Principal Investigator will perform an RCFA logic tree or what, why-and-how tree diagram based on the evidence gathered. They will do their best to reconstruct the sequence of events that led to the failure based on the evidence they have uncovered. If there are levels in the RCFA logic tree where they cannot be answered by the group, the evidence-gathering team will proceed to Step 5 once again and gather additional evidence. After completing the RCFA logic tree diagram, the principal investigator and the evidence-gathering team will consolidate and finalize the physical cause, human cause, system cause, and the organizational latent cause of the problem. This will supersede the physical, human, system, and latent cause written by each evidence-gathering team. However, the latent cause uncovered will still be tentative as it will be the Stakeholders who will finalize the latent cause of the problem for both organization and personal.

Step 8: Principal Investigator and Evidence Gathering Team to Plan for the Initial Stakeholder Meeting: After completing the physical, human, system, and organizational latent cause of the problem, the principal investigator, together with the evidence-gathering team, will name the people involved in the incident. The Principal Investigator will request a meeting with them as the evidence gathering team presents their evidence to them and why they were involved in the incident. The Principal Investigator, evidence gathering team, and the people involved in the analysis will meet for the first time. This will be called the Stakeholder meeting. Note that the stakeholder meeting will only be applicable for both medium and large-scale events. Small-scale events will not require a stakeholder meeting. The stakeholder will include any of the following; the person being accused, the person whose behavior needs to change, the person who has to spend money, the person accusing, or anyone that the Principal Investigator and evidence gathering team think is involved in the incident. More details are provided in Section 8.6 on how to proceed with the Stakeholder meetings.

The Principal Investigator and the Evidence Gathering team must understand and plan the stakeholder meeting carefully. This will be a very sensitive meeting because the person involved in the failure and the management team will be involved in most cases. It is good if the Principal Investigator can provide house rules to avoid blame, sarcasm, condemnation, and this sort of stuff during the meeting duration. The main objective of the stakeholder meeting is to find the truth about what causes the problem, learn from it, and finally do something about it so that the same cause of the problem can totally be avoided in the future. There will be a series of meetings that will take place with the stakeholder, especially if the incident will be a large-scale event. The Principal Investigator should explain to the stakeholder the importance of their presence in the meeting.

Step 9: Conduct the Initial Stakeholder Meeting: During the first meeting, the Principal Investigator will explain why they are called and included during the stakeholder meeting. The evidence-gathering team will present the summary of their investigation. The Principal Investigator will present their logic tree diagram and discuss the outcome of the investigation based on the pieces of evidence uncovered during the investigation process. It must be clear at the beginning that the Golden Rule will be followed since we want to learn from the things that go wrong. Usually, this will take about an hour or two at the most. Special care must be taken as this will be a sensitive meeting. Our reason will be to let them know their involvement in the meeting. If you recall our previous discussions that when a person who committed the error was known, there are ghosts or invincible people involved that cause the person who committed the error. These people will be included in the Stakeholder meetings. Assign someone to take the minutes of all the stakeholder meetings.

Step 10: Second Stakeholder Meeting: After all the three forms of evidence have been presented to the stakeholder, the Principal Investigator (PI) will ask the stakeholders if they agree with the physical, human, system cause of the problem presented. If not, the stakeholder will brainstorm and present what they think the problem's physical, human, and system causes. Usually, this meeting can last from 2 to 4 hours. The stakeholders will also be asked if they agree or if there is any additional information they would like to add to the physical, human, and system causes derived by the Principal Investigation and Evidence Gathering team.

Step 11: Final Stakeholder Meeting: On the last meeting with the stakeholders, two things will be asked, which include determining the latent cause and providing a SMART Countermeasure. The acronym SMART means the corrective actions should be Specific, Measureable, Actionable, Reasonable, and Time-Bounded. The stakeholder will also discuss the possible corrective actions for the physical, human, and system cause of the problem stated. The latent cause will not require corrective action. Usually, this meeting can last from 2 to 4 hours. To answer the latent cause, the stakeholder needs to ask the following questions;

- **What is it about the way WE ARE that contributed to this incident?** The question should be answered by the whole group of stakeholders, open-discussion style. Their answers must be prefaced by the word "we." Only list the answers on the flip charts that everyone agrees upon. Do not suggest or tell them "the answer." They must come up with their own answers.

- **What is it about the way we are that contributed to this incident**? Each person included as a stakeholder is required to ask and answer this question individually. Ask people to initially write their answers on flipcharts. Their answers must start with the word I.

ROOT CAUSE FAILURE ANALYSIS SUMMARY

Determining The Physical, Human and System Cause of The Problem

(P) Summary of Physical Cause	Permanent Countermeasure	Responsible	Completion	Status
(H) Summary of Human Cause	Permanent Countermeasure	Responsible	Completion	Status
(S) Summary of System Cause	Permanent Countermeasure	Responsible	Completion	Status

Bottom Line Learning : What do you think is the lesson we can learn from this failure ?

Root Cause Failure Analysis Performed by :	RCFA Reviewed by :	RCFA Approved by :

Figure 8.4: Corrective Action Form

Once the Stakeholder completed writing both the organizational and personal latencies, they will be presented to the principal investigator and evidence gathering team. The Principal Investigator will also share the latencies they wrote individually and consolidate them with the stakeholders if they agree or disagree. Suppose the stakeholder disagrees with the latencies of the Principal Investigator; in that case, the latencies written by the stakeholder will be used. Typically, the principal investigator and the evidence-gathering team will answer the physical, human, and system cause, while it will be the stakeholder that should come up with the latent cause of the problem. However, additional inputs can be added to the stakeholder's physical, human, and system causes if they think it would be necessary.

Step 12: Stakeholder to Determine a Smart Countermeasure: The Principal Investigator will ask the stakeholder what they think will solve their problem and determine a Smart Countermeasure or Corrective Action. Note that the stakeholder will only recommend, but the actual people to execute this corrective action may be other people or third parties such as Engineering, OEM, vendors, etc. Note that countermeasures will be required only for physical, human, and system causes. Both the Stakeholder and Principal Investigator will recommend who should execute the corrective action. These corrective actions can be done inside or outside the plant. The latent cause will not require a countermeasure since this is where a change in oneself and the organization needs to happen. The stakeholder needs to brainstorm what they think is the latent cause and likewise will need to write what they think is their personal latent causes that need to be changed.

Step 13: Translate the Findings: The Stakeholder, together with the Principal Investigator and evidence-gathering team, will generate a plan to translate the findings to other departments or business units that they think should also learn from this incident. The translation may be in the form of posters, comic storybooks, memos, One Point Lesson, Good to find, or anything that can easily be read and understood by the people in the plant. Lessons from the failure should be disseminated and conveyed to other plant areas where the problems are most likely to happen. This will serve as a learning and guide for the people in the industries.

Figure 8.5: Translate the Findings

Step 14: Principal Investigator and Evidence Gathering Team Conclude the Investigation: The principal investigator and the evidence gathering team finally conclude the investigation and generate a report on the incident. This report will be submitted to the head of the plant, the management, and the Root Cause Core Team. The final root cause report should include the following;

- Indicate the problem as precise as possible
- Request for a root cause investigation (figure 8.3)
- Evidences gathered on the people, physical and paper evidence
- Copies of all the interviews conducted which includes the question and answer
- Root Cause Logic Tree Diagram conducted by the Principal Investigator and Evidence Gathering Team
- Minutes of the Meeting with the Principal Investigator and Evidence Gathering Team
- Reasons of the Selection for each of the Stakeholder by the Principal Investigator
- Minutes of the Meeting with the Stakeholders
- Summary Sequence of Events based on the Evidences Uncovered
- Physical, Human, System Causes
- Summary of the Organizational and Personal Latent Causes
- Summary of Corrective Actions for the Physical Cause (figure 8.4)
- Summary of Corrective Actions for the Human Cause (figure 8.4)
- Summary of Corrective Actions for the System Cause (figure 8.4)
- Revised Procedures if any
- Lessons learned from this event and means of translating the Findings
- Other Relevant Documents

ROOT CAUSE FAILURE ANALYSIS CHECKLISTS

Validation of the Physical, Human and System Causes of The Problem

A	PHYSICAL CAUSE CHECKLIST	YES	NO	REMARKS TO QUESTION
1	Was the failure recurring or was it the first time you encountered it ?			
2	Was an instrument used to verify the physical cause ?			
3	Was an outside laboratory necessary to determine the physical cause ?			
4	Was there sufficient evidence gathered to verify each hypothesis ?			
5	Was CBM Diagnostic Instruments necessary to verify the hyphothesis ?			
6	Was a containment action necessary to implement ?			
7	Was the Physical Cause of the failure determined ?			
8	Was the countermeasure effectively implemented ?			
9	Was there a recurrence of the failure after implementing the countermeasure ?			
10	If there was a recurrence was the failure caused by the same Physical Cause			
B	**HUMAN CAUSE CHECKLIST**	YES	NO	REMARKS TO QUESTION
1	Was Operator involved in the failure ?			
2	Was Maintenance involved in the failure ?			
3	Was the failure attributed to the same person always ?			
4	Was other department involved in the failure ?			
5	Is training required and necessary to eliminate the human cause of the problem ?			
6	Was the Human Cause of the failure determined ?			
7	Was the countermeasure effectively implemented ?			
8	Was there a recurrence of the failure after implementing the countermeasure ?			
C	**SYSTEM CAUSE CHECKLIST**	YES	NO	REMARKS TO QUESTION
1	Was there an existing procedure, system or specs before ?			
2	Does the existing PM Specs or documents need to be revised ?			
3	Does the existing procedure, system or specs need to be revised ?			
4	Was other department involved in the failure well informed on the countermeasure ?			
5	Is there a way we can simplify the existing process, system or procedure ?			
6	Was the Latent Cause of the failure determined ?			
7	Was the countermeasure effectively implemented ?			
8	Was there a recurrence of the failure after implementing the countermeasure ?			

RCFA Analysis Performed By :	RCFA Analysis Noted By :	RCFA Analysis Approved By :	Department / Area	Equipment ID Number	Case No

Figure 8.6: RCFA Validation Check for Corrective Actions

Step 15: Validate the Corrective Actions   Once the root cause investigation has been concluded, someone should be assigned to check the corrective actions' validity done 3 to 6 months after implementation. Validate if the cause of the problem had been eliminated or check if the problem repeats itself or if any of the corrective actions generated did not provide any new problems on the asset or equipment. Suppose an item had been modified and kept in stock in the storeroom; in that case, the modification should be discussed with the PM group, storekeeper, and purchasing people especially if the lifespan of the part has been lengthened since the initial usage and consumption will be much less, hence, purchasing needs to adjust their quantity to be ordered.

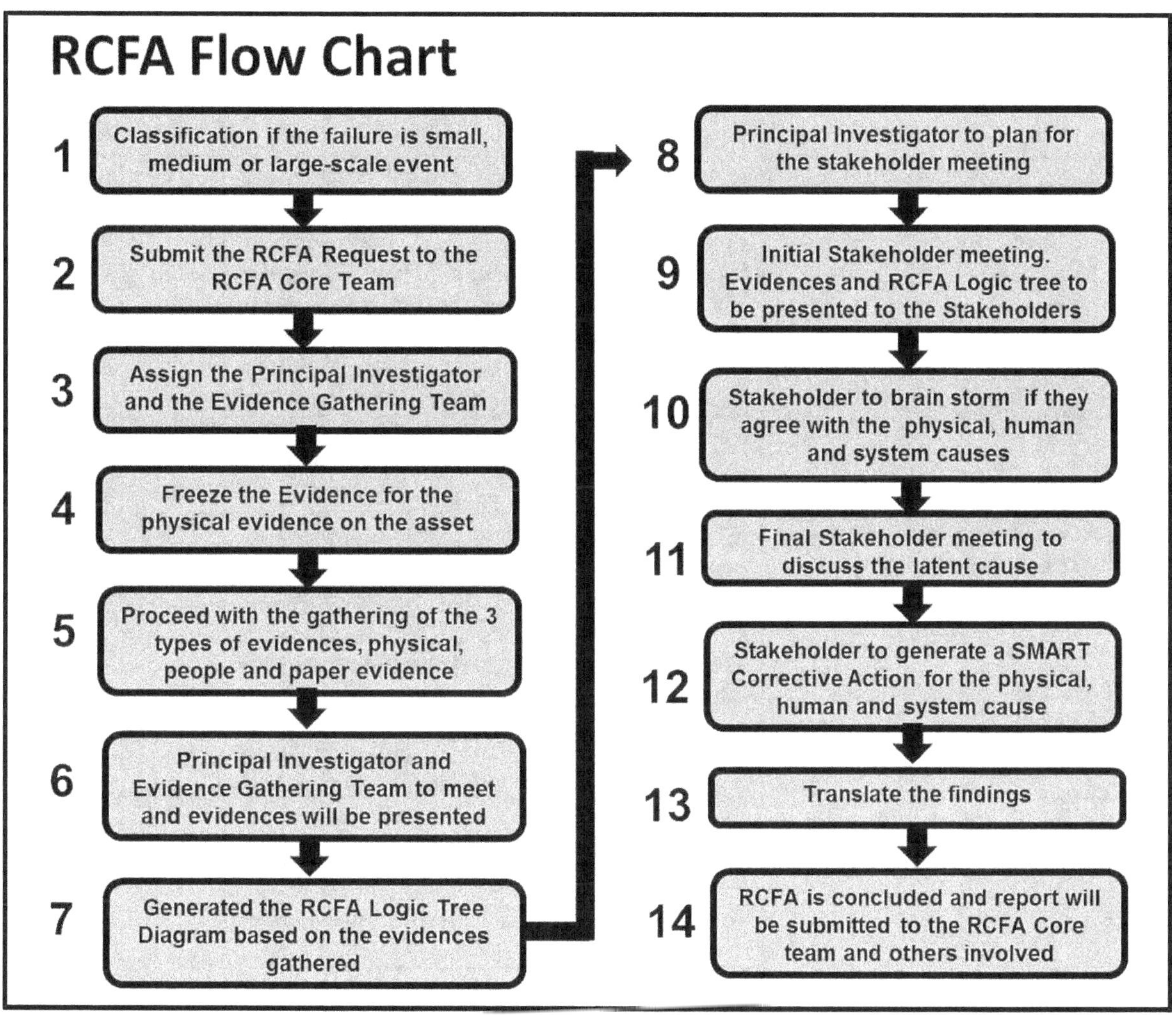

Figure 8.7: RCFA Flow Chart

No	RCFA Steps	Detailed of the RCFA Steps	Responsible
1.	Determine whether the failure is a small, medium, or large-scale event.	a. Prepare the request for root cause request of the asset or equipment that failed.	Maintenance Repair Team Manager Principal Investigator
		b. Repair team to extract all physical evidence on the equipment before commencing the repair.	
		c. Request for a Principal Investigator to investigate the failure incident.	
		d. Give the root cause request statement to the	

No	RCFA Steps	Detailed of the RCFA Steps	Responsible
		Principal Investigator assigned for the failure incident.	
2.	Submit RCFA request	a. Area manager or concerned person should submit an RCFA request to the RCFA Core Team. RCFA Core team will assign the Principal Investigator to handle the case.	Area Involved in the Failure and RCFA Core Team
3.	Set up the RCFA Investigating team.	a. Principal Investigator (PI) will select the evidence gathering team to gather evidence on the failure incident. - Small-scale event, PI to act as the evidence gathering team - Medium-scale event, PI to assign 1 or 2 evidence gathering team -Large-scale event, PI to assign 3 evidence gathering team - Note: PI means Principal Investigator	Principal Investigator Evidence Gathering Team
4.	Evidence Gathering team to be deployed to gather pieces of evidence regarding the failure incident.	a. Assuming this is a maxi event, one person will be assigned to collect all physical evidence on the failed equipment.	Principal Investigator Evidence Gathering Team
		b. One person to be assigned to conduct the interview with a minimum of 10 people, directly and indirectly, involved in the failed incident.	
		c. One person to be assigned to collect all relevant paper documents during the failure incident.	
		d. Principal Investigator should regularly communicate with the three evidence gathering team regarding their status.	
5.	Summarize the evidence gathered by the Evidence Gathering Team.	a. Each of the evidence gathering teams will summarize their evidence and provide their own physical, human, and system causes based on their evidence on hand.	Evidence Gathering Team
		b. It is important for each evidence-gathering person not to compare their evidence with one another and should remain independent.	
6.	Principal Investigator and the Evidence Gathering team to meet and prepare the RCFA Logic tree diagram.	a. Each of the Evidence Gathering teams will summarize all their evidence collected and discussed this with each other together with the Principal Investigator.	Principal Investigator Evidence Gathering Team
		b. The Principal Investigator will now proceed with the RCFA Logic Tree Diagram based on the evidence collected.	
		c. The RCFA Logic Tree Diagram will start with the Failure incident followed by the causes based on the evidence collected by the evidence gathering team.	
		d. If there are levels on the RCFA Logic Tree Diagram that are unclear to both the PI and Evidence Gathering team, they will repeat Step 4 of the RCFA Step.	
		e. Principal Investigator and the Evidence Gathering team to reconstruct the Sequence of Events that led to the failure incident.	
		f. The Principal Investigator and Evidence Gathering team will complete the RCFA Logic tree diagram and	

No	RCFA Steps	Detailed of the RCFA Steps	Responsible
		determine the physical, human, and system cause of the problem.	
7.	Principal Investigator and the Evidence Gathering team to plan for the Stakeholder Meeting.	a. PI and the Evidence Gathering team to list all concerned people involved in the failure incident.	Principal Investigator Evidence Gathering Team
		b. An email will be sent by the Principal Investigator to the Stakeholder for a special meeting regarding the failure incident.	
		c. A date will be set for a meeting between the Principal Investigator, Evidence Gathering team, and the Stakeholders involved.	
8.	First meeting with the Stakeholder	a. The Principal Investigator will debrief the Stakeholder regarding why they were included in this meeting.	Stakeholder Principal Investigator Evidence Gathering Team
		b. The Evidence Gathering team will summarize all the evidence and present them to the stakeholder.	
		c. The initial or first meeting will last for 1 to 2 hours.	
9.	Second meeting with the Stakeholder.	a. The Principal Investigator will present the RCFA Logic tree diagram to the stakeholder and present the physical, human, and the system cause of the failure incident.	Stakeholder Principal Investigator Evidence Gathering Team
		b. The stakeholder, PI, and Evidence Gathering team redefine and brainstorm what they think is the physical, human, and system cause of the failure incident.	
10.	Final meeting with the Stakeholder	a. The Principal Investigator will present once again the complete RCFA Logic Tree Diagram and Sequence of Events to the Stakeholder and present the Latent Cause of the problem.	Stakeholder Principal Investigator Evidence Gathering Team
		b. Each stakeholder will redefine and brainstorm what they think is the latent cause of the problem.	
		c. To answer the latent cause is to ask the following questions. First, what is it about the way we are as an organization that contributes to the problem? Second, what is it about the way I am that contributes to the problem?	
		d. The Principal investigator will facilitate the Stakeholder to define the corrective action for the physical, human, and system cause of the problem. The latent cause will require no corrective action. Corrective Action generated should be SMART, specific, measurable, actionable, reasonable, and time-bounded.	
11.	Generate the corrective action for the physical, human, and system cause of the problem	a. The stakeholders will generate a corrective action for the physical, human, and system cause of the problem.	Stakeholder Principal Investigator Evidence Gathering Team
		b. The stakeholders will finalize who will perform the corrective action and estimate the time of completion.	
12.	Translate the findings and share the lessons learned.	a. The Principal Investigator, together with the Evidence Gathering team, will translate the findings and share the lessons learned from this failure incident to the different areas of the plant.	Principal Investigator Evidence Gathering Team
		b. Translating the findings may come in the form of posters, One Point Lessons, Good to find so that other departments and areas will also be aware of the	

No	RCFA Steps	Detailed of the RCFA Steps	Responsible
		problem.	
13.	Principal Investigator and the Evidence Gathering team to conclude the investigation.	a. The Principal Investigator, together with the Evidence Gathering team, to conclude the investigation and generate a report about the failure incident.	Principal Investigator
		b. The report will be submitted to the head of the plant and the affected area where the failure incident had occurred.	
		c. The RCFA Investigation is now concluded.	
14.	Translate the findings	'a) Lessons learned from the incident should be translated in an easy-to-understand method and will be given to other functions, departments, or business units in which the incident or failure can occur.	RCFA Core Team Principal Investigator Evidence Gathering Team

Figure 8.8: RCFA Step by Step Investigation Process

8.6: Details in Preparing for the Stakeholder Meeting

Once the principal investigator and evidence gathering team have named the people to be included in the stakeholder meeting, it is very important to prepare for this meeting carefully. What questions to be asked, how to present the evidence so that the principal investigator must be objective at all times. It is important to take note of the following before proceeding with the meeting.

1) **Prepare the necessary things for the meeting**, such as a whiteboard, whiteboard pens, flip charts with paper, a marker for flipcharts, a multi-media projector, and a computer or laptop. Reserved the room in advance. Assign someone to take the minutes of the meeting at every stakeholder meeting. The minutes of the meeting should be given to everyone present during the Stakeholder meeting. The principal investigator will email or inform each stakeholder personally if they have no email for the preliminary meeting. If the meeting takes more than an hour, might as well prepare for some refreshments such as sandwiches and drinks for the people present in the meeting.

2) **Principal Investigator explains why the stakeholders were selected.** The stakeholder in the RCFA process will include the people involved in the incident. The stakeholder will include any of the following; the person being accused, the person whose behavior needs to change, the person who has to spend money, the person accusing, or anyone that the Principal Investigator and evidence gathering team would think that is involved in the incident.

3) **Evidence Gathering Team to Present their Evidence to the Stakeholders.** After the introduction by the Principal Investigator, each of the evidence gathering team will present their pieces of evidence to the stakeholders present during the meeting.

4) **Principal Investigator to Present the RCFA Logic Tree Diagram:** After the evidence gathering team's presentation, the principal investigator will explain the RCFA Logic Tree Diagram to the stakeholders. The meeting will then be adjourned. The principal investigator will set a date for the second meeting with the stakeholder. The stakeholder needs to confirm their presence during the succeeding stakeholder meetings.

5) **Stakeholder to Brainstorm What They Think is the Physical, Human, and System Cause.** Usually, this will be done during the second meeting. The principal investigator will ask the stakeholder to brainstorm, based on the evidence presented and the RCFA logic tree diagram, what they think is the physical, human, and system cause of the problem.

6) **Stakeholder Finalizes their Physical, Human, and System Cause of the Incident:** The Stakeholder will present to the Principal Investigator and Evidence Gathering team the result on what they think is the physical, human, and system cause of the problem. After the Stakeholder's presentation, the principal investigator will present the physical, human, and system cause of the problem based on the pieces of evidence gathered.

7) **Consolidation of the Physical, Human, and System Causes:** Once both the Stakeholders and Principal Investigators presented their physical, human, and system causes, they need to consolidate them to arrive at the final physical, human, and system cause of the problem. Once they have consolidated the different causes, the meeting will be adjourned. The Principal Investigator will set for the final meeting to discuss the latent causes. The principal investigator needs to have the commitment of the stakeholders to be present during their final meeting as this is the most important meeting of all.

8) **Stakeholder to Determine the Latencies:** Once the physical, human, and system causes had been consolidated and finalized, the principal investigator will ask the stakeholders the following questions. First, what is it about the way we are as an organization that contributed to the problem or incident? Second, what it is about the way I am that contributed to the problem. The first question can be done through brainstorming, while the latter will be answered individually by each stakeholder involved. They need to write their answer about what they think they need to change. This is about looking at themselves in the mirror. The principal investigator will collect both organizational and personal latent causes, which will be included in the Root Cause Final Report.

9) **Stakeholder to Determine a SMART Countermeasure:** Once the individual and organization latent causes have been completed, the stakeholders will again brainstorm what they think would be the corrective action for the physical, human, and system cause of the problem. They should also indicate who will be responsible for carrying on these countermeasure and corrective actions. Estimated the deadline for completion. Suppose a budget is required for a modification. In that case, the stakeholders need to convince the decision-makers to approve the budget needed for the modification and explain the reason behind it.

10) **Stakeholder Meeting is Concluded:** The principal investigator, together with the evidence gathering team, will conclude and thank the stakeholder for their commitment and cooperation in providing a solution to the crisis or incident. Anyone from the stakeholder can also speak and share their message regarding their experiences during the Stakeholder meeting. The Stakeholder meeting is now concluded. The Principal Investigator will now finalize his report regarding the Root Cause of the incident and return to his original work in the plant.

8.7: How to Make RCFA as Part of a Structured System

Step 1: Train and educate all people on RCFA: Management must set out expectations for why its people will be trained on RCFA. Start by educating your maintenance, safety, and technical people on RCFA. Conduct a management presentation and overview of the basics of RCFA and when it should be applied. Note that Root Cause Failure Analysis is done on current failures and not on failures experienced in the past. Hence, people must understand the importance of freezing the evidence, taking photographs, and preserving the part that failed.

Step 2: Set up the RCFA Team and Freeze the Evidence: The Principal Investigator, as well as the Evidence Gathering Team, will be assembled currently to investigate the failure. It is highly recommended to assign an RCFA Facilitator and third-party consultant to guide the root cause analysis's initial process. The Principal Investigator and Evidence Gathering Team must be free from bias on the problem at hand. Generate an operating context statement for both the Principal Investigator and the Evidence Gathering team to understand how the equipment is being operated. It is also vitally important for the evidence-gathering team to conduct interviews and collect as much data as possible regarding the failure to later verify every single hypothesis uncovered.

Step 3: Conduct the RCFA Investigation: Based on the evidence, the principal investigator and the evidence team will summarize their findings and determine the Physical, Human, and Latent Cause of the problem. RCFA Analysis that stops identifying the physical root causes or ends at the component level always lacks depth. Analyses that focus on people who make bad decisions are often called witch-hunting expeditions. A Root Cause Failure Analysis will take the time to understand why good people make bad decisions. Why did the person who decided think it was the right thing to do at that time? All RCFA efforts must steer away from blaming people.

Step 4: Recommend for the Improvement Plan: Management commitment should be clear at the beginning of any RCFA initiative. Teams will lose motivation and enthusiasm if their recommendations fall on management's deaf ears. Management should review and have the recommendations approved by the team who performed the analysis. When performing RCFA, getting to the causes will be the easy part; getting something done to address the causes is a whole different story. The most important thing that people must understand is that it does not matter who did something wrong. What matters is why. If we do not address the how and why of the failure, it will likely recur; therefore, if we have verified beyond a reasonable doubt that a latent cause exists, the team undergoing the analysis should end its probe on the latent cause of the problem.

Step 5: Implementation of the Improvement Plan: Once the recommendation has been approved, implement the improvement and countermeasure. Counter-measure should be two-fold: to address the physical cause, the human cause, and the system cause of the problem. Management should understand that problems will still resurface unless latent causes of the problem are being addressed by the stakeholders. We must understand that blaming one another has no room in the RCFA investigation and what is important is the how and why and not who caused the problem.

Step 6: Track Results and Measure Key Indicators on RCFA: ROI is only one measure of effectiveness. However, our Root Cause Analysis efforts should be further measured as to how they contribute to the corporate's Key Performance Indicators (KPI's). It is vital to demonstrate this linkage, as it will make the attainment of these goals dependent on conducting a Root Cause Failure Analysis. This further helps to justify the existence of the Root Cause Analysis effort. Measurement will allow us to learn from the things that go wrong and focus our efforts on improving them. We are what we measure, and if we measure what is important, our efforts will definitely be rewarded.

Step 7: Translate the Findings and Share Lessons Learned: Lessons must be spread across the plant. Do not provide each person with a one-hundred-page report. It is important to translate your findings in such an easy way that a layman or a five-year-old kid can grasp and understand. This can be done through posters or comic books that every employee will have fun reading instead of providing some thick pages of reports.

Chapter 9

RCFA Corrective Actions and Countermeasures

> *The distinction between a maintenance and a mechanic is that a mechanic use his hands often and becomes good at it. A maintenance uses a balance of the hand and the brain. They know that if the failure keeps on repeating, then its time to use their brain and start analyzing failures. Let us treat them as maintenance and not just as mere mechanics*

9.1: RCFA Corrective Measures

The goal of any RCFA or RCA is three folds:
• First is to know the truth on what caused the problem or incident.
• Second is to learn about the things that go wrong in their industry.
• Lastly is to do something about it to avoid a repeat of the cause of the failure.

The latter is the objective of this chapter. Corrective actions and countermeasures can be said as the changes or the improvements to be done to address a particular problem. In Root Cause Failure Analysis, the physical, human, and system cause will require corrective action or countermeasure. The latent cause will require no corrective action since this is where a change in oneself needs to happen. It may come in different formats where the ultimate goal of the corrective action is to mitigate or eliminate the chances of failure from recurring. Corrective Action is a list of things that need to be done to reduce, mitigate, or completely eliminate the cause of the problem. These are the changes and improvement that needs to be done on the problem. Once corrective actions are done, they should be validated and evaluated after implementation to determine their effectiveness. These corrective actions may be in the form of modification, or redesign, engineering change, poka-yoke solutions, change in the process or SOP, adding failsafe and protective devices, changes in the maintenance tasks, training, or any other form.

Corrective actions must be well documented in the RCFA Final Report, indicating the changes done for the problem's physical, human, and system cause. Once corrective actions are finalized, assign the person responsible for the design and target the dates of the completion of the redesign or modification. The person responsible may not be the one who will do the design changes but should be held responsible for providing updates regarding its completion. The Action Plan outlines what steps are needed to implement the solution, who will do them and when they will be completed. A simple solution will only need a simple action plan, while a complex solution needs more thorough planning and documentation. Corrective actions and countermeasures performed on any RCA or RCFA should be "**SMART,**" which means that they should be specific, measurable, achievable, realistic, and time-bounded. It is important to remember that these corrective action Items are defined for only the investigation's physical, human, and system causes. Action Items are usually not defined for Latent Causes. Latent causes will be addressed by asking what changes we should need to do about the organization and to ourselves. This will be addressed during the stakeholder's meetings.

Containment is a short-term action being initiated to keep the defects or failures from further damage. Containment is just a temporary solution and will not fix the problem. Recurrence prevention is a list of things and actions that need to be done to indicate that the cause of the failure has not repeated itself. This will usually be done in the case of the physical, human, and system cause only. This can also be a series of questions that need to answer several months after the corrective action had been initiated to determine if the failure had been mitigated or eliminated completely and no longer considered a threat to the industry.

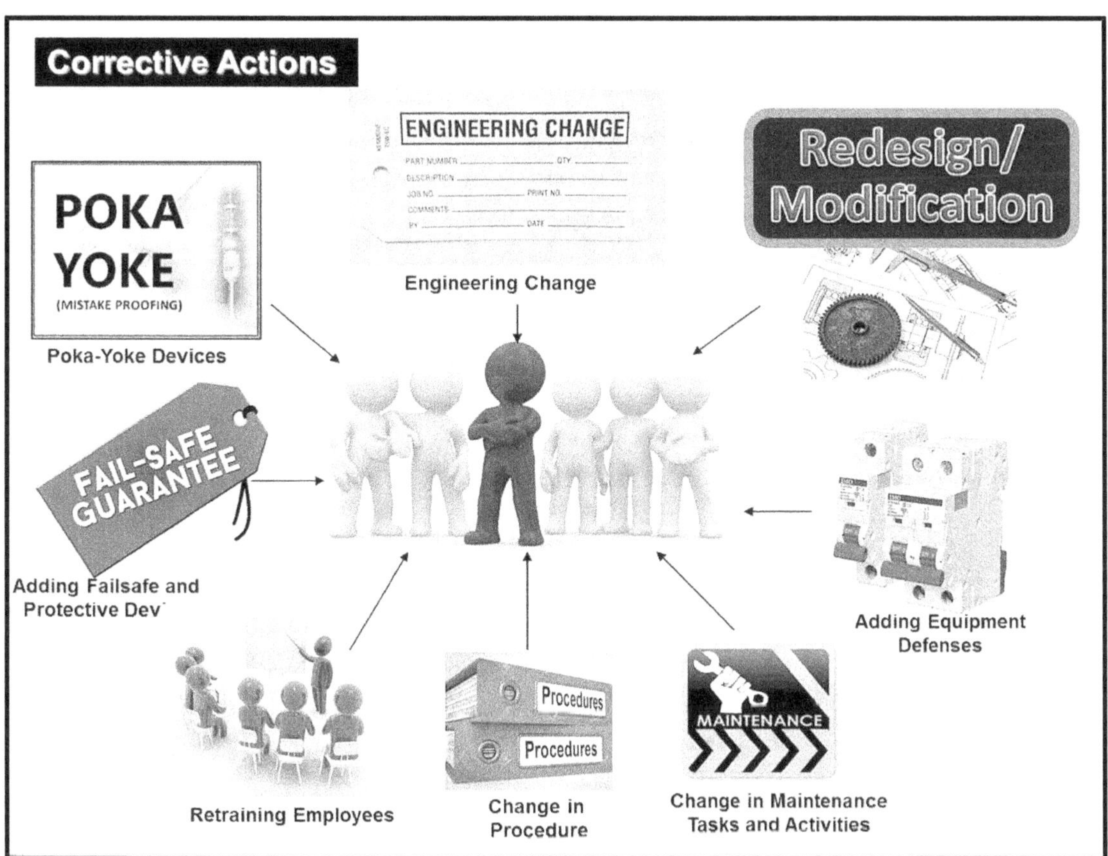

Figure 9.1: Different Types of Corrective Actions

9.2: Different Types of Corrective Actions

Once the physical, human, system causes had been identified, a list of corrective actions will be generated by the Stakeholders involved. These are changes and improvements done on the existing design, process, or system to address a particular problem or incident. The goal of any corrective action or countermeasure is to eliminate, reduce or mitigate the cause of the problem. Corrective actions can come in different forms as follows;

a) Modification or Redesign means any changes or improvements done on an existing design. The goal here is to eliminate, reduce or mitigate the consequences of failure from happening. Modifications may take several months depending on the amount of time, and at times it might be costly, therefore before proceeding with the modification or redesign, cost savings must be made and compared between the existing and modified design and projected in a year to determine if it will be feasible or not to continue. These corrective actions can be used to address the physical cause of the failure. Redesign or modification may include the following:

• Includes changing the specification of a component
• Adding a new item or relocating a machine
• Replacing an entire machine with a different type
• Changing the process or procedure
• Changing the strength of the material of a part or spare
• Adding a redundancy in place if it is not currently present
• Changing the measurements and dimensions of parts with design weaknesses
• Changing the strength of materials to a stronger one
• Increasing the lifespan of parts with inherent design weaknesses
• Improving and lengthening the life cycle of the equipment

The purpose of redesign or modification is to reduce the probability of the failure mode occurring to an acceptable level. This means changing the component or item to a stronger design making the failure no longer a threat, especially when the failure will have safety or environmental consequences. Even before proceeding with the redesign or modification, we need to answer all these questions. If we answer yes to all of them, then redesign, or modification is feasible. Figure 9.2 and 9.3 is an example of a modification done on a flywheel belt pulley.

1. Does the failure involved major operational consequences?
2. Is the cost of scheduled, or Breakdown maintenance high?
3. Are there specific costs that can be eliminated by the design change?
4. Does the design have no harmful effects which can be generated afterward?
5. Is there an economic trade-off study on expected cost savings?
6. Is the asset to stay or be used for a long time and not be decommissioned soon?

Modification of Flywheel Belt for MECO Plating Machine

Division: Central Lead Finish
PM Step: PM Step 4
Team Leader: Jojo Santos
PM Committee: Cesar dela Torre
Model Machine: Meco 2

Impact: Lengthen lifespan of flywheel belt from 5 to 24 weeks

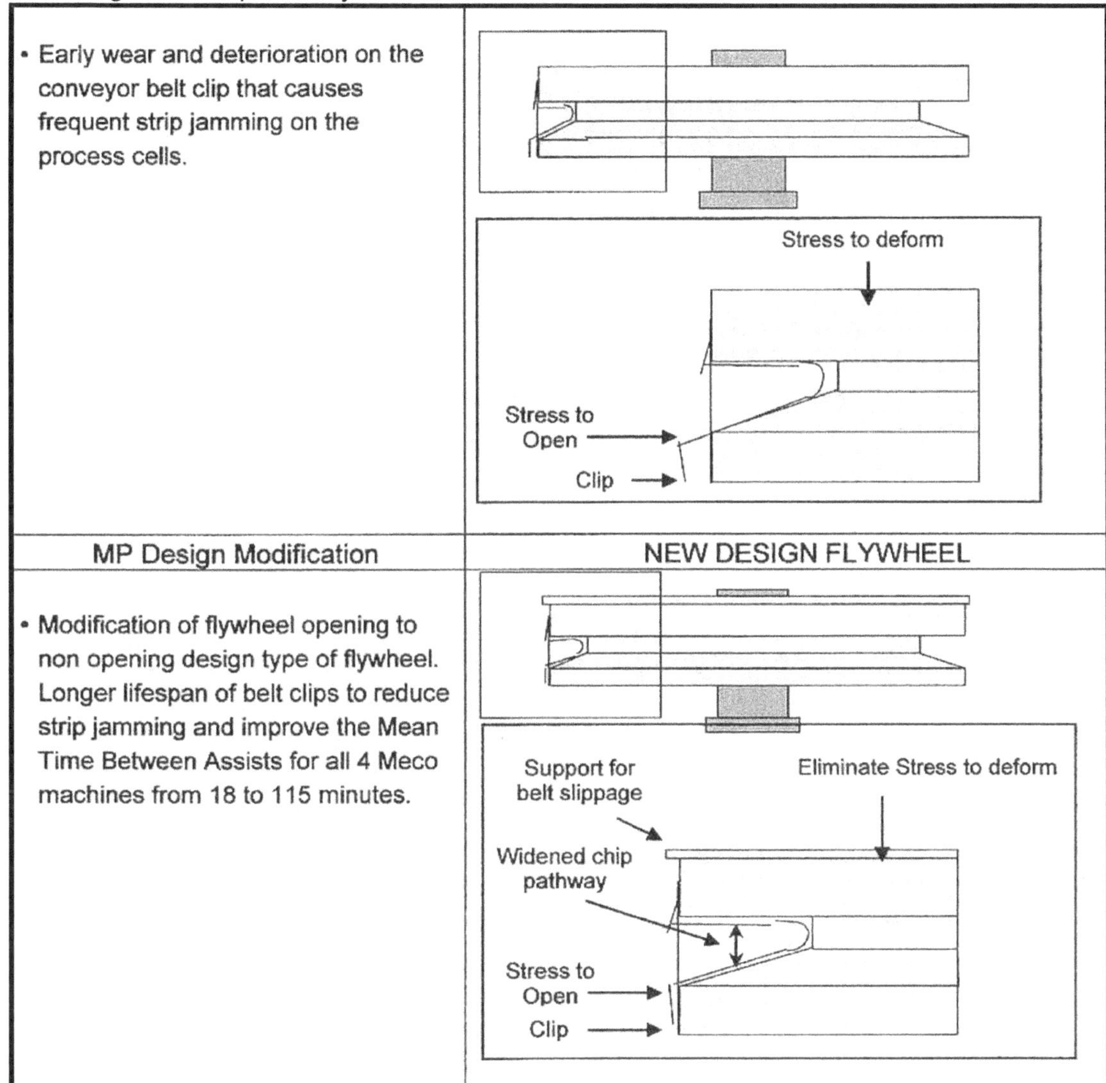

Figure 9.2: Flywheel Belt Modification

Details	Before	After	Unit
Replacement Frequency of Belts	4 to 5 weeks	24 weeks	weeks
Cost per 1 set	$ 600.00	$ 2,200.00	USD
Usage per Machine	2	2	pieces
Costs per Machine	$ 1,200.00	$ 4,400.00	USD
Total Fan Out Costs for 4 Machines	$ 4,800.00	$ 17,800.00	USD
Change in Strength of Materials from	Tool Steel	Stainless Steel	material
Average Life in Weeks	4 to 5 Weeks	24 weeks	weeks
Average Life in Years	0.0961538	0.4161538	years
Yearly Savings in USD	**$ 49,475.02 per year**		**USD**
Year Savings in PHP	**PHP 2,523,266.02 per year**		**PHP**

Figure 9.3: Cost Savings on Flywheel Belt Modification Meco Machine

If a redesign will be chosen as the corrective action where the design of part or item will be modified, it is important to evaluate this first on the equipment. If the lifespan had been prolonged, the maintenance interval should be changed. Both storeroom and purchasing should adjust their replenishment time based on the need of this item on the equipment if this part is stocked in the storeroom. If the original parts still exist in the storeroom, a decision has to be made on whether to still used the parts or proceed with the modified design. If the decision is to use the modified design, these original parts in the storeroom become part of the obsolete parts and should be removed from the storeroom.

b) Poka-Yoke is a Japanese term that means mistake-proofing. Examples of Poka-Yoke include elevator alarms, 110/220 volt or auto volt appliances, low battery indicators, fuel indicators in the car, child lock doors, electrical outlets, and so on. Originally, Shigeo Shingo termed this as Baka-Yoke, which means fool-proofing, or Idiot-Proofing, which means even the dumbest of the dumbest won't be likely to make a mistake or error. However, through the years, Shigeo Shingo changed the name to a milder one and termed it Poka-Yoke, which means Mistake, or Error Proofing. A good example of this is having both a 110 and 220-volt outlet in your house. The socket for the 110 volts can be round, and the socket for the 220 volts can be parallel. Of course, we also need to change the sockets of our appliances or place an adaptor. Poka-yoke is any mechanism used by industries, especially manufacturing, that helps an equipment operator avoid errors and mistakes. Its purpose is to eliminate product defects and failures by preventing, correcting, or drawing attention to human errors as they occur. It was developed by Shigeo Shingo as part of the Toyota Production System. The term Poka-yoke was applied by Shigeo Shingo in the 1960s to industrial processes in which its primary purpose is to prevent human errors. Poka-Yoke aims to design the process to detect and correct mistakes immediately, thereby eliminating defects at their origin. An example of Poka-Yoke for industries is placing a sensor on the equipment such that if the operator accidentally opens the door, the machine will stop automatically. This is done for safety reasons. These corrective actions can be used to address the human cause of the failure.

c) Visual Controls in the Workplace: Visual controls are placed in the workplace to expose problems and deviations easily by just looking at them. It makes deviations easily seen by operators by alerting them that something is not right. These visual controls will alert operators on any deviations from their normal operations and are one of the unique features of Autonomous Maintenance that tells you what to do and what not to do. They originated from the Japanese. They are not meant to make us dumb or look stupid. It simply entails making the problems, abnormalities, or deviation from standards much more visible to everyone in the plant. When these deviations are clearly visible and apparent to all, corrective actions can immediately correct these problems and anomalies. Visual Controls allow us to see problems and deviations more easily.   They are also meant to provide instructions and convey information.   These devices or mechanisms were designed to manage or control our operations process to meet the following purposes. They make the problems, abnormalities, and deviation from standards easily known to the operator and everyone in the entire organization. It will easily display the operating or progress status to see the format, provide instructions, convey information, and provide immediate feedback to people. Good visual control management should tell the operator immediately if a process is going beyond the specified parameters. This is also sometimes called Visual Communication. A good example of Visual Controls can be seen in an international airport. Even if this is your first time in that airport and you are looking for your gate, a place to eat, shop, smoke, or even go to the toilet, just look at the signs and symbols, and you will never be lost. These visual controls can be used to address the human cause of the failure.

d) Update and Simplify your Procedures: If we take a look at all the procedures we used in our industries, there will be many outdated or even obsolete procedures currently being used and implemented up to this very point in time. Perhaps these procedures were effective when it was written but would no longer hold true for now. It is important to make the procedures, SOP's as simple as possible, which anyone can understand and implement easily. Use diagrams, pictures if necessary. These procedures should be simple, precise, and should be easily understood by the user. Make the fonts big enough so that they can be easily read. What is important is that if there is a way of simplifying the things being done, procedures should likewise be revisited, reviewed, revised, and simplified whenever necessary. These corrective actions can be used to address the human and system cause of the failure.

e) Application of Precision Maintenance: This involves performing maintenance work in a consistent, precise, and industry-accepted way. If properly implemented, this means that maintenance should yield the exact same results, no matter who is performing the tasks. Precision Maintenance is much more than merely having procedures on PM. It must be the correct culture of the organization. The good thing about doing Precision Maintenance is that it can reduce the chances of Infant Mortality Failures seen during the start-up of assets right after a Preventive Maintenance shutdown or Major Scheduled Overhauls are performed. It involves performing maintenance work accurately and precisely, which means that the maintenance should provide the exact same results no matter who is performing the work, whether the work is done by the most or the least experienced maintenance craftsperson in the plant. When different maintenance people perform the same PM on the same equipment, variations occur. These variations cause problems since they do not meet the requirements for the process. Precision maintenance help to ensure that equipment will be maintained to the highest possible standard so that the variations that can cause defects and failures can be reduced, or eliminated enabling equipment and assets to run at their optimized and peak level of reliability. Precision maintenance rebuilds machines and equipment to the highest standards so that fewer problems can occur during operation. It is a matter of ensuring the important things for equipment and machinery health are done correctly and accurately. These corrective actions can be used to address the human cause of the failure.

f) Training and Skills Assessment: Essential to any industry are competent people who understand their equipment intimately. It is mainly the missing link ingredient in any change or continuous improvement initiative and effort. Skill is the ability to do one's job correctly. It is gained from the things we learned and our experiences in doing the job over an extended period. Skill is also the product of personal motivation and thorough training in which the end result is mastery. And to enable them to achieve this stage, companies must develop the most effective training strategies chand methods. The first step in any training program is to identify their people's level of knowledge, technology, skill, and competency. The second is to assess their skills from time to time. People are not born with skills. Skills are developed over some time. Knowledge and practice are required to develop the skills of their people. What we want to achieve are people who have mastered both theoretical and practical applications of their work. These corrective actions can be used to address the human cause of the failure. This is the goal of any training needs and skills assessment. The 4 levels of skills include:

• Level 1: A person lacks both theory and practical application. This is the starting point.
• Level 2: A person has gained theoretical knowledge but has not practiced the concept at all.
• Level 3: A person has practiced the concept through experience but no theoretical knowledge.
• Level 4: A person has both mastery of practical knowledge and theory.

RCFA Corrective Actions can Include

Physical Cause	Human Cause	System Cause
- Redesign or Modification	- Application of Poka-Yoke	- Changes in the Procedure or SOP
- Change in the Maintenance Tasks	- Adding Equipment Defenses	- Changes in the Process or System
- Engineering Change	- Adding Visual Controls	- Changes in the Policies
- Improving the Life Cycle	- Retraining and Skills Assessment	
- Reengineering	- Change in Workplace Conditions	
	- Regular Assessments and Audits	
	- Use of Precision Maintenance	

Figure 9.4: Different Types of Corrective Actions

Figure 9.5: RCFA Training Workshops

Another important point to consider in conducting the training is to provide a pre-post quiz and workshops to the students. This will determine the knowledge they gained from the classroom training. Workshops are also a fun and good way to add on the training that can benefit the students learning process. Figure 9.5 were pictures of my class during some of my RCFA training. Having a pre and post-quiz before and after the training would be one of the ways to determine the knowledge the students have gained before and after the training. Depending on the quiz being provided, whether this will be a multiple-choice, true or false, fill in the blanks, or matching quiz, what is important is that both pre and post-quiz should be identical. This means that the questions should be the same for the pre and post-quiz.

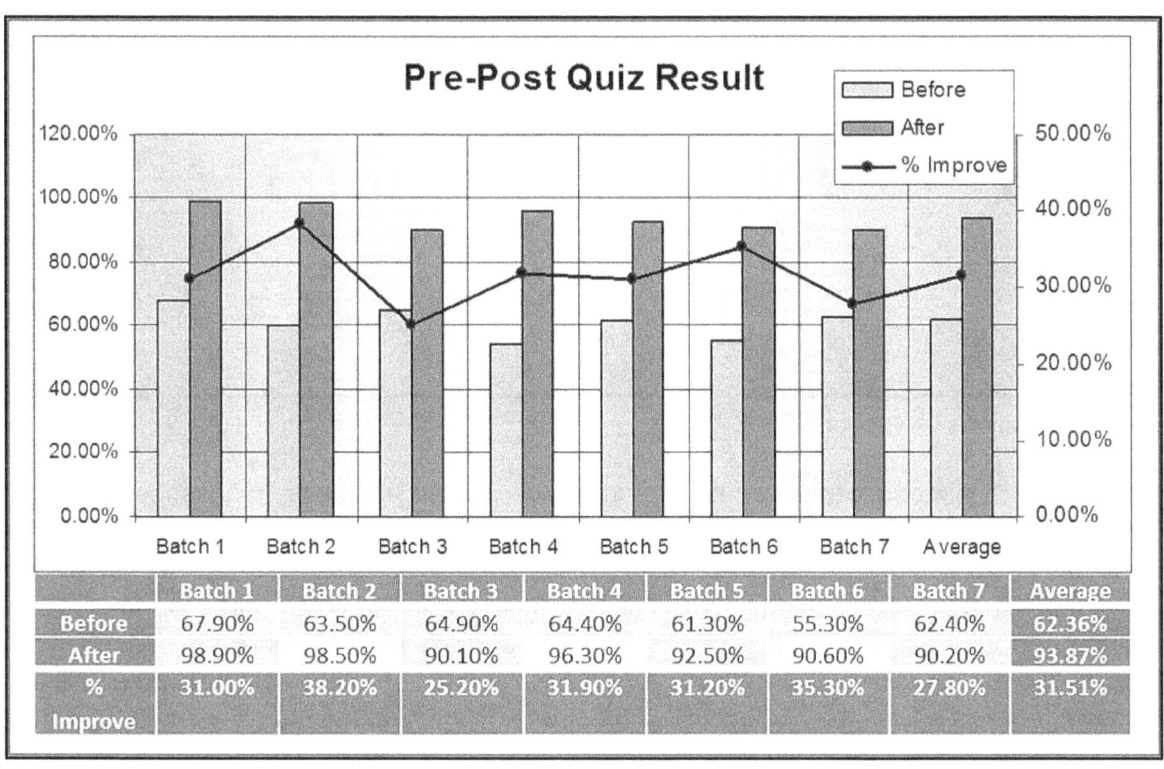

	Batch 1	Batch 2	Batch 3	Batch 4	Batch 5	Batch 6	Batch 7	Average
Before	67.90%	63.50%	64.90%	64.40%	61.30%	55.30%	62.40%	62.36%
After	98.90%	98.50%	90.10%	96.30%	92.50%	90.60%	90.20%	93.87%
% Improve	31.00%	38.20%	25.20%	31.90%	31.20%	35.30%	27.80%	31.51%

Figure 9.6: Pre and Post Quiz Training

g) Adding Equipment Defenses: These machinery defenses can be in the form of protective devices, functionality tests, or maintenance tasks used to detect that a failure mode from occurring. Identification of these machinery defenses and controls should begin with those failure mode combinations with the highest severity, impact, and consequences. These controls must detect the failure before it happens and not on the incident already happening. These controls and defenses can alert operators and maintenance that a failure mode is on the verge of occurring. They should be in place, especially when the failure's impact and consequences have very high criticality. An example may be a protective device in the form of an alarm. A low-level alarm of the main fuel tank will sound if the level in the tank is at 2000 liters. When the fuel ran out of the tank, an alarm will sound for everyone to realize that the tank is already emptied. These corrective actions can be used to address the human cause of the failure. What is important for the maintenance function is that if defenses and protective devices will be added to the equipment, a functionality inspection or failure finding tasks must be done to check if they are still working or not.

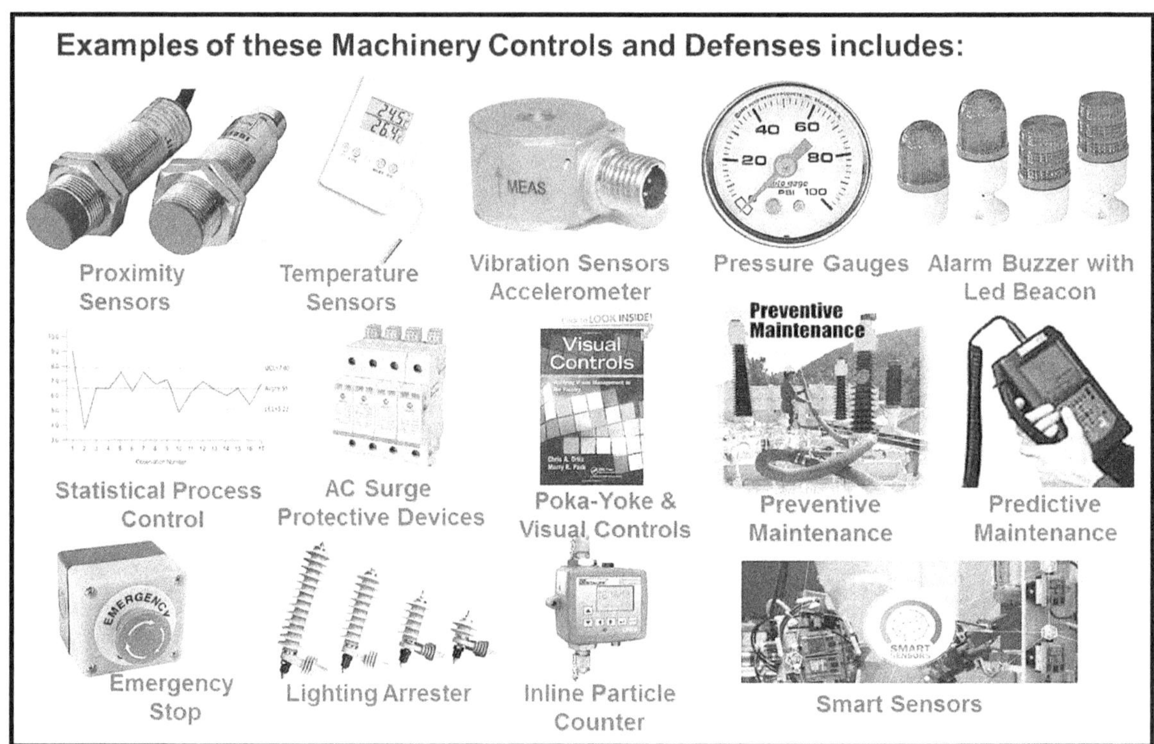

Figure 9.7: Machinery Defenses and Controls

h) Audits and Assessments can be performed on people, methods, processes, or systems. They almost have the same objective but have a slight difference. An audit is a verification of the actual results or performance to verify their accuracy conducted by an auditor to gauge their compliance to a particular standard. It can evaluate work products or processes compared to their actual specifications, standards, processes, or other forms. Audits may be done based on checklists, interviews, actual line audits, or both. On the other hand, an assessment results from the final judgment based on the actual results. It goes further than an audit as it involves the determination of actions necessary for compliance. It can also be said as a formal evaluation of the process of an organizational unit against a reference model. The standard criteria are used to determine the final result of the assessment. Typically a scoring system will be used for a pass, fail, or conditional depending on the criteria used by the assessor. It is also possible to include the audit as one of the focal criteria in assessing something. These corrective actions can be used to address the physical, human, or system cause of the failure.

i) Engineering Change: This is a document approved by the design activity that describes and authorizes an engineering change to the product, defect, or equipment failure and has been approved for changes or improvement. Correction of a design error that doesn't become evident until the problem reveals itself. There are several reasons for the engineering change, but the main goal is to correct a problem. These corrective actions can be used to address the physical, human, or system cause of the failure. Engineering change will include the following;

• Control or Reference Number. This is used for documentation control processes.

• Identification of what needs to be changed. This should include the part number and name of the component, part, item, and the drawing or photo of the change.
• Reasons for the change.
• Description of the change includes drawing the component, part, or item before and after the change.
• List of areas in which the engineering change will be reflected.
• Estimated Budget of the change, the person or team initiating the change should estimate the cost of the engineering change if any.
• Approval of the change. The engineering change must be approved only by those authorized to approve the document.
• Evaluation refers to the result after testing it on the equipment.
• Implementation Time refers to when this change will take effect. If the stock of this item is present in the storeroom, a decision has to be made if the original stocks will be scrapped or will still be used.
• Responsible person to perform the engineering change and the date of completion.

j) Change in Procedures: Usually, changes in the procedure are implemented to address the system cause. What is important in these changes is to indicate the reference number from the existing procedure that will be changed or superseded and the reason for the change. The changes should be easy and simple to read, understand and implement. Use visuals, or photos whenever necessary. Changes in workplace procedures should produce positive results, compared to the existing procedure.

One last thing to consider after any of these changes had been made to the equipment is to check the maintenance tasks and intervals. If a lifespan and MTBF of an item, part, or spare had been prolonged due to the changes made, then maintenance should adjust their maintenance interval. Purchasing and storeroom people should also be informed as their replenishment time for ordering these parts will also be lengthened.

9.3: Validation of Corrective Actions

Whatever corrective actions in section 9.2 that have been initiated should be subject to validation to determine if the changes or improvements done are effective, especially for modifications that require changes in any equipment spare, items, parts, or process. A prototype should be validated on the equipment to determine if the problem is no longer a threat or if any unforeseen problems emerged after the changes took place. It must not pose any new problems on the equipment or asset where the root cause is conducted. These are some of the questions that need to be answered during the validation period.

• Was there a recurrence of the failure after implementing the countermeasure?
• Where there new problems that emerged as a result of the corrective action?
• Was the corrective action for the Physical Cause effective?
• Was the corrective action for the Human Cause effective?
• Was the corrective action for the System Cause effective?
• Did the industry learned from the things that go wrong?

Maintenance KPIs must also be monitored after validating the corrective actions to determine if the corrective actions generated were effective or not. These maintenance KPIs can include the number of breakdowns in frequency, downtime due to breakdowns or failures, maintenance costs, MTBF, or any other KPI that has a direct impact on the incident being investigated. These KPIs will spell the difference between the success or failure of any Root Cause Failure Analysis investigation.

9.4: Horizontal Replication of Improvements

In manufacturing, one of the things Quality people do is to fan out improvements and modifications to any similar equipment, assets, or machines. They called this process improvement fan-out or horizontal replication. My point is that if we have around 100 similar equipment in the plant and modification has been done in one equipment with a problem. My question is if only 3 out of the 100 emit the same problem and 97 of them possess no irregularity or problem of the same kind, should the modification be done on all 100 equipment or only on the 3 equipment?

HORIZONTAL REPLICATION FOR CENTRIFUGAL FAN

DETERIORATIONS UNCOVERED	1	2	3	4	5	6
• Air leak in the system	●	●	X	X	○	○
• Excessive belt tension	●	●	●	○	●	○
• Excessive vibration on the fan wheel	●	●	●	○	X	○
• Soft foot on the base foundation	●	●	●	X	X	○
• Worn out bearing	●	X	●	X	●	○
• Worn out drive coupling	●	X	○	○	●	○
• Excessive grease in ball bearings	●	○	●	●	○	○
• Excessive misalignment of shaft	●	○	●	○	●	○
• Abnormal end thrust load	●	X	X	X	X	X
• Incorrect direction of rotation	●	X	X	X	X	X

○ ⟶ Restoration still ongoing   ● ⟶ Restoration completed   X ⟶ Deterioration not present

Figure 9.8: Horizontal Replication of Improvement

During my employment days when I was still working in the semiconductor industry, there was an improvement group called B.K.M (Best Known Method). The objective of this group is that all improvements, modifications, redesign, or changes made on any equipment should be replicated or fan-out with all equipment of similar types and makers whether the problem exists or not on other equipment and assets. I was totally against this group (since I previously attended RCM and learned about the equipment operating context). I used to argue with the implementers of this group a lot. My argument is that if I have 100 similar equipment in the plant of the same type and if only 3 of them emit the same problem, my point is that only the three equipment should be modified and not all 100 equipment. In this case, the B.K.M. group insists on carrying the improvement to all 100 equipment. I posed this question to some of my connections on LinkedIn who are likewise heavyweights, practitioners, and book authors on maintenance, and here are their comments.

• Hello Rolly, I agree with you that it is not always a good idea to spend money and make changes just because a similar piece of equipment needs it. Equipment are like children; they need individual attention but to be held to a standard that meets the needs of the whole. Thank you for your email. *From Nathan Wright (Maintenance Book Author)*

• Rolly, your point is well taken, but consider the following, if the identified problem is an inherent design deficiency that would be common to all 100 assets. In this case, the modification should be made to all 100 assets. However, suppose it is not an inherent design deficiency, and the failures are limited to one or two of the assets in specific areas or applications; in that case, there is no value and a high potential for creating more problems in modifying all 100. As a general rule, when problem-solving, you determine whether the problem is unique to one asset or all common to similar assets. In the former, the problem is unique to that asset, and if all similar exhibit problems, then the problem is systemic, e.g., caused by something common to all similar assets. I would suggest that you not apply one rule for all cases. Instead, determine to the best of your ability whether the problem and modification are unique to one or a few assets or systemic and exhibited by all. Quality's rollout practice is common, and many consider its best practices. Still, only in cases where the modification is typically warranted, the root cause is an inherent design weakness common to all assets in the family or a systemic (infrastructure) problem. *From R. Keith Mobley, Book Author.*

• Rolly, I have never thought about this deep question. I think of the airplane makers, and when they find an improvement, they spread it out to the entire fleet. Besides, when an RCM analysis turns up a failure mode with a dire consequence, all the units of the same type should have the modification. I am happy being wrong on this but would side with quality. *From Joel Levitt, Maintenance and Reliability Book Author*

Although Joel Levitt's point here disagrees with my thoughts, I fully agree with him since he talked about the airline industry, and we all know the consequences of failure if a plane crash. I have no argument with this but only for manufacturing equipment and assets where there will be no media, crooked old politicians, insurance people, and lawyers mingling around after your equipment failed. It is just you and the boss. Many years ago, I had a friend and mentor. We used to call him Mang Tibo. In the Philippines, the word Mang means elderly. This guy is from the hard knocks. I asked him this question a long time ago: if I made an improvement on this equipment to address a specific problem, should I replicate it with the other equipment too. He looked at me for a while and told me a story. He said that he was the bread earner of his family, and he had seven sons. One day, his youngest son got the flu and a high fever, and he went to a drugstore to get some medication for his son, who was sick. When he went home, he gave his son the medicine. He called all his sons and gave them the same medicine even if they were not sick. The eldest son said that he was not sick, and he said just to follow his orders. After an hour, his second eldest son got some rashes on his skin and has difficulty breathing. He was allergic to the medication given to him. Mang Tibo rushed him to the doctor immediately and was confined for a couple of days, which added to his bills and expenses. He said, from that moment forward, he learned his lesson the hard way. Mang Tibo never answered my question directly, but his story tells it all.

Usually, Quality people do these kinds of things in industries, especially manufacturing. Their reason is quite obvious as they adhere to the concept of JIC or Just in Case. My question is, what if it is a case of JICIDNH? You might be wondering, what on this planet is that ugly acronym? It means, what if "Just in Case it did not happen? Here is my concern with the B.K.M. group. Any two machines can be identified if they are of the same type or maker. However, even if the equipment were identical, the operating context may not be exactly and precisely identical in every sense. Different operators can operate the equipment unless they are identical twins with identical minds and skills. But even if they are twins, their character might not also be 100% identical. The equipment may not run with the same operating hours and so on. This is what I meant by not exactly having the same operating context statement even if the machines are identical. The second is the cost of the modification. This means that if the modification will be carried out to the entire 100 equipment and the cost of modification is $ 1,000 per equipment, then the total costs of modification will be $ 100,000 instead of only $ 3,000.00. I rest my case on this, and I believe that only equipment with the same problem should be included in the modification; for those that are stable, the best thing to do is to leave it alone. Besides the cost of modification, what we are avoiding is for the equipment to generate new problems in the future. This means that if the equipment is stable and not emitting any problems, then let it be and I think that's all I have to say about that.

Chapter **10**

Actual RCFA Case Study

> *We can learn from failures if we take the time to analyze them. It is an opportunity for us to begin more intelligently. Therefore, whether from the lessons of life, or problems that an industry encounters, failures tells us something. Failure can be our greatest teacher if we can learn from it.*

10.1: How Root Cause is Performed

Root Cause Failure Analysis is performed on three levels. The first level is to determine the Physical Cause of the problem. This refers to the metallurgical aspect of why the failure occurred. The physical cause is similar to performing failure analysis on the failed part. The analysis would only be limited to the physical cause of the failure itself. The physical cause can be explained as the scientific reasoning or technical explanation of why the failure occurred.

Once the physical cause has been determined, the principal investigator together with the evidence-gathering team should determine the human cause that led to the physical cause. This means that someone either did something wrong or did the wrong thing. It is believed that the majority of the human cause is not the fault of the person or beyond the control of the person who committed the mistake, as there is a lot of factors to consider. It is important to understand why the person committed that mistake so that we all can learn from it. Human errors can be classified into slips and lapses. A slip occurs when somebody does something incorrectly. Once the human cause had been determined, we now proceed and determine the system and latent cause of the problem.

10.2: Performing RCFA on a Personal Level

Root Cause Analysis can also be done not only for industries but for personal problems in our life as well. Here is a personal RCFA case I performed for several decades. This is just a very simple case, but the personal latencies I have uncovered help me realized my weakness in which up to this point in time I have applied in my day-to-day activities.

During the early 90s, only a few rich people can afford to have a cable, while others have satellite TV, majority of Filipinos rely on Aerial Antenna for their TV reception.

- Who (did it happen to)? Rolly Angeles
- What (was the undesired actual/potential consequence)? Wrong orientation of aerial antenna at home causing bad TV reception on almost all VHF and UHF channels.
- Where (did it happen)? At my home
- When (did it happen)? Can't recall but it has been this way for many years since the year 1995.

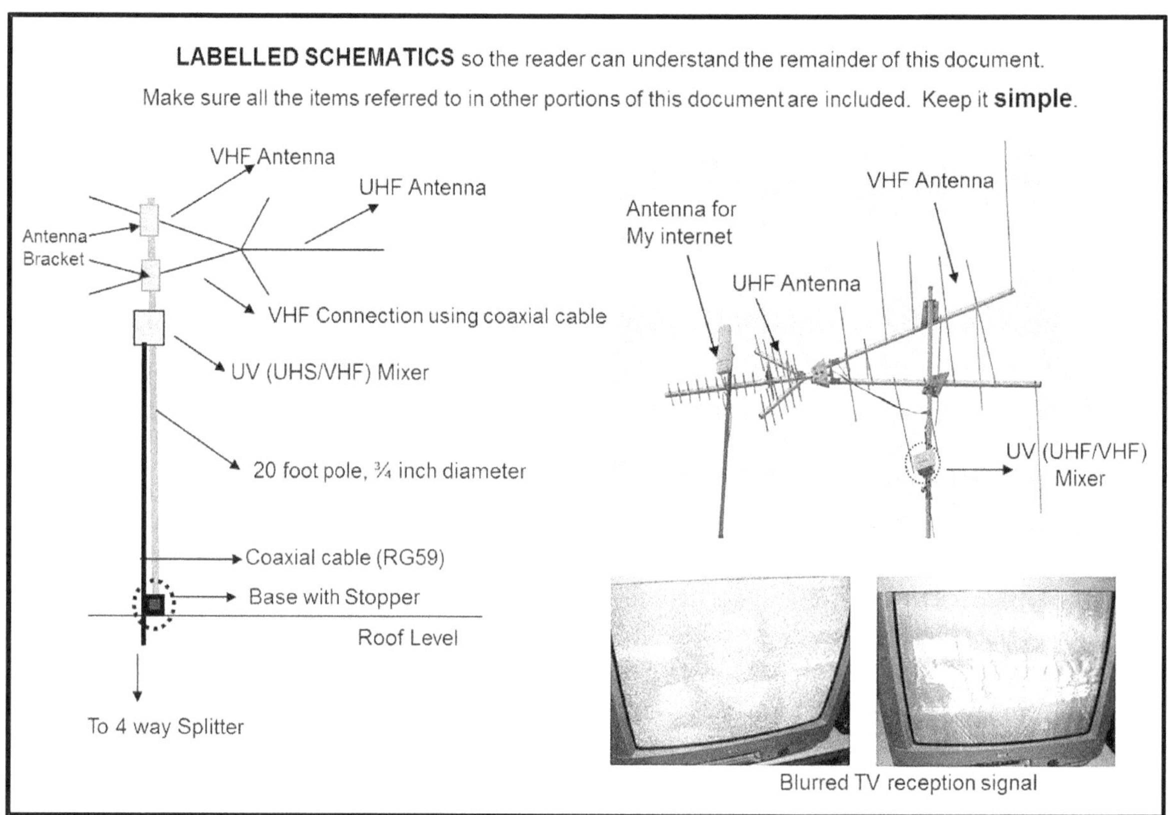

Figure 10.1: Aerial Antenna Set-up in my Residence

This aerial antenna had been with me and specifically designed by me and my friend since the 1990s as far as my poor memory recalls. We call it the Pyramidal Logarithmic or Pyralog Antenna in short because of its V-Shape. It is composed of a VHF antenna, for lower channels 2, 4, 5, 7, 9, 11, and 13 and UHF channels, 21, 23, 27, 31, 39, 45, 49. U stands for UHF or Ultra-high frequency and V for VHF or very high-frequency channels. Our country has an NTSC system for broadcasting and we don't use PAL or SECAM. Me and a friend of mine designed this aerial antenna which can accommodate 4 TV sets with good quality reception. Note that all television sets during this time were made of CRT and not LCD. I use both a coaxial cable (RG59) for the VHF and 300 ohms Feeder wire for the UHF antenna. Both wires will pass through the UV mixer, to separate the signal for both UHF and VHF broadcast on the TV set. A coaxial cable (RG59) will be the output of this UV mixer and will go through a 4-way splitter since I have 2 television sets at home at that time. This antenna is supported by a 20-foot ¾ inch diameter pole and is located on our roof. It will take around two people to install this antenna. As far as I can recall this antenna had been in my home from 1992 onwards. The summary of the events includes;

- The antenna had been in place since 1992, as far as I can recall.
- It had served our TV reception well as the signal is strong.
- During stormy and windy seasons, the antenna becomes disoriented.
- TV reception becomes blurred. My wife and kids always complain they can't watch TV.
- I went to the roof to fix the problem by rotating the pole, after a while, the antenna becomes disoriented once more and I need to climb the ladder and rotate the pole once again.

Problem: The aerial antenna becomes disoriented creating a blurred screen on my television set which can hardly be watched. This incident happens when it is windy or when a storm hits the Philippines.

Physical Evidence

- Wooden bracket. The bracket is made of ordinary plywood which I retrofit.
- Original design of the bracket was made from Acrylic Plastic (0.75 inches thick).
- There is only one bolt instead of 2 which was being attached to the antenna
- Metal bracket is loose and you can rotate the antenna easily by hand
- Feeder wire (300-ohm brown wire) for UHF antenna had become brittle over time
- Original brackets design was made of ¾ inch plastic which became brittle and cracked
- G.I wires were loosened

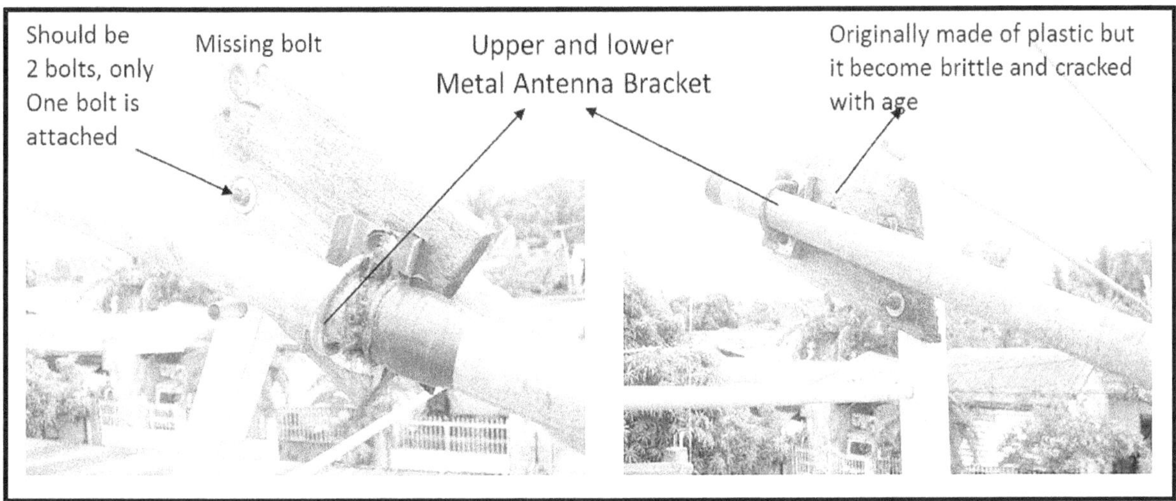

Figure 10.2: Understanding the Physical Problem on My Aerial Antenna

People Evidence: My kids and wife called me from time to time to rotate the antenna or find someone since they would want to watch their favorite TV show when they arrive home from school. Most of the time, I ignored them. Every time there is a live feed boxing in our country, (Manny Pacman Pacquiao), this was the time I go to the roof to fix the problem by rotating the pole manually. My wife inquired from a cable company and advised me to use a cable instead of an aerial-type antenna which I have to pay monthly which is way out of my budget as cable during those times was way too expensive. The cable man provided me a list of channels I can watch with their services. My thoughts on this were that I got two television sets and one is an old model and was not cable ready. This means that besides the costs if I allowed the cable to be installed, this will only be good for one television in which I still need to fix the antenna for the other TV set. Hence I decline the cable man's offer, and, I refused to install a cable TV.

Investigating Equipment Failures Through Root Cause Failure Analysis

My wife and kids were disappointed in me.

Paper Evidence: Revisited Pyralog Antenna instruction manual which I made a long time ago. Figure 10.3 serves as the paper evidence on how the antenna should be installed properly.

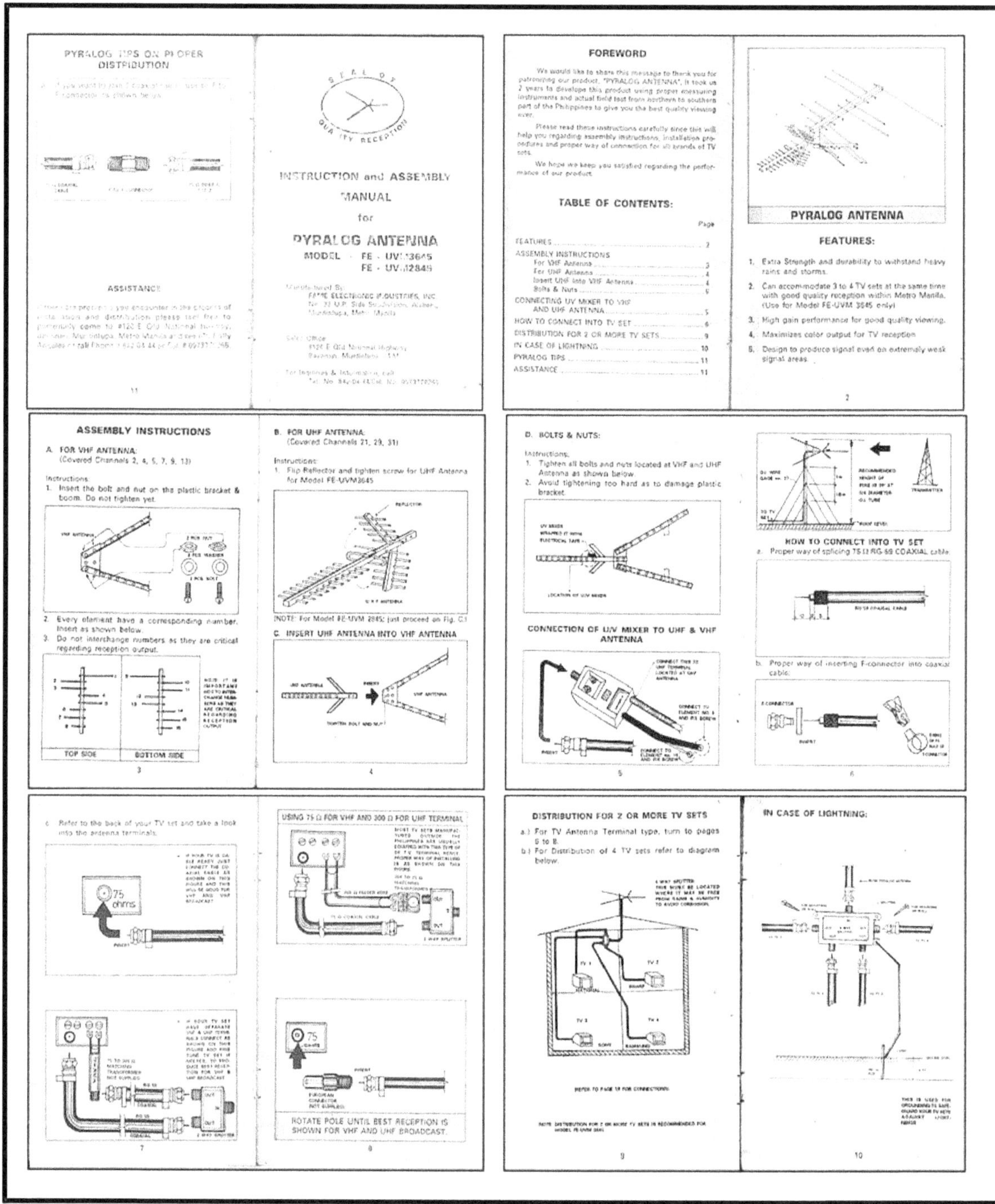

Figure 10.3: Aerial Antenna Instruction Manual

Physical Cause of Failure: The strength of the material of the plywood tends to deteriorate over time as this was exposed to sun and rain. As the wooden bracket material softens, metal brackets tend to loosen their grip on the pole which caused the antenna to rotate on its pole during a storm or when it's windy.

Human Cause of Failure: The owner (myself) did not have the right tools when he self-fabricated the wooden brackets such as the hand drill, hence, the hole on the bolt was not aligned on the antenna. That is why only one bolt instead of 2 is attached to the antenna, and the pole.

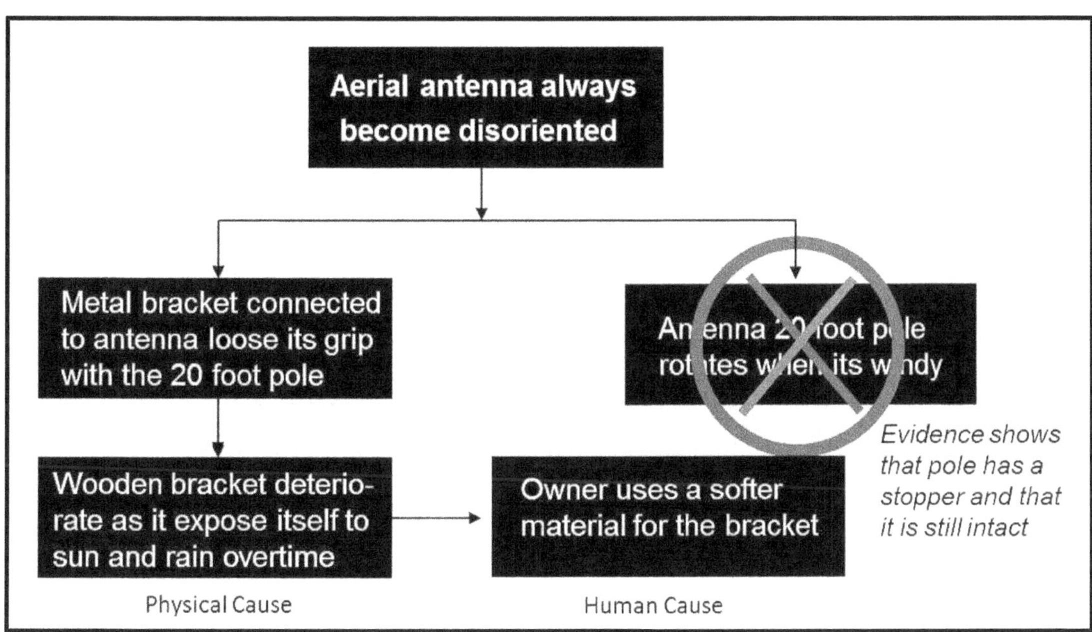

Figure 10.4: RCFA Logic Tree Diagram on Aerial Antenna

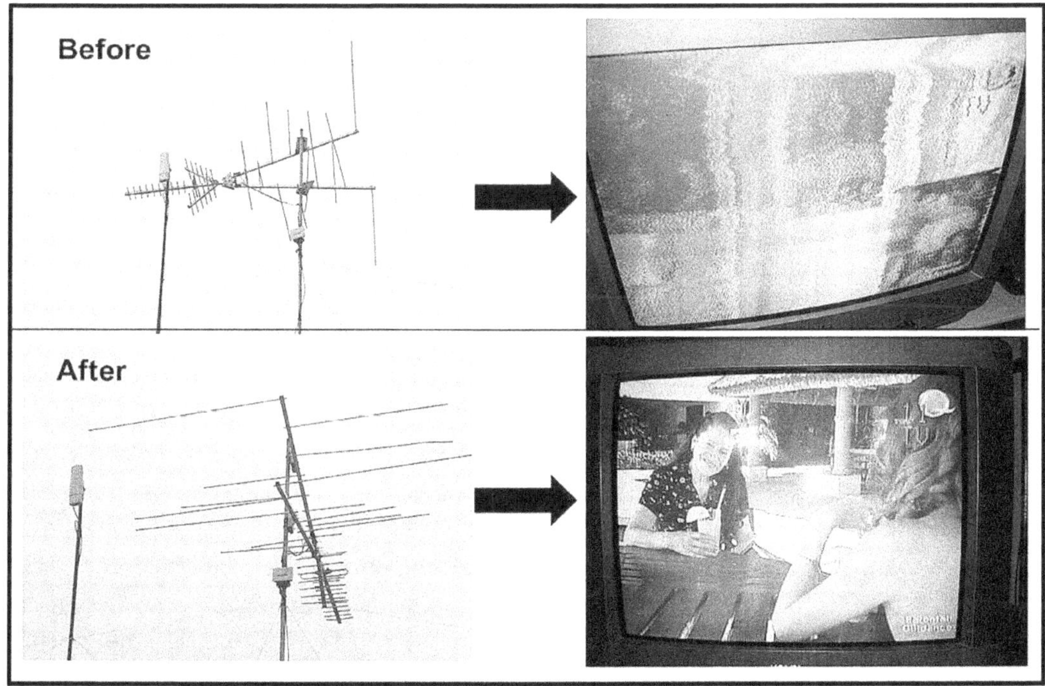

Figure 10.5: Before and After Corrective Action

Personal Latent Causes

• I tend to procrastinate a lot when the job should be done a long time ago. I always say tomorrow to my family when we need to keep things done today.
• I always have a way of fixing problems temporarily and not permanently.
• I tend to talk too much about what I cannot deliver.
• I tend to teach about the need and value to perform and analyze the problem through Root Cause yet, I cannot even solve and find a solution to this simple problem.
• I don't give time to analyze simple home problems such as this one.

Corrective Actions Generated

• I finally decided to bring the whole antenna down and replace the bracket with a kiln dry solid wood.
• The clamp brackets were designed according to the original specs.
• The antenna metal brackets had also been replaced with new ones.

Life Lessons Learned from this Small-Scale Personal RCFA: Although I've retired my aerial antenna a long time ago and using cable today, this simple RCFA taught me a very valuable lesson in life about procrastination. I procrastinate a lot. I always told myself that I will fix this tomorrow. What can be done today should be done today, hence, as the weeks gone by, I have a list of things to do and accomplish. If I have a deadline to make like writing this book, then I have to commit to that deadline with no excuses. Although these aerial antennas are a thing of the past, all I can say is that this is just an example of a small personal problem that meant a large impact on my life which is never procrastinate. This is also the time when I started to make a list of things to do and crossed out all the things I have completed monthly until this point of writing.

• I need to refrain from procrastination and do things on schedule.
• It took me more than 5 years to finally resolve this simple problem.

10.3: Actual Case Study on Root Cause Failure Analysis

Here is an actual RCFA case. A centrifugal pump is used to pump cooling water to several equipment in the Operation process for ABC company. The pump is a stand-alone and currently has no backup. On the night of August 10, 2009, at around 1900 hours, the production of ABC company stopped since the pump was declared failed since it was not discharging water for their cooling system. A work order request was handed to the maintenance. Charlie who was the maintenance manager decided to perform a Root Cause Failure Analysis on why the discharge of the pump stopped to determine the real cause of the problem as this is the third time it has happened this year. The maintenance manager submitted a request for an RCFA to investigate the incident and called the plant's RCFA Core Team. The RCFA Core Team immediately contacted several principal investigators they have in the plant on who was available to take the case. Finally, Mr. Stones (not his real name) from the process department agreed to take the case, the plant's RCFA core team emailed Mr. Stones regarding the RCFA request. Mr. Stones immediately called Charlie the maintenance manager to ask the maintenance people in charge of repair to preserve the parts affected which Charlie agreed and complied.

Request for RCA/RCFA Investigation

Control Number: *To be updated by RCFA Core Team*   Date: August 21, 2007
Time: 1915 hours

Who: (This will be the department or function requesting an RCA/RCFA Investigation)

Mr. Charlie Maintenance Department Manager

What: (State the Problem in its precise terms. Specify the asset to be investigated)

Pump failure - No discharge of cooling water from the pump

When: (State when did the failure occur? Specify the date and time)

The failure occurred at 1915 on August 21, 2007 during Night Shift·

Who: (Specify the people involved in the problem, no names just indicate the position)

Operator, and maintenance on duty during the shift

Oddities: (Specify if there are any irregularities, deviations before the failure)

Strange noise heard several days before the failure· Reduction in flow rate was experienced in the past

Person Requesting for the Investiation   Received by RCFA Core Team

Signed by Maintenance Manager   *Signed by RCFA Core Team*
(Indicate Position)   (Indicate Position)

Figure 10.6: Request for RCFA Investigation Form

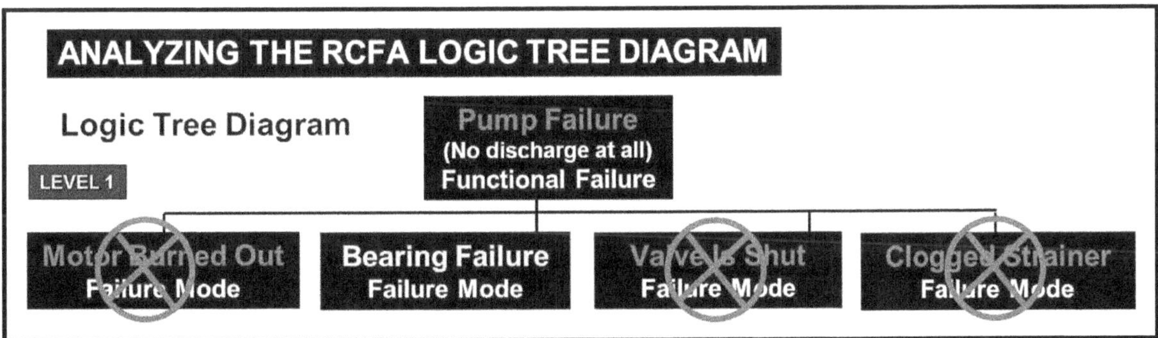

Figure 10.7: Level 1 of the RCFA Logic Tree Diagram

Mr. Stones the Principal Investigator was assigned to handle the RCFA investigation. He organized a two-man independent evidence-gathering team to probe the failure of the pump. Mr. Stones himself handled the physical evidence. He assigned Liza to conduct the people's evidence and Mr. Jones was assigned to collect the paper evidence. Their mission is to determine the root cause of the problem. Usually, a typical maintenance job will be just to replace the bearing with a new one since the part had evidently failed, and production must be up and running, but the question we ask is if we just repair and replace the bearing, will the problem be gone away for good? No, it will recur again at a given time. When our engineers look at the failed bearing, they then look at the failure history and data of the pump's previous failures and conclude that a different type of bearing or a more heavy-duty should be installed. In short, engineers love to modify without performing a root cause investigation.

Level 1 on RCFA Analysis: Mr. Stones (the Principal Investigator) got his notes and lists all the possible causes why there was no discharge of water from the pump for him to check. At this point all Mr. Stones know was there was no discharge on the pump. After a few minutes, Mr. Stones went to the place where the incident occurred and Charlie, the maintenance manager provided Mr. Stones with the resources he needed. The equipment was not yet repaired at this moment and so Mr. Stones gathered his possible lists of causes and check them one by one on the equipment. All that Mr. Stones listed are still considered as a hypothesis and must be proven if they exist or not. Hence, the problem stated in the RCFA request was that there was no discharge of water from the pump. Mr. Stones started to write some possible causes such as burnt motor, the valve is closed, strainer was clogged, and a bearing seizure.

- Mr. Stones went to the motor to check it, and it was working; hence, the motor burned out has been disregarded.
- The valve was open; therefore, the valve shut had also been disregarded.
- Mr. Stones asked the technician to open the strainer and it was clean. Hence, the strainer clogged was disregarded.
- Mr. Stones tried to rotate the shaft but he was exerting a tremendous force in rotating it. He asked the technician to take out the bearing and concluded that the bearing was damaged.
- Other possible causes were also checked. Lubrication in the bearing was checked and found out it was sufficient.
- Finally, Mr. Stones checked and noted all the readings, gauges and took several photos of the components and the bearing. He asked the technician to wrap the bearing and took it with him and finally, he called it a day and went home at around 1145 in the evening.

At this point, the bearing can fail for several reasons. All we know is that it was the bearing that causes the pump not to discharge. Again there were many possible causes on why the bearing seizes. Therefore to distinguish the facts from hearsay, the bearing was sent out to an independent third-party metallurgical lab for further analysis to determine how the bearing failed to fulfill its function. On the other part of the investigation, both Liza and Mr. Jones started gathering the people and paper evidence respectively. Lubrication in the bearing was checked, and it was found out that it was sufficient. The only possible hypothesis left was dirt, debris, and premature wear, and so the team decided to have the bearing tested in a metallurgical laboratory. The management approved the budget on sending the failed bearing to a metallurgical laboratory to determine its physical cause as the plant has insufficient tools and knowledge about mechanical failures. Hence, level 1 on the RCFA logic tree will represent the different possible causes of why there was no discharge of water from the pump.

Level 2 on RCFA Analysis: At this point, we speculate why and how the bearing prematurely seizes. It was learned that the bearing installed has a life of L_{10}, however, it was still less than a year since their last replacement of the bearing since January of 2007, in which the bearing just lasted for 8 months which was quite unusual. Although the metallurgical report was still ongoing and has not yet arrived. Mr. Stones listed some hypotheses that caused the bearing to fail. Although all that Mr. Stones listed are just hypotheses, they are still waiting for the results of the metallurgical lab. It was evident that the bearing did not achieve its designed life obviously.

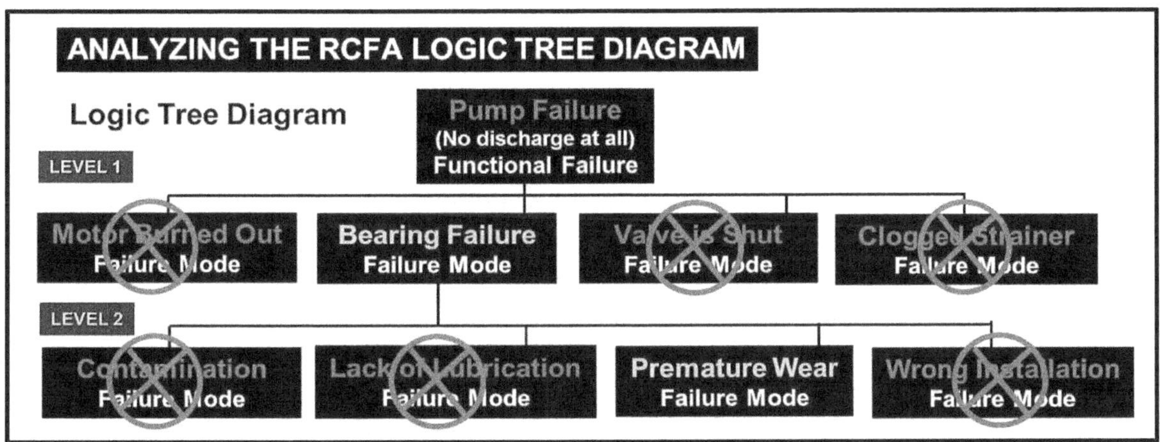

Figure 10.8: Level 2 of the RCFA Logic Tree Diagram

Level 3 on RCFA Analysis: Finally after a week the report from the metallurgical laboratory had been concluded that there was strong evidence of spalling on the raceway. This means that the bearing had fatigued prematurely. Several small fractures on the raceway, which creep to the surface, resulted in particle pitting. The other probable causes had been therefore eliminated, and we asked ourselves how fatigue can occur on the bearing. The physical cause of the bearing had finally been determined, that the bearing failed due to fatigue. Both Liza and Mr. Jones had already completed the interview and the collection of the paper evidence. They finally reported to Mr. Stones that the pieces of evidence for both paper and people evidence are now at hand and so Mr. Stones, who was the principal investigator called Liza and Mr. Jones for a preliminary meeting to complete the sequence of events on the RCFA logic tree diagram.

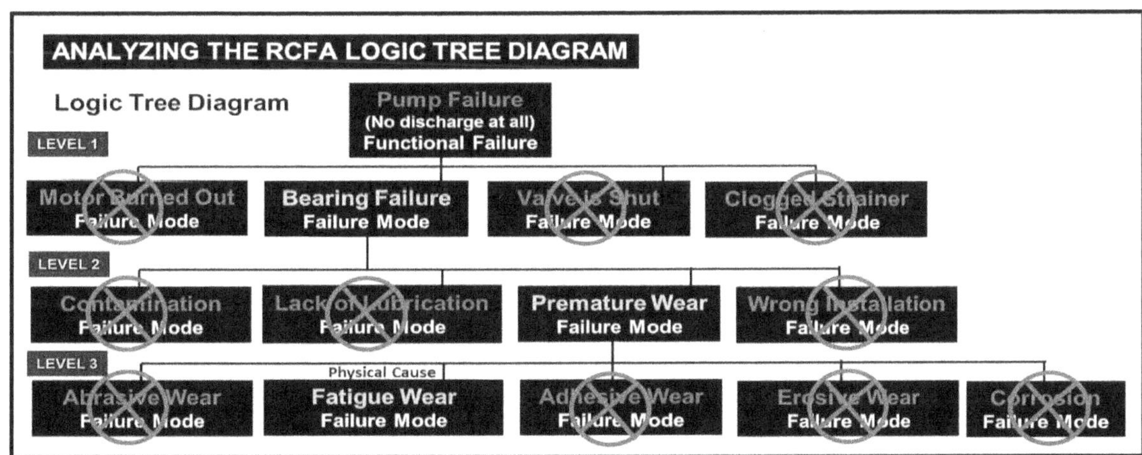

Figure 10.9: Level 3 of the RCFA Logic Tree Diagram

Level 4 in RCFA Analysis: At this point, both the Principal Investigator and evidence gathering team knew that the bearing's physical cause was fatigue. They now move on to determine both the human and system cause of the problem. And so the question was raised how could fatigue occur on the bearing? The team hypothesized that it could come from a high vibration. Many factors can contribute to fatigue, such as contamination, incorrect loading, excessive loads, inadequate lubrication, improper handling or fitting, faulty installation, and excessive vibration. Hence, each of the hypotheses was drawn and checked if any pieces of evidence supported any of them. However, several causes have already been ruled out such as lack of lubrication and contamination. Liza also confirmed during her interview with one of the technicians that they regreased the bearing just a few days before the failure when the operator reported the unusual sound. She also noted that the unusual sound again recurred just a few hours after regreasing. While on the other side, both Mr. Jones and Liza confirmed that there was evidence of high vibration based on the Vibration data. The other hypothesis on incorrect installation and incorrect loading as possible causes for high vibration was ruled out since there was no evidence to support it.

The vibration monitoring records confirmed strong evidence of excessive vibration. Excessive amplitude from our vibration data supports our hypothesis that fatigue that occurred on the bearing was due to high or excessive vibration; hence, contamination and incorrect loading had been disregarded. Evidence on the bearing raceway showed strong evidence of fatigue and spalling on the raceway of the bearing.

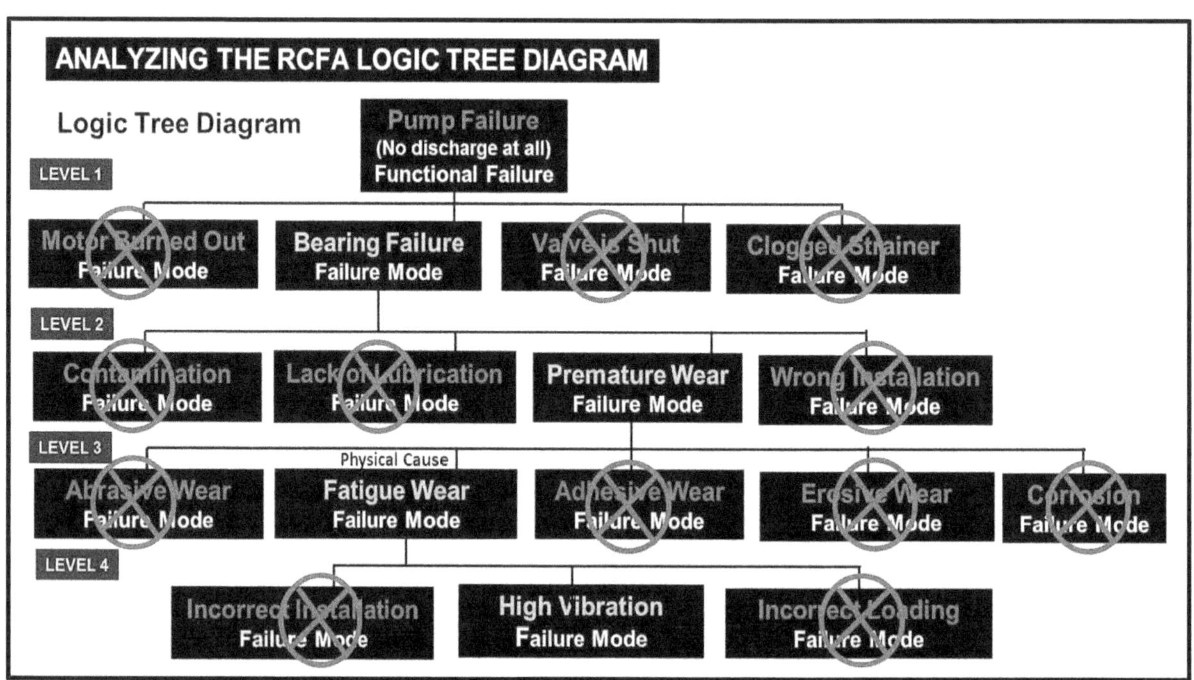

Figure 10.10: Level 4 of the RCFA Logic Tree Diagram

Level 5 Analysis: As both Mr. Stones, Liza and Mr. Jones dug deeper into the root cause, again, they hypothesized how could we have excessive vibration. The possibility is that it could come from imbalance, resonance, and misalignment.

According to both Liza and Mr. Jones, the vibration analyst records show that both resonance and imbalance were inconclusive but confirmed that and supports his data that misalignment was the cause of the high vibration. Vibration data indicates strong axial vibration at 1x RPM and is accompanied by harmonics of the shaft rotating speed with low amplitudes. Mr. Stones asked Liza to call the maintenance who conduct the alignment if he can provide them how the alignment was conducted. The evidence gathering team, together with the principal investigator and the person in charge of alignment all went to the equipment to reenact on how he performed his alignment. At this stage, evidence shows that fatigue was caused by excessive vibration.

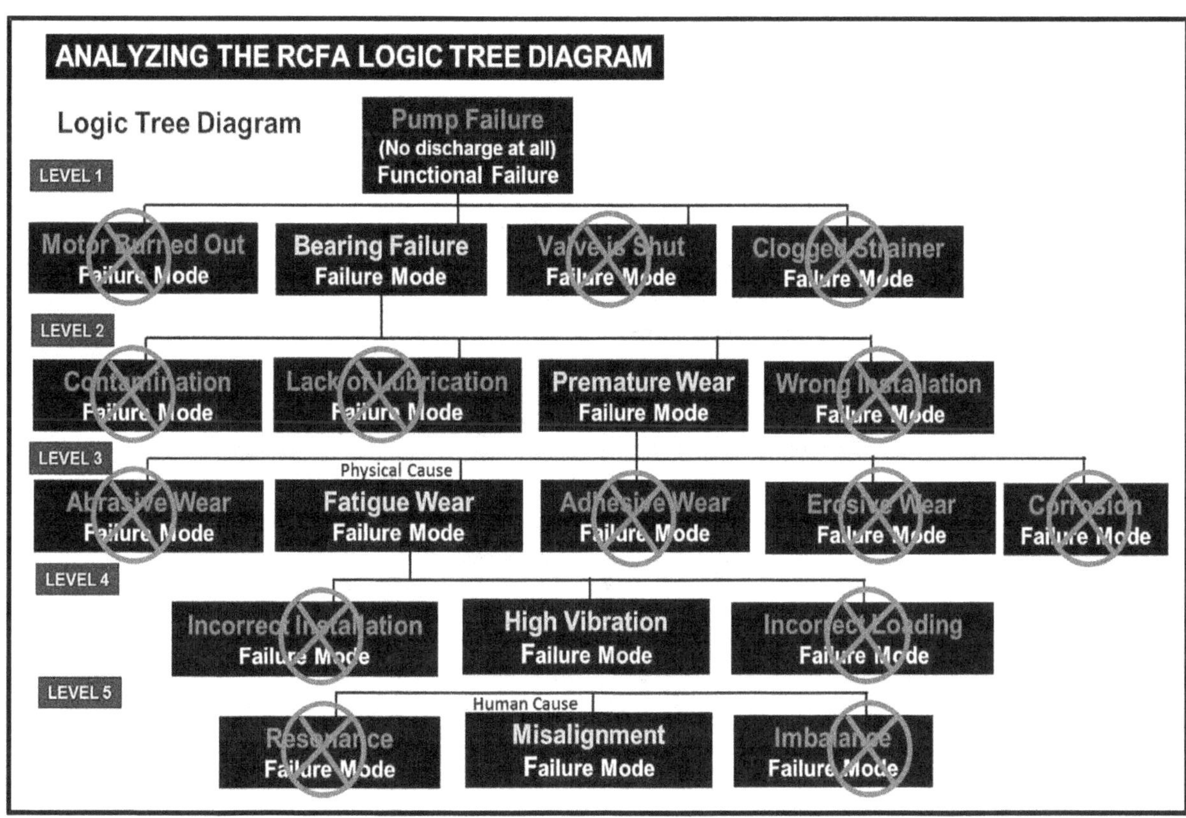

Figure 10.11: Level 5 of the RCFA Logic Tree Diagram

Level 6 on RCFA Analysis: Liza, Mr. Jones, and Mr. Stones, the principal investigator observed the maintenance person who was aligning the motor and pump. From their observation, they were certain that the maintenance did not know how to align the pump properly because he was just using his eyesight. The person was not using any instruments in conducting the alignment procedure on the motor and pump, and it was how he had done it since the beginning. This is the human cause of the problem. The maintenance does not really understand how to align the pump and motor itself. After studying the evidence that leads to this misalignment problem, the team found out that the person performing the alignment does not possess the skills and tools to perform such practices. He was only using his eyesight. Liza asked the maintenance if he was trained in alignment and the maintenance responded negatively. Liza's follow-up question was if any procedures or instruments were being used in alignment, and the maintenance replied that this is just what the previous maintenance did who had already retired. Hence, probing further, the evidence-gathering team and the principal investigator concluded that there was no training, procedures, and instruments to perform the correct alignment, not only on this particular motor and pump that failed but on every single rotating equipment they have in the plant.

People often misalign because they were never trained in proper alignment practices. No procedure exists outlining alignment as a required practice with the specification, or the current alignment equipment used was already worn out or inadequate for the application. This is the start of probing into the depths of the latencies of the problem. Investigation shows no instrument (Laser Alignment) and no training on why the equipment needs to be aligned. And when we probe on with depths of the Latent Causes, we understand that each of us is also part of the problem. The Principal Investigator, Mr. Stones, Liza, and Mr. Jones went back to the room to complete their RCFA Logic tree diagram. Mr. Stones, the principal investigator finalized the physical, human, and system cause of the problem as shown in figure 10.12.

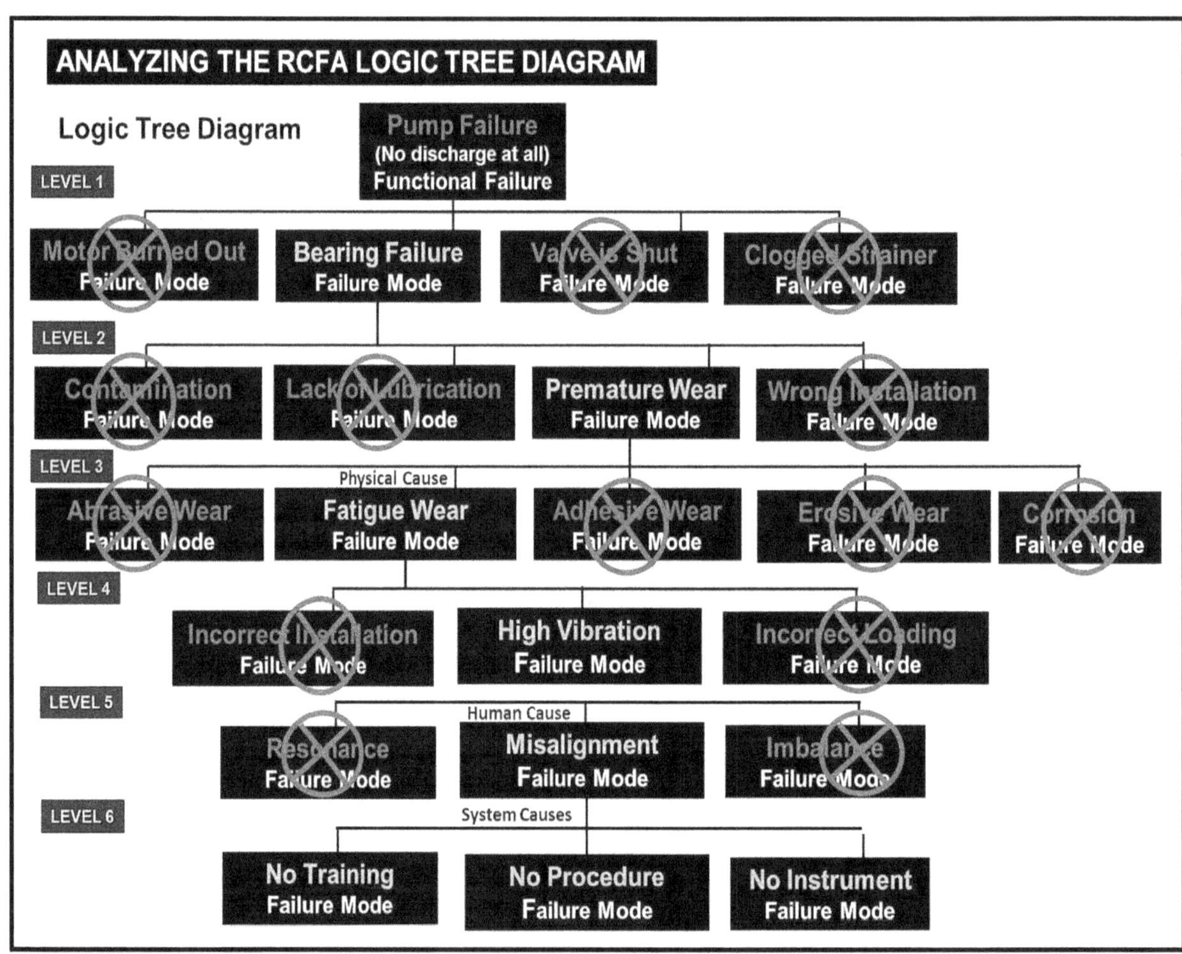

Figure 10.12: Level 6 of the RCFA Logic Tree Diagram

10.3.1: Identifying the Stakeholders, and Deriving the Latent Cause

Once the RCFA Logic Tree Diagram has been completed, the principal investigator together with the evidence gathering team should identify the people involved in the RCFA logic tree diagram. In this case, the following group of people has been identified. They will be called for a series of meetings which is called the stakeholder meeting:

• Maintenance person who performed the alignment
• Maintenance manager in charge of the department where the incident took place.
• Training Manager
• Head of Engineering
• Operations Manager
• Plant Manager

Figure 10.13: The RCFA Stakeholder Meeting

During the initial meeting the Principal Investigator, Mr. Stones, should give a brief introduction and explained to the Stakeholders why they were called in the meeting and would be seeking their support and cooperation in finding a resolution to the problem. After the brief introduction, the principal investigator will present the logic tree in detail which should be supported by the pieces of evidence gathered until they reached the physical cause of the failure. Once the physical cause has been presented, the principal investigator will ask the stakeholder if they agree with the physical evidence or if there are any inputs from the Stakeholder team. The initial meeting usually lasts for an hour or two at most and would be resumed at a date agreed upon by the stakeholders.

During the second meeting with the stakeholders, the principal investigator will continue with the RCFA Logic Tree diagram until the human cause of the problem had been determined based on the pieces of evidence presented. After the human cause of the failure has been presented, the principal investigator will ask the stakeholder their opinion if they agree or disagree or if they have additional thoughts that need to be considered. Also, take note that this is a sensitive meeting since the person who committed the error is also part of the Stakeholder team. The goal of this meeting is not to defend nor condemn the person who commits the mistake but to learn why the mistake or error was committed and what can be done about it in the future to avoid a similar occurrence. Once the human cause has been completed, the principal investigator will continue presenting the RCFA Logic Tree diagram until the system cause of the failure has been determined. After the presentation, the principal investigator will ask the stakeholders if they agree or disagree with the system cause and the meeting will be adjourned.

During the last meeting with the stakeholders, the principal investigator will discuss the concept of the latent cause. There are two questions about the latent cause that will be answered by the stakeholders. The first question in the latency is about asking the stakeholders about the way we are as an organization that has contributed to the problem. This will be brainstorm by the stakeholders as a team and determine the organizational latent cause. The last question is about what is it about the way I am as an individual that contributed to the problem. This will be answered independently by each stakeholder present during the meeting. These two questions should expose the latencies or hidden causes and we want the stakeholders to reflect on what changes do the organization needs as well as myself to address the latent cause of the problem. Latencies must be addressed by looking at the man in the mirror and admitting to the fact that we too as an organization and individual have contributed to the problem and we want to change that. Summarizing the causes;

Physical Cause: According to the metallurgical laboratory report, strong evidence of fracture on the outer raceway, balls, and cages caused the bearing to fail due to fatigue. Small traces of metal fracture exists on the raceway due to excessive vibration on the motor and pump.

Human Cause: Although many human errors from each function contributed to the problem, the human error committed was because the high vibration was caused by misalignment as the person conducting the misalignment was only basing it with his eyesight.

System Cause: The human error was committed since there was no training, no instruments, and no procedure on how to perform proper alignment on our rotating components.

Latent Cause: This will be answered by the stakeholders, and according to them these are the latencies they have learned from the failure.

- **Training Department Head:** After this incident occurred, we at the training department realized that a course on misalignment should be provided for all our maintenance people to understand the importance of misalignment. We have not provided this course because we do not understand the problem induced in our operations. Besides this, we will come up with a training needs analysis for both our operations and maintenance people.

- **Maintenance Department Head:** It is normal for management to budget and reduce operating costs, but this instrument is what we need to justify and improve our equipment uptime. We requisitioned this instrument a year ago, but it was denied by Top Management. We admit that we did not justify this instrument, and we believe that if we have prepared some justification and feasibility study for this instrument, management will provide us with this instrument. I believe we can justify this instrument if we really want to by performing a cost study in the first place and presenting an ROI to management. We simply forgot about it because we were too busy putting out fires daily. There are just too many failures in the line.

- **Engineering Head:** We at engineering always resort to modification. We based our decision on the data of the failed bearing, history records and recommend a solution by redesigning without really understanding the root cause of the problem. We have to admit that some of our previous redesign and modification resulted in new problems with the machine. We agree that before we redesign, we need to know the root cause of the failure.

- **Operations Department Head:** We focus on providing productivity and do not really understand the value of shutting down the equipment for performing an alignment or Preventive Maintenance. We often waive the equipment for production to commence since there is pressure to ship these products to the customer based on the schedule.

- **Plant Manager:** We should have provided the budget for training and the alignment instrument a long time ago. We at management finally learn to realize after this severe downtime that this instrument is needed by maintenance. We are cutting costs on the wrong grounds.

ROOT CAUSE FAILURE ANALYSIS SUMMARY

Determining The Physical, Human and System Cause of The Problem

(P) Summary of Physical Cause	Permanent Countermeasure	Responsible	Completion	Status
According the metallurgical laboratory report, strong evidence of fracture on the outer raceway, balls and cages. Small traces of metal fracture exists on the raceway due to excessive vibration on the motor and pump.	*a) A dial indicator will be used to correct the alignment not only for this equipment but also to all rotating assets.*	*Engineering*	*Start: 10/01/07*	*Completed*

(H) Summary of Human Cause	Permanent Countermeasure	Responsible	Completion	Status
The high vibration was caused by the misalignment as the person conducting the misalignment was only basing it with his eyesight.	*a) The training department will source a 3rd party consultant to conduct a training on the correct alignment practices for those responsible for operating and maintenance rotating equipment in the plant.*	*Training*	*To be scheduled for the month of November 2007*	*Ongoing*

(S) Summary of System Cause	Permanent Countermeasure	Responsible	Completion	Status
- No training on alignment *- No procedures on alignment* *- No instrument on alignment*	*a) After the training on alignment is completed, both the maintenance and engineering will provide a procedure for alignment.* *b) Budget will be provided for both training and purchase of Lazer Alignment.*	*Training Engineering Maintenance* *Plant Manager*	*To be scheduled for the month of November 2007*	*Ongoing*

Bottom Line Learning : What do you think is the lesson we can learn from this failure ?
- Training will provide a training needs analysis not only for maintenance but to all employees
- Engineering and Maintenance will allocate budget for instruments needed
- The stakeholder meeting provides us an opportunity for the different people involved to work for a solution on the failure.
- We will build an RCFA culture on our plant just like this for us to learn from the things that go wrong in our plant.

Signed by the Principal Investigator	*Signed by the Stakeholders*	*Signed by the Plant Manager*
Root Cause Failure Analysis Performed by :	RCFA Reviewed by :	RCFA Approved by :

Figure 10.14: Summary of the Physical, Human, and System Cause

And again, tell me if we are all part of the problem, or is the person who performed the misalignment be condemned alone? He that is without sin among you let him first cast a stone at maintenance. Let us all be in the shoes of this person who made a mistake and ask ourselves if we are certain that we would have done it differently. Latency is not about system causes but rather understanding how we people, in a way, contributed to the problem since we are taught that we are serving in the best interest of the company. I consider latency to be a higher form of human cause. Again, exposing these Latent Causes is the only way to understand what a true and meaningful Root Cause Analysis is all about. Understanding the latencies is the way to learn from the things that go wrong with our industries.

Once the latencies have been uncovered from the stakeholders, they will finalize them and submit them to the principal investigator to be included in the RCFA report. The stakeholder will generate a SMART Corrective Action for the physical, human, and system cause of the problem. The designated corrective action will have a deadline for completion as well as the person responsible for implementing it. It is also possible that neither the stakeholders, the principal investigator, or the evidence-gathering team may not be in a position to recommend the corrective action, in this case, they will recommend who will perform the corrective action and they can be invited to a separate meeting. The stakeholders may or may not be present during this meeting.

After the stakeholder meeting had been completed, the principal investigator together with the evidence gathering team will generate a report which will include the RCFA request, the summary of the evidence gathered which includes the physical, people, and paper evidence, the RCFA logic tree diagram, Stakeholder meeting minutes, summary of the Stakeholders response on the two latent cause questions and their respective SMART corrective action. The report will be submitted to the RCFA Core Team and will indicate an RCFA Control number and will be provided to the plant manager and others involved. The lessons learned from this event will be translated into an easy-to-understand format and will be provided not only to the area or function where the failure occurred but to other areas, functions, and other business units of the organization in which this same failure is possible to occur. The RCFA report is now concluded. Mr. Stones, Liza, and Mr. Jones can now go back to their original responsibilities. After the corrective actions have been implemented, Charlie, the maintenance manager will be responsible for verifying if the same cause of the problem recurred or not on their component.

Chapter 11

FAQs, Tips, and Don'ts on RCFA

> *Big problems start with little ones. It is just an accumulation of small problems that have been left unattended and neglected. Taking care of these big problems will just assure us of a continuance of bigger problems. Start addressing these small problems, and it will take care of the bigger problems in our industries.*

11.1: Frequently Asked Questions on Root Cause Failure Analysis

Here are lists of questions and answers on Root Cause Failure Analysis. Some of these questions have been raised in my previous training classes on RCFA.

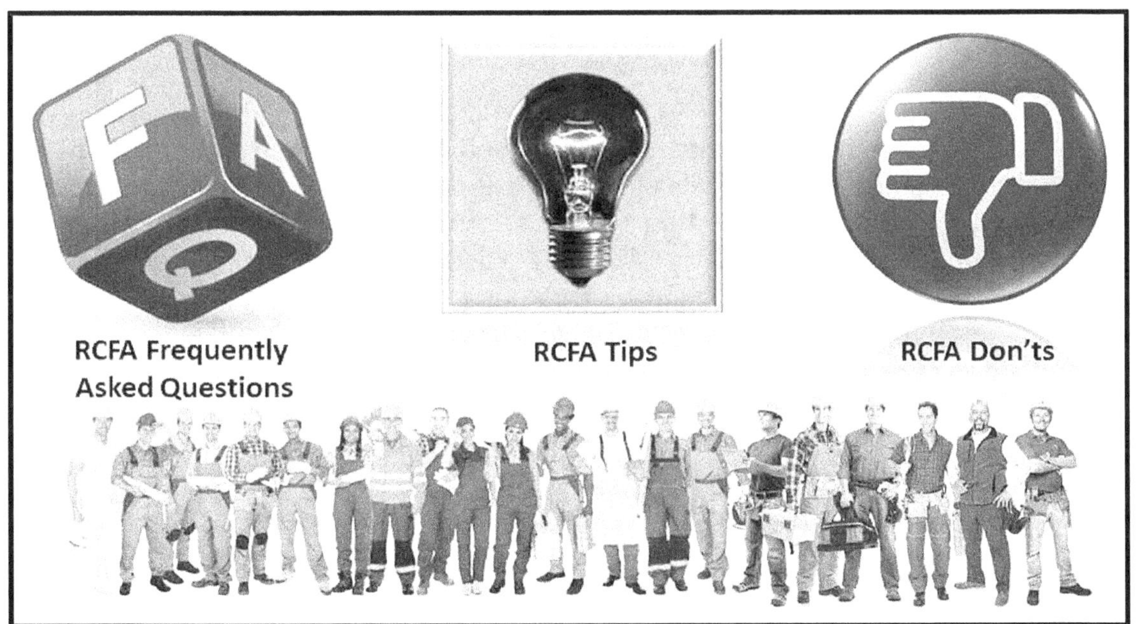

Figure11.1: RCFA FAQs, Tips and Don'ts

1. **Who will be eligible to conduct the Root Cause Investigation?** A Root Cause Analysis or Root Cause Failure Analysis investigation should be conducted by a Principal Investigator. The Principal Investigator has the freedom to choose their evidence-gathering team. All

principal investigators and evidence gathering teams should be fully trained in root cause by a qualified third-party firm or consultant as a preliminary requirement.

2. **How deep should the probe or investigation be, or when do we end our probe in root cause?** Both Root Cause Analysis and Root Cause Failure Analysis should end on the latent cause of the failure. The latent cause will answer two questions. First, what is it about the way we are as an organization that contributed to the problem? This is the organizational latent cause. Second, what is it about the way I am that contributed to the problem? This will be the personal latent cause of the problem. Hence, to answer the question, where do we end our probe on Root Cause Failure Analysis? The root cause investigation process ends when we reach the Latent Cause of the problem.

3. **How should the team decide if they want to perform a root cause or an analytical problem-solving tool?** Root Cause can only be done when the failure is fresh or the incident just recently happened and the physical pieces of evidence have been well preserved. An analytical problem-solving tool can be done on existing failures that happened in the past and will rely mostly on data of the failure or this can be a product of brainstorming.

4. **Why-why or 5-why states that if the team finds it difficult to answer the why, does this mean that the root cause has been known?** The answer is no. As the word implies a 5-why analysis will only answer the question on why, while conducting an RCFA or RCA investigation will answer the question on what, why, and how the incident occurred. Other books will use why-why as a tool to determine the root cause of failures, but I tend to disagree with this. Even if the 5-why was based on evidence, it will not explain what and how the failure occurred. To understand the root cause is to answer the what, how, and why the failure occurred in the asset.

5. **Can Root Cause Failure Analysis totally eliminate equipment failures 100%?** Root Cause Failure Analysis will not completely eliminate the failure but it will just eliminate the particular cause of the failure based on the evidence uncovered. For example, the physical cause of the mechanical seal derived from the evidence was a shaft misalignment. This means that after corrective action is implemented, the mechanical seal may again fail but not for the same reason. It may fail due to low or high pressure, internal line plugged, line pressure too high or the seal is not compatible. Note that conducting a Root Cause Failure Analysis is not designed to address all the possible failure modes that can occur on the equipment but only on that specific cause as based on the pieces of evidence gathered.

6. **Why do latent causes do not require corrective action?** Latent causes do not require a corrective action because no amount of corrective action can address a latent cause. What is needed to address a latent cause is for the people to change. What I believe is that performing a successful Root Cause Failure Analysis investigation not only requires corrective action but also requires people to change. Once people and the organization accept change then only can we conclude that the industry has finally learned from the thing that went wrong.

7. **Is system and latent cause the same?** No, these two are different. The system causes will come first, the last cause in any root cause investigation will be probing the depths of the latent causes. A system cause will mostly reflect on the system, procedures, and SOP. The latent cause is much deeper than this. A system cause may indicate that there was no procedure to follow or no training provided to the maintenance function. This means that it is unlikely that maintenance can write a procedure if there was no training. A latent cause will go deeper and probe why there was no training. Perhaps a response to this is that management is implementing a cost-cutting scheme due to a decrease in revenue due to this **Covid 19** pandemic. The question is, is cost-cutting the only means to cope up with the company's losses? Do cost-cutting measures improve the reliability and the way we maintain our equipment? Why did management make that decision? Is cost-cutting the only alternative to reduce the plant's operating costs?

8. **Why do we need to perform an RCFA and why not just call our OEMs Consultant?** The problem is that if you provide the details to your OEM or Consultant, in most cases, they already got the answer even without going through and evaluating the evidence. This is what we are trying to avoid in any root cause investigation, which is jumping to conclusions. What we need is for the Principal Investigator and the Evidence Gathering Team to explore the reason behind the failure based on the pieces of evidence gathered.

9. **If you are assigned as the person to gather the people's evidence, how would we know if what we are conducting is an interview or interrogation?** Use the 80/20 rule. In conducting an interview 80% of the time, the interviewee should be the one talking, if this will be reversed and 80% of the time, the interviewer is the one talking, then it now becomes an interrogation. For the interviewer just ask one question, and let the interviewee talk. If the interviewee finished talking, a follow-up question will be, and then what happened next?

10. **Can we apply Root Cause Failure Analysis to all failures?** Technically yes, but I would not advise it unless you have the manpower to do so. I would recommend performing a Root Cause Failure Analysis on repeating failures, failures with high consequences and impact, as well as those with high cost and consequences or when the industry wants to learn from the things that go wrong.

11. **Can Root Cause Failure Analysis be done in less than 24 hours?** Perhaps this will be possible for a small-scale, or mini-event, but not on medium or large-scale events. Both the gathering of pieces of evidence, summarizing the evidence, performing the RCFA Logic Tree Diagram, and the stakeholder meeting will not be possible to be completed in 24 hours, not even 48 hours. Sending a part that failed to an independent metallurgical laboratory may already take several days or a week to determine the physical cause. Major catastrophic failures, such as explosions, fires, or fatalities, may take weeks, or even several months.

12. **What are the qualifications of a Principal Investigator on Root Cause Failure Analysis?** As discussed in Chapter 8.3 of this book, the Principal Investigator should be unbiased, objective, have good interview and communication skills, be capable of balancing his time despite other workloads, be capable of thinking outside the box, willing to seek outside assistance when needed, Neutral and taking no sides during the investigation process, should

not jump easily to conclusions, capable of handling the stakeholder meeting, and dedicated to the completion of the Root Cause investigation.

13. What are the preliminary requirements in conducting a Root Cause Failure Analysis Investigation? If a failure warrants a root cause investigation, the first thing to do is to preserve the evidence. This means that whatever failed on the equipment must be extracted with care, the positions, photographed and all physical evidence should be well preserved. Maintenance people responsible for repair should surrender all physical evidence to the person responsible for handling the physical evidence.

14. Is conducting an analytical Problem-Solving Tools and Root Cause Failure Analysis the same or different? In some books on Root Cause Analysis, these problem-solving tools such as 5-why, fishbone analysis, Pareto, 8-Disciplines, quality tools will address the root cause of the problem. My personal view on this is that these tools will not address the root cause but only the most likely or probable cause of the problem. Any problem-solving tool that is not based on evidence cannot be considered a tool to address the root cause. Data alone won't get to the root cause, these data should be conclusive and based on evidence since root causes are evidence-driven endeavors. Another distinction between these two is that you can use an Analytical Problem Solving tool for failures that happened in the past, while a Root Cause Failure Analysis can only be done for fresh failures that happened recently for as long as the evidence is preserved. However, if the evidence is not preserved, then a Focused Improvement team can be deployed using conventional problem-solving tools. The distinction between a root cause and an analytical problem-solving tool is that the root cause will always end up in the latent cause of the problem.

15. Can we perform RCFA if only 2 of the 3 types of evidence are known, for example, if we only have people and paper evidence but not the physical evidence? My response is no, you need the three types of evidence in conducting a root cause investigation otherwise we will just end up in the probable or most likely cause of the failure.

16. Is it really required to have a third party conduct the root cause investigation if this is a large-scale event where the lawyers and insurance people are involved? Although I have mentioned several times in this book that it is highly recommended that a qualified outsider or third party should perform the investigation to avoid any form of biased and cover-ups. However, the industry involved can perform their own internal investigation independent from the third party which can be consolidated with the third-party investigator at the end of the investigation.

17. Is it possible to conduct a Root Cause Failure Analysis when the equipment is already repaired? It is difficult or impossible to conduct a root cause investigation if the equipment has already been repaired since the physical evidence had already been destroyed. A preliminary requirement in any root cause investigation is to preserve the physical evidence, however, the team can perform a conventional analytical problem-solving tool which may only result in the probable cause of the failure.

18. What's the main difference between a Root Cause Analysis (RCA), and a Root Cause Failure Analysis (RCFA)? RCFA is the term used to address the root cause for equipment-related problems and incidents, while RCA will be used for non-equipment-related problems for industries such as safety incidents, accidents, near misses, quality defects, or even administrative problems. RCA is also the term used for non-related industry problems such as in hospitals, IT, psychology, performing autopsies, and others.

19. Is Root Cause Failure Analysis a Top-Down or Bottom-Up approach? My response to this is that it is both. Anyone can perform a Root Cause Failure Analysis investigation as long as they know the process and are supported by management. However, if the industry wants this process to be part of their structured system, then it should start from the top so that industries can have the right mindset and culture on dealing with problems and failures. If a Root Cause Core team will be established at the very beginning, then this will be a top-down approach.

20. What if the Principal Investigator and Evidence Gathering Team lack evidence in proceeding with the RCFA Logic Tree Diagram? In this case, both the principal investigator and the evidence gathering team need to source additional pieces of evidence to find the sequence in their RCFA logic tree diagram. They may also seek help from others who experienced failure before. In figure 8.7, RCFA Flow Chart on page 231, the three evidence-gathering teams will once again proceed to step 5 and gather additional pieces of evidence.

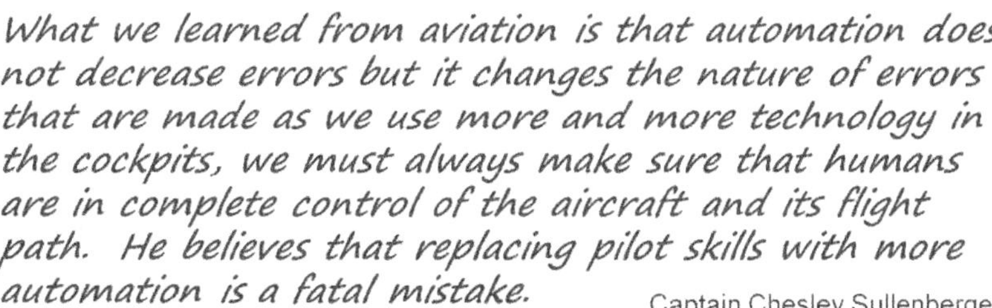

Figure 11.2: Auto Pilot Error and Not Pilot Error

21. **Can the use of automation such as robotics, artificial intelligence, automation, or similar means eliminate the use of root cause, as these robots definitely will not commit human errors?** Automation is not a foolproof solution for industries especially when used in operations but as Captain Chesley Sullenberger said, it only changes the nature of the kind of errors that might be produced in the future. What if your system is hacked despite all your security measures? As technology upgrades itself, so as these hackers. This means as we use these technologies today in our industries, we can expect a new breed of problems that had not yet been anticipated.

22. **How many hours should the principal investigator allot for the RCFA case when he has also other responsibilities in the plant?** Regarding how many hours will the Principal Investigator will work on the RCFA case and his existing job will depend on the policies developed by the RCFA Core Team. If it requires the Principal Investigator to work full time on the investigation, then he or she must inform their management.

23. **What is the difference between implementing RCM from performing Root Cause Failure Analysis, meaning, is it still necessary to perform RCFA if the industry is already implementing RCM?** The main difference between conducting an RCM and RCFA is that in RCM, we accept failures by assigning maintenance tasks to address a particular failure mode, while in RCFA, we challenge the failure by understanding its causes and doing some corrective action so that the same cause of the failure will not repeat itself. However, these two can be integrated respectively. In the RCM decision diagram, the last question on the algorithm states that if we can redesign or allow the failure to occur. We can just add one more question in case if this is a run to fail, will this failure warrant an RCFA investigation, and for redesign if it was not successful, will this warrant an RCFA investigation.

24. **Is performing a Root Cause Failure Analysis Reactive or Proactive?** Root Cause Failure Analysis is both reactive and proactive. This means that we need a failure to occur first before we can conduct a root cause investigation which makes it reactive. However, when we implement these SMART corrective actions in which the same cause of the failure will no longer repeat itself, then this is when RCFA becomes proactive.

25. **Can we apply root cause investigation for failures that happened in the past?** Root Cause Failure Analysis can only be done when the failure is fresh or just recently happened and the physical evidence on the equipment has been well preserved. However, we can use an analytical problem-solving tool for failures that happened in the past.

26. **What is the number one obstacle or hindrance in performing a root cause investigation?** Actually, there are two, the first is management commitment. I would prefer to use the term commitment than support because management can support but never commit to the initiative especially if cost will be involved in addressing the root cause. If these corrective actions will fall on management's deaf ears, then everything will be useless. The second is addressing the personal latent cause as we need to admit our fault on how we contributed to the problem, and as human nature states, no one will ever admit to their mistakes.

27. **Will the golden rule on amnesty apply to all root cause investigations?** Before answering this question, humans are humans. Human error is part of being human. Ask yourself, if given the following conditions and you were in the shoes of the person who committed the mistake, will you do the same thing or different? If the error or mistake is unintentional, then amnesty should be in place. If this is a clear case of sabotage to hurt someone or damage something, then this is a different story and amnesty may not be applied, however, it is still important for the industry to know why the person committed the acts of sabotage in the plant.

28. **What is the main difference between a Focused Improvement team and a Root Cause Failure Analysis Team?** The main difference will be the composition of the team since a root cause will include a fact-finding team composed of the principal investigator and evidence gathering team. Their analysis is based on facts supported by the pieces of evidence collected. A Focused improvement team will compose of a cross-selection of people with knowledge of the problem. Their analysis will be based on data and a product of brainstorming in the majority of cases.

29. **Who will be the responsible person who will validate the effectiveness of the RCFA corrective action?** The best person to validate the corrective actions will be the department or area that requested the RCFA investigation as they will be the first people who will detect if the problem resurfaced once again.

30. **What is the key to a successful Root Cause Failure Analysis investigation?** The key to a successful RCFA investigation is for both operations and maintenance to understand that the preservation of evidence is the key to Root Cause Failure Analysis which means that the equipment should not be repaired immediately if a failure will warrant a root cause investigation

31. **What if the person being interviewed refused to use a tape recorder to record the conversation?** As a general rule, before we can use a tape recorder to record the conversation during the interview, the interviewer should seek the permission of the person being interviewed. It will be unethical to hide and record the conversation without the consent of the person being interviewed. If the person being interviewed refused to record the conversation, then the person in charge of the people's evidence should manually write the key points of the conversation.

11.2: Tips on Conducting a Root Cause Failure Analysis

1) It is highly recommended that the Principal Investigator to be assigned to investigate the incident should not come from the area where the failure incident occurred but should come from other areas in the organization to avoid being one-sided and biased.

2) It is recommended that Root Cause Failure Analysis only be conducted on recurring failures or failures with devastating consequences and not on every failure or breakdown experienced in the plant.

3) It is highly recommended that each area in the plant should have a list of their own Principal investigators, which can be called if there is a failure incident in the plant. The minimum requirement for a Principal Investigator and the evidence gathering team is to be trained initially on Root Cause Failure Analysis. The principal investigator should possess the traits and criteria mentioned in Chapter 8.3 of this book.

4) If a breakdown or failure warrants an investigation, the repair team must preserve the part that failed, take pictures, collect metal debris, check the position, and the like before repairing the failure. What is important is to preserve the evidence. This might slightly prolong the repair time, but this will be for the industry's benefit so that they can finally have a chance to learn from the things that go wrong.

5) It is important for the Principal Investigator initially not to let the people involved in the evidence gathering know who is collecting the paper evidence, conducting the physical evidence, and people's evidence so that they will not compare their evidence with each other. This means that if Bob was assigned to collect the physical evidence, he did not know that Charlie is assigned to conduct the people's evidence, and Mike was assigned to collect the paper evidence. Likewise, Charlie does not know that Bob was assigned to collect the physical evidence so that they can be independent of each other as much as possible. At this stage, we are avoiding them from comparing their evidence, which can compromise the integrity of the evidence they are carrying.

6) Top Management should clarify to all people in the organization that no one should be punished due to the incident or failure if the human error committed was unintentional so that people will contribute what they know regarding the failure. If people know that someone will be blamed and take the fall, then the people involved will be defiant, remain silent, and will not share what they know regarding the incident. They will be hesitant to speak, and little information can be gained from the people's evidence.

7) For the area that the failure incident occurred, a request for root cause form should be provided to the Root Cause Core Team. They will be the ones who will be assigning the Principal Investigator to handle the case. The root cause request statement should include the following information.

• State the actual problem as detail as possible
• Where did this event actually occurred?
• Who did this happened to?
• When did this event actually happen?
• Were there any oddities or irregularities detected before the failure?
• Were there any signs or symptoms before the failure occurred?
• Did this failure already happened in the past?
• Were there any parts or spare that failed?
• Were the parts affected been preserved?

8) For the person in charge of conducting the people's evidence, have at least two sets of questionnaires that will be used during the interview process. One set will be used for those people directly involved in the problem, such as operators, maintenance, their supervisors, while the other set of questionnaires will be used for people indirectly involved in the problem. These are the people who have experienced the failure before. Ensure that the same questionnaire will be used for all directly, and another set of questionnaires will be used for indirect people involved in the incident. It is best to conduct an interview with one person at a time.

9) When the Root Cause Failure Analysis investigation is concluded, share the lessons learned on the affected area and on all areas in the plant and other business units. Translate the findings so that a 10-year-old boy will understand the lesson learned from this incident. Our goal is to learn from the failure and generate awareness for all people in the organization.

10) Each area in the plant should have its own Root Cause Failure Analysis kit, which will include the following: gloves, safety tapes, pen and notes, zip-lock plastic bag for collecting evidence and debris, flashlight, tweezers, camera, tape recorder, and magnet. This will serve as the Principal Investigator's initial kit in conducting the investigation. Another option is that the RCFA kit will be kept by the RCFA Core team and will be given to the principal investigator to handle the case. Once the case is concluded, it will be the responsibility of the principal investigator to return the RCFA kit back to the RCFA Core team.

11) It is important for the Principal Investigator not to discuss their initial findings with anyone in the organization for as long as the investigation is still ongoing and should remain confidential for the time being.

12) When the Principal Investigator and evidence gathering team is conducting the RCFA logic tree diagram, there is a possibility that there will be more than one cause that leads to the problem based on the evidence gathered. If this is the case, separate these causes by providing their own RCFA logic tree diagram. Do not combine all the causes in a single logic tree diagram since this will only complicate the investigation. Treat each of the causes separately.

13) Both operations and maintenance need to understand that the preservation of evidence is the key to conducting a Root Cause Failure Analysis which means that the equipment should not be repaired immediately. The evidence should be extracted before repairing the equipment. If the equipment will be repaired, then it is most unlikely to perform a root cause investigation.

14) Instruct maintenance, technicians, or the people involved in repairing the failure on how to preserve the evidence before repairing the equipment. This means extracting the part or item that failed, recording all positions, taking photos, collecting debris, recording all readings from the equipment during the time the equipment or asset failed. These items will be surrendered to the person in charge of gathering the physical evidence.

15) For the person handling the people's evidence, know the distinction between an interview and an interrogation. Use the 80/20 percent rule, which means that the interviewee should be talking most of the time, if this will be reversed, then it becomes an interrogation.

16) In Conducting the 5-Why, do not assume that the 5th Why is the Root Cause: As discussed previously, determining the root cause will not only answer the question on why, but it should also answer the question of what and how. The root cause analysis investigation should always end on the latent cause of the failure.

17) The stakeholder meeting is a very sensitive as this will be composed mostly of the person who committed the mistake and management people. Our main focused is not to put the blame on anyone but to let them know that they too are involved in the problem. It is best to set house rules during the beginning or at the very start of the stakeholder meeting. An example perhaps is that when you enter this room, kindly leave your positions and ego behind as our main focus is to learn from the failure and avoid the same incident in the future. Avoid derogatory or sarcastic remarks during the stakeholder meetings. The principal investigator will also act as the facilitator and mediator. If someone is being accused during the stakeholder meeting, the principal investigator should intervene and remind them of the rules of the meeting.

11.3: Don'ts on Root Cause Failure Analysis

1) **Do Not Rush the Investigation:** Depending on the magnitude and consequences of the failure, conducting a root cause investigation will take several weeks to months especially for a large-scale event. Remember that in most cases the principal investigator and evidence gathering teams are employees of the plant which means that they also have other tasks and responsibilities in the plant. This means that the time these people will spend on the investigation will depend on the time allotted to them by the RCFA Core team.

2) **Principal Investigator Should Not Jump at Once on the Conclusion:** Although this is still related to item 1, rushing the investigation may pressure the RCFA principal investigator and evidence gathering team to rush with the conclusion or skip the stakeholder meeting which may result in the investigation to become inconclusive, or being compromised.

3) **Do Not Perform RCFA without Management Commitment:** It will be a complete waste of time, effort, and resources if management will not support and will not be committed to the RCFA investigation. Remember there will be an investment to be made such as sending the failed item to a metallurgical lab and generating correcting actions especially modifications. I may be wrong on this but I have seen many RCFA efforts end up in waste since the RCFA recommendations will just fall on deaf ears of the management.

4) **Do Not Assign the Principal Investigator from the Area Where the Failure Occurred:** It is highly recommended that the principal investigator should come from other areas, departments, or functions and not from the area where the failure or incident happened. Hence, each industry implementing root cause investigation should have a list of qualified

principal investigators. The reason behind this is for the principal investigator to remain unbiased and to avoid any cover-up during the investigation.

5) **Never Punish People for any Unintentional Mistakes and Errors**: Human error is part of being human. Even the smartest and the most intelligent person commit errors and mistakes. Industries must adopt the golden law on root cause which is amnesty especially if the error committed is unintentional. The exception to this rule is if the failure or incident is a case of sabotage, corruption, or similar acts.

6) **Never Compare the Pieces of Evidence During the RCFA Investigation:** The people handing the physical, people, and paper evidence should avoid comparing their pieces of evidence from one another. We want them to be as independent as possible. What we are avoiding is comparing and consolidating each of their evidence which can compromise the integrity of the investigation.

Chapter **12**

The Conclusion

12.1: All Failures have a Reason for Failing

Majority of failures that industries experience can be classified as random and infant mortality. According to Stanley Nowlan and Howard Heap, their research on the six failure patterns shows that 11% of the failure will wear out while 89% will be caused by either infant or random failure. A bearing manufacturer states that their bearing has a life of L_{10}. If the life of the bearing is 10 years, we multiply this by 90%, which can be said as the reliability of the bearing. Hence, the probability that the bearing will last will be for 9 consecutive years. However, despite the reliability stated by the OEM, this bearing had been replaced 5 times in 5 years which average or an MTBF of around one-year lifespan of the bearing. So the question is the OEM is lying about the true life of the bearing. The answer here is inconclusive and will depend on several factors such as the environment, load, temperature, and the amount of lubricant the bearing received. Although we can conclude that the failure of the bearing is random.

If you have some books on TPM written by Japanese authors such as Suzuki, Nakajima, Gotoh, and the like, they seldom used the word random failures, but rather they would prefer to use the word accelerated deterioration. When we speak that the part had deteriorated, it means that the part had eventually worn out. Therefore, originally, the part should have a wear-out pattern, but the part's life had not been reached. In the Jost Report, friction, wear, corrosion, and lubrication cost the United Kingdom over 500 million pounds per year or roughly around 639,749,500.00 USD. He reasoned out that significant education and research are required to reduce these losses by 30% of its primary energy consumed by friction throughout the world, while 60% of the machine failures are due to premature wear. This means that the life had not been reached.

When we compare these two methodologies, RCM will declare the bearing is a case of random failure, while TPM will state otherwise that the bearing did not reach its life and experienced accelerated deterioration. Although both of them are correct, the point I would like to emphasize is that whether this is a case of random failure or accelerated deterioration, one thing for sure is that there is a reason why the bearing failed. The problem is that the industry failed to learn from the reason behind the failure. They accept failure as it comes, repair the equipment and repeat the same process over and over. Every failure whether random or not will have a reason as to why and how it happened. The problem with industries is that they accept failures and never challenge them so nothing was learned from the failure and since we do not know when the failure will happen again, then we called it a random failure and the rest is history. This means that when we just accept failures, then we just called them random, chance failures, or accelerated deterioration, but what we have failed to do is to understand the reason behind the failure. We accept failures by preventing, predicting, allowing them to happen, or simply creating a redundancy so that operations can still continue. These are the maintenance tasks we do to sustain and preserve our equipment and assets. The question I would like to ask the reader is have we actually learned from the failure by doing these maintenance tasks. I am not saying that doing Preventive, Predictive, Run to Fail, or having redundancies is wrong, all I am saying is that if the failure keeps repeating, why not understand its causes by doing a root cause investigation.

Predictive Maintenance will propose to use some high-tech instruments such as Vibration Analysis, Ultrasonic Monitoring, Oil Analysis, or other instruments to predict when a particular item or part will fail. As we said, if the life of the bearing is L_{10}, Predictive Maintenance will not guarantee that the bearing will fully reach its life of 9 years. The bearing can be predicted to fail in 6 months, 1 year, 2 years, and so on. These instruments will just tell us if the symptom of the bearing has started which is the potential failure. It can be monitored and provide a recommendation when to replace it. The question is why did the bearing not reached its lifespan of 9 years? This is where Root Cause Failure Analysis will come in.

A company implementing RCM will default and place a task on Predictive Maintenance to monitor the vibration pattern of the bearing and replaced it if the reading reached its peak or maximum limit. An ailing bearing will produce symptoms or potential failure that it is in the process of failing. Therefore, the maintenance department will use Condition-Based or Predictive Maintenance Techniques such as Ultrasonic Monitoring or Vibration Analysis. Imminent failures can be predicted, production can be notified, spares can be delivered as planned, the amount of downtime can be well estimated, operations can be advised regarding the downtime, but still, we need to replace the bearing. The thing is, Predictive Maintenance will not guarantee that the life of the bearing will be reached and maximized. It will only tell us if there is a fault on the bearing. Predictive Maintenance will not address the root cause but only the symptoms of the problem, but they can be used to determine the evidence if the root cause investigation will be done. However, we just stop when the symptoms are visible. Stretching this a little bit further, both Predictive and Condition Based Maintenance is just one more step on identifying the physical cause of the problem. Although it is good enough that both operations and maintenance will not be caught by surprise on the failure of the bearing. For industries implementing TPM, they will address the basics of the equipment, which will be done by both Autonomous and Planned Maintenance pillars. These basic equipment condition includes completing the bolts on the equipment with the correct amount of torque, making the equipment clean by addressing contamination sources, conducting the correct lubrication practices and addressing leaks of all kinds. The goal of TPM is to allow the part to reach its natural deterioration.

Although TPM and RCM are both good methodologies to adopt, one question lingers: Did we ever know why the bearing fails? I guess not. Root Cause Failure Analysis believes that whether the part wears out, failed randomly, or the failure is considered infant mortality, there is a cause that needs to be uncovered to understand the true reason behind the failure.

• Why did the bearing failed?
• How did the bearing failed
• What is the physical cause of the bearing?
• Did we apply the correct lubrication?
• Is it about our process?
• What human errors were committed that led to the bearing's failure?
• What was the latent cause of the failure?
• What corrective actions did we undertake?

What will the maintenance do? A typical job of a technician is to replace the bearing with a new one since the part had evidently failed and production must be up and running, but the question is, did the problem go away? No, if we just replace it, then this will just repeat again in the future. Although Engineering will have a different approach in the sense that if the bearing failed, engineers will often result in modification or redesign without understanding the root cause and may at times provide new problems on the equipment. In short, they will redesign or modify the bearing without understanding the physical, human and latent cause of the failure. Again, if we ask the question, did we solved the case of the bearing failure? My response is, I really doubt that. The problem with the management people at times is that their thinking is that we have always done this before and when a failure of unfathomable consequences happens, then that is the only time they realized that they should have listened to the recommendation of their engineers. This means that if nothing happens and they escape these horrible consequences, then expect nothing to change from the management people.

What I believe is that random and infant mortality failures exist because we accept them. And by accepting them, then we do not learn from the failure. What I am saying is that for every failure, there is a cause that needs to be uncovered. Industries hire technicians, and their job is to repair the failure, but have you ever heard of any industries that hire rooticians? By definition, **rooticians** are people who are knowledgeable in conducting a Root Cause Analysis or Root Cause Failure Analysis investigation. And even if we have qualified rooticians in our plant, we have another problem, and that is the process of how we do things in our plant. Meaning when a failure occurs in our equipment, machines, systems or asset, what do we do. The answer, we repair them. So what's the problem? When we repair the failure, then we destroy the physical evidence, and when asked, the maintenance will once again guess the cause of the failure and the rest is history and the failure just repeats itself. Industries must understand that something can be learned from failure if we just take the time and resources in addressing the cause of failures.

12.2: Integrating RCFA Into the RCM Process

The main objective of the RCM (Reliability-Centered Maintenance) process is to determine the maintenance tasks of the asset in its present operating context. This is done by identifying all the possible failure modes that can possibly occur on the asset. Once these failure modes have been listed, the RCM team will use an RCM Algorithm or Decision Diagram to determine the most feasible maintenance tasks to address every single possible failure mode. These maintenance tasks include Preventive Maintenance, Predictive Maintenance, Run to Fail, Failure-Finding Task, Modification, and having redundancy in place which I also include as tasks in my book on Decoding Reliability-Centered Maintenance Process for Manufacturing Industries. But thinking about the RCM process, I believe that RCFA can be integrated into the different maintenance tasks.

For Preventive Maintenance Tasks: As discussed in my previous books, these maintenance tasks will be feasible for failure modes that are considered age-related or have a wear-out pattern. This means that majority of the failure is expected to fail in that particular period. In this case, the part, item, or component will eventually wear out or have an age-related pattern. RCFA can be applied if we want to increase the lifespan or MTBF of that particular part or item. This means that if the lifespan of the part or item is consistent to reach 1 year, an RCFA investigation can be conducted on some of these parts before modifying redesigning these parts so that their lifespan can be lengthened.

For Predictive Maintenance Tasks: Predictive Maintenance is a maintenance task that is based on the actual condition of the equipment with the use of precision instruments. What is important for the user is to determine the P-F Curve. The P-F Curve is the emergence or start of the potential failure until its final descent into the functional failure. This is also known as the failure development period. Predictive Maintenance can be used to detect random failures for as long as the failure will provide signs or symptoms that it is on the verge of failing. Although Predictive Maintenance is a good strategy, there is still one problem with Predictive Maintenance, which is it will detect when the failure will occur, but it will not guarantee that the lifespan of the part or item will be reached. This means that imminent failures can be predicted, production can be notified but still, we need to replace the bearings and the life still had not been reached. Although it is good enough that both operations and maintenance are not caught by surprise by the failure. In this case, corrective maintenance is usually performed when the part or item is nearing its functional failure. RCFA can be performed on failures under the watch of Predictive Maintenance especially for those with a short inherent lifespan or those parts or items that do not reach their dictated lifespan for as long as the part or item that failed had been well preserved. The information and measurements gathered from Predictive Maintenance can be used as pieces of evidence to determine the cause of the problem for further analysis. An example will be a hot spot taken from an infrared thermography image that can indicate the precise part where the heat generated, which can provide concrete evidence in an RCFA investigation.

For Failure Finding Tasks: These maintenance tasks will only be applicable for protective devices and redundant functions. RCFA can be used to investigate the failure of protective devices and those duty components that failed for as long as these protective devices that failed are preserved.

For Other Default RCM Tasks: Other default maintenance tasks besides Failure Finding Tasks include, Run to Fail and Modification. These maintenance tasks are usually used when the failure mode cannot be Prevented, Predicted or there is no standby or redundancy in place. These tasks are usually located at the bottom of the RCM Decision Diagram or Algorithm as indicated in Figures 12.1 and 12.2. In these maintenance options, we can add additional an questions for both Run to Fail and Modification as indicated in figure 12.2.

Although the RCFA can only be done when the RCM process has already been implemented to the equipment or asset. The question to raise on this will be if the RCM team performing the analysis should be the one to perform the RCFA? My response to this is that the RCFA will be done by a separate team after the RCM had been completed and second, we need to let the failure happen first before we can conduct an actual RCFA investigation.

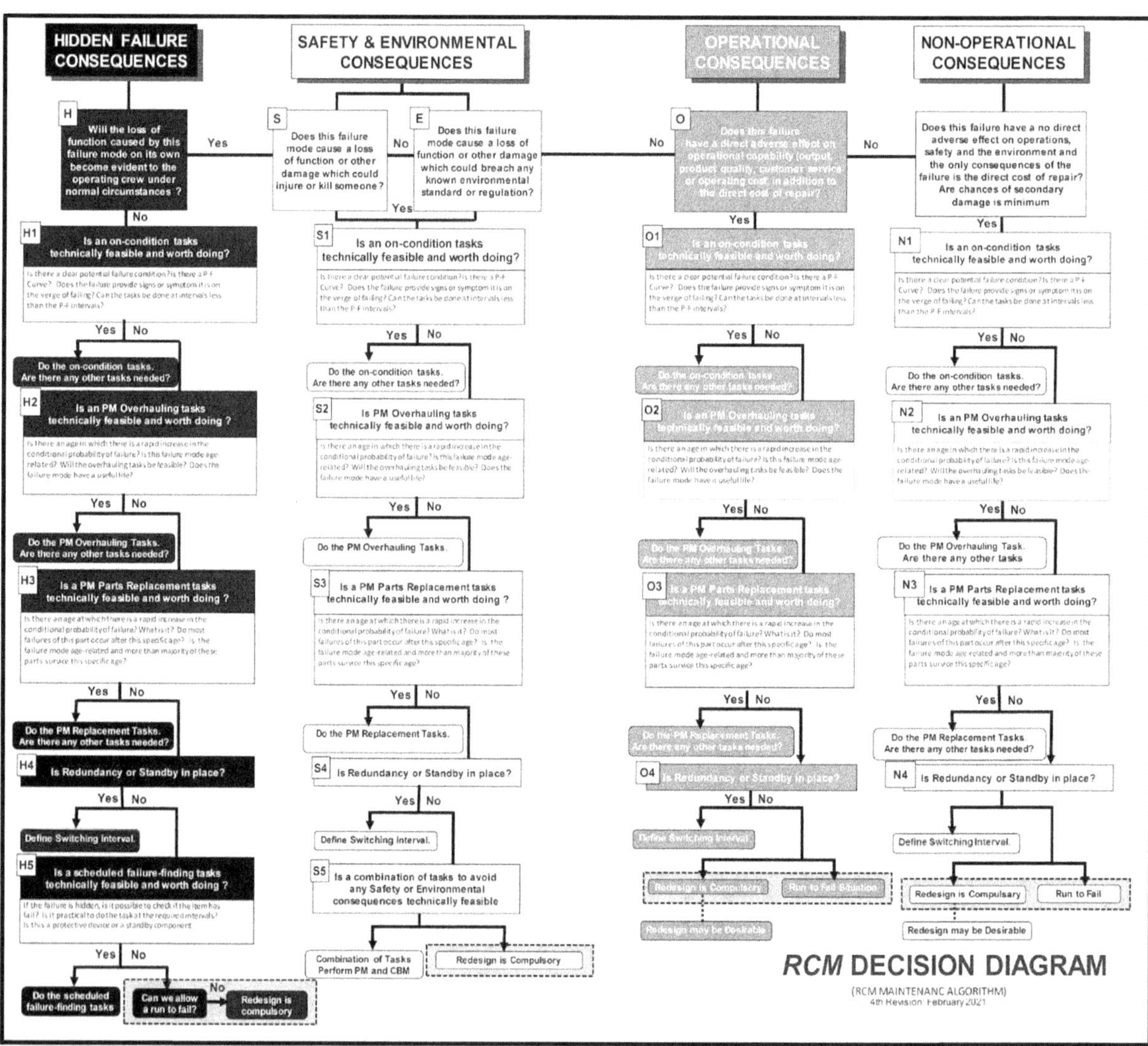

Figure 12.1: RCM Decision Diagram

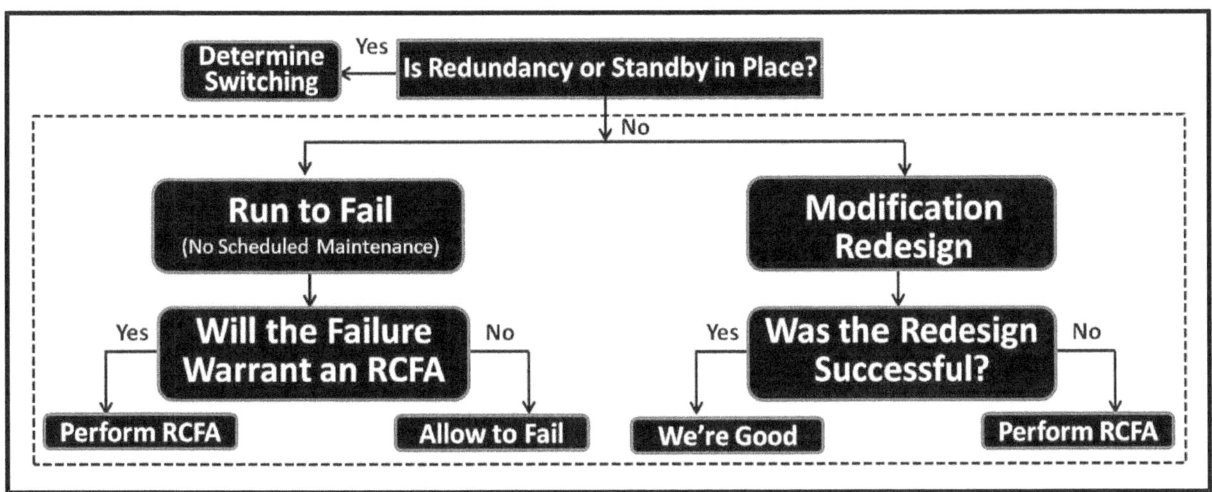

Figure 12.2: Integrating RCFA Into the Run to Fail and Redesign Tasks

12.3: <u>Root Cause Wil Not Eliminate the Failure 100%</u>

There are two parts to conducting a Root Cause Failure Analysis investigation. The first is to determine the physical, human, system, and the latent cause of the problem, and the second is to generate a corrective action to eliminate, reduce or mitigate the chances of recurrence of the failure with the same cause. A Root Cause Failure Analysis will not specifically eliminate the problem if the problem has several causes, but what it does is to avoid a repeat of that particular cause of the problem.

Failures cannot be totally eliminated by conducting a root cause investigation since we are only dealing with the cause of that particular failure based on the evidence that has been gathered. We must also consider various causes that are possible to occur on the equipment, and every single failure has its own unique causes. When we treat a single cause, then there is a likelihood of that same part failing again due to a different cause but most unlikely on the cause where a root cause investigation had been done. Remember, when we speak about Root Cause Failure Analysis, we are dealing with the pieces of evidence as to what really caused the part to fail and not all the probable causes that might have to cause the part to fail. In performing root cause, we are addressing the causes of the failure one at a time.

As my good and dear friend, Bob Nelms from Failsafe Network, once told me that you cannot move an ocean with a fork (not even a spoon), simply concurs that it is not possible to remove or eliminate all the possible causes of the failure. This means that if a root cause was done on the bearing failure and the cause was misalignment, the bearing can still fail due to other causes such as lubrication, brinelling, fatigue, electrical pitting, and so on. In this case, the root cause will only address that particular cause based on the evidence gathered. RCM was founded on the belief that its purpose was to reduce the accidents per million take-offs by studying every single part and its behavior from the airline industry. Today, the plane still crashes for other reasons, but it is unlikely from the same cause. The good thing about the airline industry is they have their own investigating team which is the NTSB or National Transportation Safety Board. The NTSB is an independent U.S. federal agency responsible for investigating and determining the probable cause of every U.S. civil aviation accident. The US

Congress established the NTSB in 1967 as an independent agency whose role is to investigate aviation accidents, identify the causes and remedies of accidents to prevent similar incidents in the future. Although NTSB is for US jurisdiction only, several countries, such as Australia, Canada, and the Netherlands, have established their own NTSBs.

Having a similar group for industries will be beneficial in the form of an RCFA Core Team. Performing root cause is a continuous process, the more we perform root cause, the more we learn from the causes of the failure. Once these causes have been addressed, then we can have lesser failures and breakdowns.

12.4: The Message of Root Cause Failure Analysis

The saying goes that if our maintenance people become so good at fixing failures, then something is definitely wrong with our organization. And what is wrong? It seems that nobody is performing any Root Cause Failure Analysis when a failure happens, that is why the failure just repeats itself. Try to assess the condition of your organization and answer the survey in figure 12.2 where 1 is the lowest and 5 points the highest. This assessment questionnaire is an extract from the overall assessment on World Class Maintenance Management, The 12 Disciplines, where RCFA is discipline number 9. You can use this RCFA assessment with your people to determine the need for performing a root cause in your plant.

• What is hindering your people from conducting a root cause failure investigation?
• Does your people know how to conduct a root cause investigation?
• Does your industry have a criterion on when to repair and when to conduct a root cause investigation?
• Are maintenance pressured by operations to bring the equipment up and running?
• Does both operations and maintenance understand the value of conducting a thorough Root Cause Failure Investigation?
• Are there qualified people in your organization that can handle the role of the principal investigator and evidence gathering team?
• Is top management committed to the completion of root cause especially if a budget is required?
• Is it about your culture?
• Are you waiting for the insurance people and lawyers to be involved in the root cause?

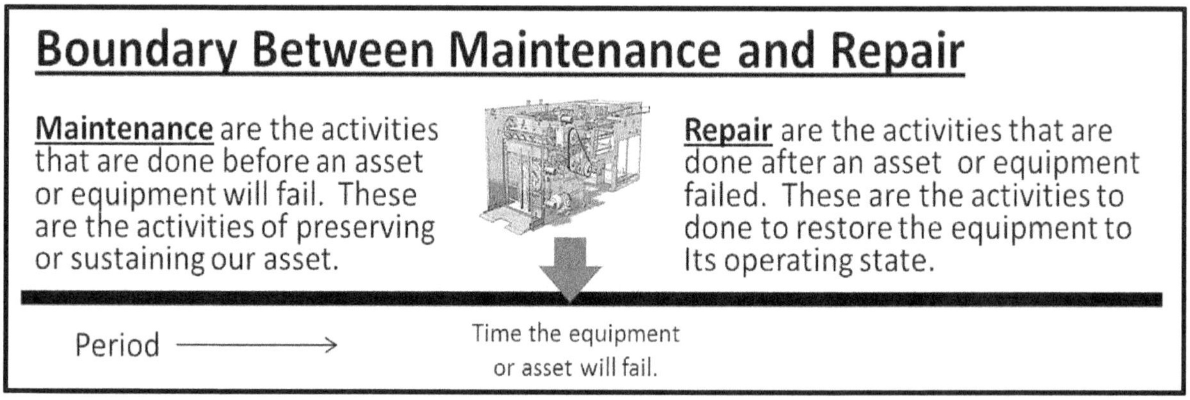

Boundary Between Maintenance and Repair

Maintenance are the activities that are done before an asset or equipment will fail. These are the activities of preserving or sustaining our asset.

Repair are the activities that are done after an asset or equipment failed. These are the activities to done to restore the equipment to Its operating state.

Period ⟶ Time the equipment or asset will fail.

Figure 12.3: Distinction Between Maintenance and Repair

<u>**Chapter 12: The Conclusion**</u>

All I can say is that Root Cause Failure Analysis can be considered as maintenance Ultimate Weapon in finally challenging failures. This should be considered as one of the responsibilities of our maintenance people in industries. Maintenance is measured not by how fast they repair, but rather by how they were able to finally analyze the problem to prevent a recurrence of the cause of the failure. Here are some messages on why we need to perform root cause in our industry.

RCFA is about taking things Slowly: There are way too many reasons why industries are reluctant to perform a Root Cause Failure Analysis. The main reason is obvious, the industry is reactive. In a reactive industry, repairing the equipment will be the main priority. My thoughts on this may contradict other consultants or practitioners but this is what I believe with all my heart. Maintenance and repair are not the same. Figure 12.1 shows the boundary between maintenance and repair. Maintenance tasks are the activities we perform on our equipment and assets before a failure. On the other hand, performing a repair on the equipment are the activities we do after the failure occurs. Maintenance activities include Preventive Maintenance, Predictive Maintenance, Condition-Based Maintenance, Failure Finding Tasks, or Routine Maintenance. How about Run to Fail? As indicated by the definition, maintenances are activities and tasks. In a run to fail situation, there are no maintenance tasks done on the equipment. It is a matter of decision on allowing the equipment to fail. When the equipment fails, maintenance has two options, first is to repair the equipment, and the second is to perform a Root Cause Failure Analysis to understand what caused the equipment to fail. However, when we decide to conduct a root cause investigation, we delay the repair activity as we need to extract the physical evidence on the equipment. RCFA is about taking things slowly. We investigate the situation so that we can finally learn what went wrong with our equipment and assets. When we finally understand the cause and implement corrective action, then we finally learn from the problem, while on the other hand, just repairing the equipment will just allow a repeat of the failure and will improve the MTTR of the maintenance which is not always a good thing.

Human Errors Cannot Be Eliminated: Industries must understand that humans are fallible, and will continue to commit errors and mistakes in the future. It is part of our human nature. The truth is we cannot eliminate human errors, they are given, but the good thing is that we can manage these errors. We simply cannot change the human nature of committing errors since it is part of being human, but we can change the conditions in which humans work. Only by understanding the reasons for the error can we do something to avoid recurrence. There is no one best way to address human error. They occur at different levels of the organization. In fact, the higher your position in the industry is, the higher will be the consequences of that particular error. Even with the best standard and procedures, people will still commit errors and mistakes, but what is truly important is that admitting our mistakes and learning from them definitely makes us a better person. Implementing effective Error Management will require an understanding of the different varieties of human error and how it is promoted. Depending on the human error, there is no one-stop solution as different errors require different corrective actions. People become successful in life by failing. The truth is, we cannot achieve success if we do not fail. Failure is inevitable. It will happen. Success comes after failure. The thing is, we need to have the guts to accept it, learn from it, and move forward. Whether from the lessons of life or from equipment failures, the process is the same. Accept it, learn from it, challenge it and finally overcome the failure.

No.	Discipline 9: Root Cause Failure Analysis Assessment	A 1 point	B 2 points	C 3 points	D 4 points	E 5 points
1.	Does your maintenance team perform Root Cause Failure Analysis on recurring failures, especially those with high impact and consequences?					
2.	Is your team familiar with the most basic types of wear, as well as how to distinguish them, such as abrasion, erosion, adhesion, and fatigue, and how to control them?					
3.	Is your maintenance team equipped with the knowledge and technique on how to investigate equipment breakdowns and failures?					
4.	Is your management team aware that the reasons for conducting RCFA are not to blame or punish anyone but to learn from the things that go wrong with our equipment in the plant?					
5.	Do your people understand that Root Cause Failure Analysis can only be done when the evidence is preserved, meaning that it is not technically possible to perform a root cause when the equipment is already repaired.					
6.	Is your maintenance team practicing any of these analytical problem-solving tools such as brainstorming, Pareto Analysis, FMEA or FMECA, why-why analysis, fault tree, and fishbone diagram to find the possible cause of the problem?					
7.	Do your people understand that performing root cause failure analysis is not only about performing countermeasures and corrective actions but also addressing the latent cause of the problem?					
8.	Is your organization no longer on the blame culture that someone will always be blamed for the failure when something goes wrong in the plant?					
9.	Are your people in your industry well trained in conducting a proper Root Cause Failure Analysis investigation on how to probe in dealing with equipment failures and breakdowns?					
10.	Is there any strategy or training program in your plant design to deal with human errors and reduce or manage them for both operations, and maintenance?					
Please write your feedback or recommendation on how to improve this discipline in your industry.						

Figure 12.4: Root Cause Failure Analysis Assessment

Addressing the Latencies Requires People to Change: The root cause is not only about performing a corrective action, implementing them, validating its effectiveness so that there will no longer be a repeat of the problem. No amount of corrective action can address a latent cause. What is needed in this case is for people to change. Latencies are generic and are usually deal with the present condition of the organization. Latencies are not just about the current failure we experience, but learning from our latencies can avoid failures in the future. Changing the system or writing a detailed 50-page procedure will not solve latencies. Addressing the latency is all about looking at the man in the mirror and asking themselves the following:

• What did I do wrong?
• Why did I do that?
• What was the reason I did that?
• How did it connect to the problem?
• What am I going to do about it?
• What do I need to change?

Small Things Matters Most: Whether from the lessons of life or from the industry's failures, there is always a pattern. Almost all failures started small. In fact, the majority if not all of them are just simply an accumulation of small things that are left unattended. The problem is we are blind to see them even if they are just in front of our face. TPM (Total Productive Maintenance) termed this as addressing the basic equipment condition. These include dirt, leaks, missing bolts, incorrect lubrication on our equipment and assets. The thing is we do not address these small problems because we always get away with them. The only time we address them is when the equipment is fed up with these small problems which end in a catastrophic failure. I remember what our JIPM Consultant once said that taking care of the basics will address 60% or more of your breakdowns and failures. I believe him since we actually experienced it when we were implementing TPM. One missing bolt can increase the overall vibration of the equipment which can lead to cracks and fractures of mechanical parts, not to mention secondary problems that are possible to occur. A vibration monitoring instrument can surely detect that, but the problem is we are too blind to see that single missing bolt. Take care of the small problems, and it will take care of the bigger problems.

Performing Root Cause Allows Industries to Learn from the Things That Go Wrong: Failures can be our greatest teacher if we can learn from them. It tells us what to do and what not to do. Industries will always face risks, losses, wastes, problems, incidents, and failures. Just like human errors, we cannot eliminate them, but we can learn from them if we analyze them. It's an opportunity for industries to learn from the cause of the problem so we can avoid a repeat of that particular cause of the problem. Just like the stories of these inspiring people in chapter 1, they face their failures, overcome them, learn from them and become better. The same goes for industries. We have a weapon in the form of Root Cause Failure Analysis to deal with these problems and incidents. This message perhaps will be the key for industries to survive and remain in business in the future. It means that when an incident happens, we have two options, either we learn from them or just ignore them. The message of Root Cause Failure Analysis is simply to take the first option and learn from the failure.

Funny as it seems, but as I recollect the times, I was often told many times that you need to look dirty at all times so the boss knows that you're doing your job, but as the decades passed by, I wonder and asked myself if that is who we really are. And my feeling is that it should not be the case, as we need to understand the failure and learn from it. All I can say is that Root Cause Failure Analysis is maintenance' Ultimate Weapon in dealing with equipment-related breakdowns and failures. What is important for industries is for the maintenance and technical people to know how to investigate equipment-related failures on their equipment and assets so that we can finally put them to an end and provide corrective measures on failures that affect our operations. Fixing failures will just allow the failure to repeat itself. It's a thing of the past. What we need are people who can analyze failures in their equipment and assets. Just like the lessons of life, learning from failure is the only way to move forward, and I think that's all I have to say about that.

Appendix A: Answers on RCFA IQ Quiz

RCFA IQ Quiz Part 1 (Multiple Choice)

1.	b	11.	b	21.	b
2.	c	12.	a	22.	a
3.	b	13.	Bonus	23.	a
4.	c	14.	a	24.	d
5.	b	15.	c	25.	b
6.	a	16.	b		
7.	c	17.	Bonus		
8.	d	18.	d		
9.	a	19.	d		
10.	b	20.	b		

Note: Bonus means any of the letters is correct

RCFA IQ Quiz Part 2 (True or False)

1.	b	11.	a
2.	a	12.	b
3.	b	13.	b
4.	b	14.	b
5.	a	15.	a
6.	b	16.	b
7.	a	17.	b
8.	b	18.	a
9.	b	19.	a
10.	b	20.	a

Chapter 4 Quiz

1.	a	11.	b
2.	b	12.	a
3.	d	13.	b
4.	b	14.	b
5.	c	15.	a
6.	a	16.	d
7.	d	17.	d
8.	b	18.	c
9.	a	19.	a
10.	c	20.	b

Chapter 6.20: How many letter E

Answer: 67

Appendix B: RSA Maintenance Courses

RSA Reliability and Maintenance Consultancy Firm have been around for 17 years, and through these years, we have developed more courses suited for our reliability and maintenance people in industries. Here is a complete list of maintenance courses and services that we offer in-house, public or online training in your plant.

RSA Courses on Total Productive Maintenance

1. Total Productive Maintenance (3 days)
2. Planned Maintenance 4 Phases to Zero Unplanned Breakdown (3 days)
3. Understanding Autonomous Maintenance, Operators 7 Steps to Empowerment (3 days)
4. Understanding Focused Improvement-Kobetsu Kaizen (1 day)
5. Relationship between OEE and Equipment Losses (1 day)
6. Advance Maintenance Strategies on Planned and Autonomous Maintenance (2 days)

RSA Courses on Reliability and Maintenance Strategies

7. Lubrication Strategy-Understanding Tribology, the Importance of Oil Contamination Control (2 days)
8. Reliability-Centered Maintenance for Industries (3 days)
9. Condition-Based Maintenance, Total Approach to Failure Prediction and Analysis (2 days)
10. Root Cause Failure Analysis-Understanding Equipment Failure (3 days)
11. Optimizing Equipment Reliability-Streamline RCM Approach (3 days)
12. Optimizing Preventive Maintenance Strategy (2 days)
13. World Class Maintenance Management - The 12 Disciplines (3 days)
14. Understanding MRO Spare Parts and Storeroom Management (3 days)
15. Failure Mode and Effects Analysis (FMEA/FMECA) (1 day)
16. Practical Best Maintenance Practices (3 days)
17. Advance Maintenance Leadership in TPM, RCM, LUB, and RCFA (5 days)
18. Advance Maintenance Strategies on RCM and RCFA (2 days)
19. Advance Maintenance Strategies on Lubrication and CBM (2 days)
20. Cutting-Edge Maintenance Management Strategies (3 days)

RSA Courses on Reliability and Maintenance Concepts

21. Meaningful Measures of Equipment Performance - Understanding MTBF, MTTF, MTBA, MTTR, Failure Rate, OEE, and Weibull Overview (2 days)
22. Basic Maintenance Concept, Understanding Reactive, Preventive, Predictive and Proactive Maintenance (1 day)
23. Understanding Proactive Maintenance (1 day)
24. Maintenance Best Practices on LUB and CBM (2 days)
25. Advance Maintenance Strategies on RCFA and RCM (2 days)
26. Preventive and Predictive Maintenance Strategies (1 day)
27. Proactive and Precision Maintenance Strategies (1 day)

<u>**Appendix B: RSA Maintenance Courses**</u>

<u>**RSA Facilitation, Guidance, and Consultation Services Includes**</u>

• Facilitation and consultation on Total Productive Maintenance implementation
• Facilitation and consultation on Root Cause Failure Analysis
• Provide guidance on starting up a Predictive Maintenance strategy in the plant
• Facilitation and consultation on Reliability-Centered Maintenance
• Conduct initial assessment on maintenance Technical Training Needs Analysis
• Facilitation on Strategic Planning for Maintenance and Reliability professionals
• Provides assessment on World Class Maintenance-The 12 Disciplines
• Consultation on setting up an Oil Contamination Control in your plant
• Implementation of TPM Planned Maintenance Pillar
• Implementation of TPM Autonomous Maintenance Pillar
• Implementation of World Class Maintenance Management-The 12 Disciplines
• Maintenance Assessment to determine where your industry currently stands

RSA Reliability and Maintenance Consultancy Firm accept in-house training services for industries for local and international overseas countries. Special arrangements can be offered for overseas countries on any of our maintenance courses selected. These courses are what industries need to improve how we maintain and sustain our equipment and assets in the plant.

The following are lists of training I offered to plants and industries concerning reliability and maintenance courses. My mission is to uplift our maintenance human resources' technical competence in industries searching for ways to achieve maintenance excellence by capturing the industry's reliability and maintenance best practices. These courses provide in-depth details and a wealth of information regarding maintenance best practices from the most basic to the most advanced strategies. These powerful courses have been proven by industries that the best way to reduce their maintenance cost is to sustain their equipment's reliability. With consistent focus on output and productivity and secondary to maintenance and reliability, the latter results in frequent failures, costly unscheduled repairs and unexpected downtime, and an inevitable high cost of maintenance that calls for the adoption of a more rigorous and more effective maintenance strategy that is truly world-class.

Finally, it is also undisputed that every maintenance manager's challenge is maximizing the equipment reliability through a traditional and often self-designed Preventive Maintenance system. This practice is why the approach to maintenance management seems to remain reactive rather than proactive. Truly, these courses are designed for all maintenance managers, engineers, and professionals whose mandate is to optimize their equipment capacity and reliability at the lowest possible cost. In the preceding light, we designed the courses that can be tailored fit and made available to be conducted in-house in your plant for your people's wider participation. Should you be interested in any of these courses, you may reach me at www.rsareliability.com or email me at rollyangeles@rsareliability.com.

Appendix C: RCFA Training Course Details (3 Days)

Training Package Includes

- Morning / Afternoon meals and Lunch
- Printed Hard Copy of the RCFA Training
- Important Articles on RCFA
- Detailed Step by Step Guide on How to Conduct the RCFA Process
- Exercises and Workshop on RCFA
- RCFA Case Study Video Showing on the Challenger Disaster
- Actual Case Studies on RCFA
- USB on RCFA which Contains the Training Package
- Video Showing on Challenger Explosion
- WCM Book in pdf copy (314 pages)
- Certificate of Attendance
- Chocolates for the Workshop
- Good Old Rolling Stones Music

We learned how to conduct the root cause and failure analysis of a certain breakdown. We also learned that in order for the RCFA to be successful, all the evidences must be preserved. The principal investigator must select the right people that will gather the physical, human as evidence. Operations people must attend the training on maintenance to know the heart and soul of maintenance.
From Marnold Celis, Diagnostic Manager, Manila Water

About the Resource Speaker

Rolly, is a seasoned international reliability and maintenance consultant with 30 years of solid experience in the field. He had been invited in different countries and have conducted reliability and maintenance trainings in United Arab Emirates, India, Malaysia, Indonesia, Brunei, Thailand, Nigeria, Bangladesh, South Africa, China and Botswana. His portfolio of maintenance trainings include Maintenance Management courses on TPM, Lubrication, Tribology, Condition-Based Maintenance, RCM, RCFA, Planned Maintenance, World Class Maintenance Management, The 12 Disciplines, Oil Contamination Control, Maintenance Indices and KPI's, Maintenance Management Strategies and much more. Rolly previously worked with Amkor Technology Philippines, as a TPM Senior Engineer, an industry engaged in the manufacture of Integrated Circuit products and spearheaded their Planned Maintenance organization compose of maintenance managers and engineers. He was also responsible for the dramatic reduction of unplanned breakdowns in their TPM Journey as well as RCM implementation on their Facilities AHU units and as well as their substation equipment. Rolly is currently working as an independent reliability and maintenance consultant. Rolly had released his three books title World Class Maintenance Management – The Twelve Disciplines, Maintenance – Roadmap to Reliability and Reliability – A Shared responsibility for operators and maintenance.

Who Should Attend

- Maintenance and Reliability Managers
- TPM Office, Facilitators and Coordinators
- Facilities/Utilities Managers
- Preventive/Predictive Maintenance Group
- Reliability Engineers and Managers
- Operations and Production Managers
- Top Management and Decision Makers
- Continuous Improvement Groups Plant and Facility
- Maintenance Planners
- Mechanical and Electrical People
- Sustaining group, Technicians and involved in repair
- CBM and PdM Personnel
- Maintenance and Reliability Managers
- Maintenance and Technical Support Group
- People in Charge of Planning and Scheduling
- People Involved in Repairs and Troubleshooting
- Head of Maintenance Organization
- Maintenance Engineers
- People Involved in Reliability and Maintenance
- Reliability Engineers and Maintenance Supervisors
- Safety and Quality People are welcome
- People involved in Reliability and Maintenance
- Other Departments are also Welcome in this Training

Figure App C1: RCFA Training Module 1

About Root Cause Failure Analysis

- It is a common practice due to lack of time and information that equipment and system failures are often investigated at a superficial level. As a result both operators and maintenance keep running unreliable plant, which cause repeated losses and become experts at fixing failures rather than understanding the its causes. This training will help delegates to learn and understand what a true and meaningful Root Cause Failure Analysis and how deep the probe should be to finally uncover the root causes that will eventually end recurring problem making us proactive.

RCFA Course Objective

- Understand the importance of Root Cause Failure Analysis as a tool in achieving a Proactive environment in our workplace

- Provide an understanding on why we need to perform Root Cause Failure Analysis

- Provide an opportunity to practice Root Cause Failure Analysis and the different tools on a real workplace problem situation

- Provide participants a working knowledge of Failure Analysis case study which will enable participants to correctly pinpoint design, installation, operating factors that contribute to short equipment lifespan

- Study the different types of wear and why do mechanical component fails in the first place and how to control them the different wear.

- Understand how deep our probe or when do we stop with our analysis on Root Cause

- Understand the different levels of Root Cause Failure Analysis and addressing the latent cause of the problem

- Finally understand what it takes to address the latent Cause of the problem

RCFA DAY 1 : Morning Session

Module 1 : Understanding Failures – Why Failures are Important
- Failure defined
- What Famous People Say About Failures
- Lessons from Failures

Module 2 : Understanding Root Cause Failure Analysis
- Root Cause Failure Analysis Defined
- Simple Root Cause Case Study – Missing Money
- Difference Between Failure Analysis, RCFA and RCA
- When to use RCFA
- Workshop on sequence of events
- Basic Rule on Root Cause Failure Analysis

Module 3 : Top Reasons Why We Need RCFA
- The Need to Analyze Failures
- Management Role on Root Cause

RCFA DAY 1 : Afternoon Session

Module 4 : Physical, Human and Latent Cause of Failures
- Understanding the Physical, Human and Latent Cause
- Root Cause Failure Analysis Logic Tree Diagram
- Difference between RCFA, RCA and Failure Analysis
- Difference Between Root Cause and Problem Solving Tools
- Mini RCFA Case Study

Module 5 : Discovering the Life Blood of Root Cause Failure Analysis - Evidence
- Evidence Defined
- RCFA Tool Kit
- Types of Evidences – Physical, People and Paper Evidence
- Role of Principal Investigator & Evidence Gathering Teams
- How the RCFA Interview Should Be Conducted
- Organizing the Stakeholder Meeting

Module 6 : Understanding the Physical Side of Failure
- Why Do Mechanical Components Fail
- Understanding Brittle and Ductile Fracture
- Understanding the Physical Cause of Failures

END OF DAY 1

Figure App C2: RCFA Training Module 2

RCFA DAY 2 : Morning Session

Continuation on Module 6 : Understanding the
Physical Side of Failure
- Understanding Common Types of Wear Abrasive,
  Adhesive, Erosive, Fatigue and Corrosive Wear
- Understanding how to address these types of wear

Module 7 : Understanding Common types of Bearing Failures
- Common Causes of Bearing Failures
- Lubricant Failure, fatigue, contamination, misalignment and false brinelling

Module 8 : Understanding The Human Side of Failure
- Why Do People Commit Mistakes – Study of Human Errors
- Causes of Human Errors
- 4 Categories of Human Errors
- Understanding Anthropometric Factor
- Understanding Human Sensory Perception
- Understanding Physiological Errors
- Understanding Psychological Errors
- Understanding Slips and Lapses
- Personal Encounter on Human Error
- Is it possible to Eliminate Human Error
- The "E – Experiment on Human Error

RCFA DAY 2 : Afternoon Session

Module 9 : Case Study on Human Error – Why Did the Titanic Sunk
- Details of the RMS Titanic and Structure
- Key People Involved in the RMS Titanic
- Actual Photos of RMS Titanic Beneath the Atlantic Ocean
- Failure Analysis and Physical Cause on Why the
  RMS Titanic Sunk
- Human Errors Committed on the RMS Titanic
- Design problems on the RMS Titanic
- Human Cause on Why the RMS Titanic Sunk
- Video Showing on RMS Titanic (Approximately 20 mins.)
- Workshop on RMS Titanic
- Answer to workshop
- Remembering the RMS Titanic
- Lessons about human error

END OF DAY 2

8:15 to 8:30 am – Registration
8:30 am – Start of Training
10:15 to 10:30 am – AM Break
12:00 to 1:00 pm – Lunch Break
3:15 to 3:30 pm – PM Break
5:00 pm – End of Training

Excellent content and very knowledgeable facilitator/trainer. The training provided great opportunity to gain knowledge/new learning's in identifying the root cause of equipment breakdowns. The topic Root Cause Failure Analysis paved a way in giving importance to Predictive Maintenance to effectively practice good maintenance. We were able to classify the difference between physical cause, human cause, system cause and the latent cause. We were able to identify the different kinds of wear which is beneficial in our field as we are handling various equipment in the factory. Good training exercises and workshops. The use of videos that are related to the topic make the training interesting and fun. Mr. Angeles is World Class Trainer. *From Krista /Leslie Cars, Area Supervisor, URC Passi*

Figure App C3: RCFA Training Module 3

RCFA DAY 3 : Morning Session

Module 10 : Understanding The Latent Cause of Failure
The Challenger Explosion – The Untold Story
- Video Showing on Challenger (1 hour and 21 minutes)
- Compose of 8 Episodes approximately 9 to 10 minutes each
- Each Episode will be explained in detailed
- Workshop To Explain if Latent Cause Really Exists and
- The Need to Expose them in any RCFA Analysis
- Team Presentation
- Lessons Learned From The Challenger Disaster

RCFA DAY 3 : Afternoon Session

Module 11 : RCFA Workshop – Determine the Physical,
Human and Latent Cause of the Problem
- Case Study : Pump Failure
- Determine the correct RCFA Logic Tree Diagram
- Determine the Physical, Human Cause and Latent
  Cause of the Problem
- Team Presentation on RCFA Workshop
- Answer to Case Study on Pump Failure
Module 12 : Building RCFA as a Part of a Structures System
- Step by Step Activities on How to Make RCFA as
  Part of our Structured System
- Delegates to Take Pre-post Quiz on RCFA
- Closing Remarks
- Awarding of Certificates to Delegates

END OF RCFA MASTER CLASS

8:15 to 8:30 am – Registration
8:30 am – Start of Training
10:15 to 10:30 am – AM Break
12:00 to 1:00 pm – Lunch Break
3:15 to 3:30 pm – PM Break
5:00 pm – End of Training

Understand the values of components in regards to the reliability of the equipment. The pattern of analyzing failure enables me to show more effort on how to eliminate and reduce cost of parts involve after equipment failed. This training helps me identify details of Root Cause Failure Analysis and to share to fellow mechanics the principles of why-why. From Elijah Loctonagan, Mechanical Technician, Team Energy Corporation, Sual, Philippines

Testimonies and Feedback

New ideas about RCFA. Very comprehensive course. Step by step learnings and development. Additional Key learnings on RCFA. From Kris Lingat, Shift Supervisor, URC, Passi

Root Cause Failure Analysis is the key to perform in our company URC PASSI on how to investigate equipment breakdowns and analyze the problem that will occur as well as solve the problem in our factory. From Niel Palmares Supervisor, Civil Works, URC Passi

Figure App C4: RCFA Training Module 4

Appendix D: Previous RCFA Training Classes Conducted

Figure App. D1: May 29 to 31, 2019, DMCI, Masbate Benguet Philippines

Figure App. D2: September 29 to October 1, 2010, Public Traning, Mumbai, India

Figure App. D3: February 16 to 17, 2015, Funai Electric, Cebu Philippines

Figure App. D4: July 4 to 6, 2018, Universal Robina Corporation, Philippines

Figure App. D5: October 11 to 12, 2019, Allegro MicroSystems, Sucat, Philippines

Figure App. D6: August 29 to 31, 2012, Petron Corporation, Bataan, Philippines

Figure App. D7: November 7 to 9, 2018, Manila Waters, Philippines

Figure App. D8: August 26 to 28, 2014, Davao, Philippines

Appendix E: Feedback from Root Cause Failure Analysis Training

- I have learned the proper procedure in conducting RCFA. I have already identified the difference between problem-solving and RCFA, which is a good start before conducting an RCFA. *From Ralph Baguio, Petron Bataan Refinery*

- It serves as a guide for lead investigators in the defect elimination team on knowing where to start and where to stop. This training also reminds us people in the industry that it is important to take things slow when it comes to the restoration of problems and failures to avoid destroying evidence to be able to pinpoint the root cause, thus avoiding future recurrence of failures. *From John Michael Calupas*

- The training is a perfect combination of technical and inspirational training. Mr. Angeles uses very practical and real-life situations to explain his points in videos regarding Michael Jackson which creates a relaxing atmosphere. Mr. Angeles is very patient in answering all our questions. We see his passion and expertise very much with the subject matter. He has a broad knowledge in other fields like history when he discussed JFK's life story. I personally enjoyed and learned a lot from this training. I will truly keep all my notes for reference in the future. Mr. Angeles is indeed a world-class trainer. *From Lisaer Pineda, Process Engineer, Petron*

- The use of videos related to the topic is very helpful and made the training very interesting and easy to understand. *From D. S. Bernabe*

- Clarifies the difference between RCFA, RCA, and problem-solving tools. Introduction of latent causes that will further improve human error reduction. The training is unbiased, frank, and direct to the point. The facilitator diverts the topic a bit for the attendees to understand the topic fully. *From Jeffery Babasoro*

- Clarified the RCFA process will be very useful. It made me realize that it would be very difficult to know the root cause without knowing the mechanisms of failure. Learned that shreds of evidence must never be manipulated. *Name withheld*

- Enumerates practical application of training subject. Dig deeper on how to properly apply the learning's acquired. The trainer teaches learning that can also be applied in real life, not only with the technical aspect. He deals with the subject matter on a personal level. *From Madison Carpio*

- Understood that RCFA is about using the word "HOW and WHY" and not "WHO" to arrive at the Latent Cause of the problem, and as a member of the management, we are part of the Latent Cause. *From Ramonchito C. Bringel, Coal Plant Superintendent*

- Excellent presentation materials. Excellent knowledge of seminar topics. An excellent example related to RCFA. Trainer good sense of humor. *From J.R. Villanueva, Mechanical Supervisor*

- The course is well designed for maintenance personnel. The learning from this course applies to all plants with recurring failures. If these tools are used in finding the course of equipment failure, the possibility of recurrence in the future will be avoided. *From Fernando Cabula, Electrical Supervisor*

- You make the training lively and inject good points and standards, highlighting the importance of every topic and its application. You are a very excellent presenter! *From Emmanuel Tayson*

- The resource speaker undoubtedly delivered the training by citing examples based on facts and experience. *From Marianne Canba, Associate Engineer, Analog Devices*

- It is useful in doing troubleshooting and determining the root cause of the problem. As part of the maintenance group, it should be cascaded to other equipment personnel to solve equipment problems. The resource speaker is very familiar with this topic. His technical background and knowledge are a big help to us in understanding the topic. *Name withheld.*

- The training course itself was very much applicable to the participants. Live discussions and exercises create a good training ambiance that makes the participants at ease, lively, and eager to learn. *From Fritzie Magno, Test Equipment Engineer, Training Supervisor, Analog Devices*

- Most of the topics discussed were relevant to our job assignment. The entire training was very alive and active. Participants were enjoying themselves, and the visual aids were good. *From Marlon Mararagan, Associate Equipment Engineer, Analog Devices*

- The seminar is concise and useful both for our profession and actual lives. Educational as well as very entertaining indeed. Very professional and informative course. *Name withheld*

- The case study, especially on the shuttle challenger explosion, is a very good example of the conduct of RCFA. The Root Cause Failure Analysis as presented helped us to understand the concept of the training. *From Elvie Orin, Power Generation, EDC*

- The seminar is quite informative. It has made some points understandable with regards to my misconception regarding RCFA. The things I learned from this seminar will definitely be taken in my line of work. Mang Tibo should come to the next training. *Name withheld, EDC*

- Informative, interesting, and new revelation. This training is full of information new to my knowledge. Root Cause Failure Analysis becomes a thing that needs to be given priority whenever there is equipment or parts that fail. By ignoring it, a fatal incident could happen. This training teaches participants how to conduct a Root Cause Failure Analysis investigation in the workplace. It is full of relevant workshops and real situations in life. Rolly was somehow able to elevate every worth not just because of the diploma but this principle I can apply in every aspect of life. *From Ernesto Salarda, Chemist, Energy Development Corporation*

- It helped me understand and provide me with best practices and ideas in conducting RCFA, especially on our bearing failures related to our plant site. This training provides technical ideas on different bearing failure modes and better understand some mechanical failures. *Name withheld*

- Previously what I been taught to get the root cause, you must do the why-why analysis then you will choose the high possible cause, and then, from the high possible cause, you will get the answer to the root cause, but from this training, it is something new and interesting. To get to the root cause is actually in a different way, and it shows and proven that this training works. I will apply this knowledge to my daily work. *From Mohd Fathi Bin Elias, Maintenance Engineer, Aerospace Industry*

- The training opens the idea of looking at ourselves to make changes and prevent failure from happening again instead of finger-pointing or guessing the root cause. *From Mohd Asha, Aerospace*

- I think this training can be shared with all among staff to increase productivity and also a good way to find the root cause of problem or failure. This also and always can change the mindset of the management team. Good presentation by Mr. Rolly. *From Mohd Zulazmi Bin Zoolbefely Manager*

- Learned the proper way of conducting Root Cause Failure Analysis. The course has deepened my knowledge of the cause of the problem. Learned the techniques in identifying the probable and root cause of each failure. *From Val Sakamancu, Process Engineer, OPM*

- It gives us clear knowledge of how to apply Root Cause Failure Analysis on our equipment. This course motivates us to learn how to handle the problem without pin-pointing individuals for the fault and instead go for the more effective way of solving problems using RCFA. The way we treat the problem on the equipment, the more effective way it will increase the efficiency of our plant if we can operate our equipment for a long time without failures. *From Danilo Reyes, Area Shift Supervisor, KILN Production*

- The topics are clearly explained. The facilitator is good at capturing the crowd's attention and shares a lot of information about the subject and topic. Uses good presentation materials. *From Raymond Lazaro, KILN, Production*

- Identifying the Root Cause of the problem in a more systematic procedure. Learned the true essence of performing a Root Cause Failure Analysis. Analyzing and solving the problem to improve the equipment and not to blame others. *From Leonardo Pagsuyoin, Maintenance Planning Manager, Power Generation*

- I have learned how to conduct and facilitate the RCFA precisely to uncover the physical, human and latent cause, which is very helpful in discharging our duties and responsibilities and improving the plant's operation. *From Domingo Delfin Jr. Reliability Engineer, Maintenance*

- Understand the values of components in regards to the reliability of the equipment. The pattern of analyzing failure enables me to show more effort on how to eliminate and reduce the cost of parts involve after equipment failed. This training helps me identify the details of Root Cause Failure Analysis and share with fellow mechanics the principles of conducting a root cause failure analysis. *From Elijah Loctonagan, Mechanical Technician, Team Energy Corporation, Sual, Philippines*

- With this training, I had gained knowledge about solving problems that I can apply not only in my place of work but also which I can apply to my day-to-day living and even our house. *From Edwin Barnes Utility, Power Generation, Team Energy Corporation, Sual, Philippines*

- This training is very much applicable to my job, especially in evaluating the root cause of why the equipment failed. There are plenty of failures in our equipment, and with this training, we can apply what we have learned from Rolly. *From Chito Ulanday, Supervisor, Team Energy Corp. Sual, Philippines*

- RCFA is very important in our daily works, not only in our equipment failure but also in our personal lives. This training has more benefits in our company to solve and reduce the problem and minimize costs. *Name withheld, Team Energy Corporation, Sual, Philippines*

- The training is prepared professionally. The trainer is an expert in the field presented. The training sessions were lively, and the training is presented in a manner that is easy to understand. *From Felipe Agcawili, Senior CNI Tech. Power Generation, Team Energy Corporation, Sual, Philippines*

- The facilitator's ability to explain was excellent, and the presentation of modules was excellent too. The program materials were easy to understand and improved my knowledge and skills. *From Michael Zarris, Senior Mechanic, Team Energy Corporation, Sual, Philippines*

- A well-planned, well-organized, cultured event useful at every level in the industrial field. Very practical oriented, knowledgeable and useful training that provides guidance at every level. *From A. B. Udavant, Mumbai Port Trust, Superintendent Engineer, Mumbai, India*

- The training covers both theoretical parts as well as sample case studies. The overall training program was excellent. *From A.K. Das, National Aluminum Company, Ltd.,*

- Excellent. It helps us to find a systematic way of Machinery Failure Analysis. The video film showing details and discussion in it was very good. Analyzing the bearing failure logic tree was excellent. *From Rajendra Adhyapak, Assistant Manager, L&T Komatsu, Ltd., India*

- Presentation about seminar subject, and presentation on videos and explain the truth about life. Case studies and explaining the situation were good. Time management was good. An interactive session and practical workshop were good. *From M. Dhana Sekar, Manager, L.G. Balakrishnan Ltd., India*

- Good presentation skills. Have an insight into human errors. Looking at machine failures with a human touch to reach the root cause rather than accusing the person working. *From V. S.Anand, Chief Manager, NAICO, India*

- This program is good and covered overall all groups of industries. Types of failure cause like physical causes, human causes, and latent causes. Analysis of major failure events was good, like the Titanic and Challenger. *From T. Soundamrajan, Manager, Sundaran Clayton, India*

- Mr. Rolly, an experienced trainer with perfect presentation skills and lots of humor. This training is very relevant to plant maintenance personnel in any industry. He is simply leading the respectable journey of maintenance professionals. *From Sanjay Panchial, Assistant Manager, Rockman Industries Ltd.,*

- The training focused more on RCFA, which we generally knew but did not know how to apply it and its benefits. The training program also emphasizes lubrication and its benefits. The interesting fact I learned is that oil does not wear out. The overall training program was interesting. *From Rahul Singh, Engineer, Hindustan Zinc Ltd.,*

- It has given basic inputs on the analysis of failure being observed and necessary steps for analysis. This training will further help in information sharing with my colleagues and better analysis of problems. *From S.P. Saxena, Chief Maintenance Manager, Indian Oil Corporation, India*

- Really good examples and videos. A good and detailed presentation on lubrication. Very good antenna example. Workshops are really good. The exercise on the pump motor was really good. *From M.V. Srikanth, Reliability Engineer, Eaton Corporation, Mumbai, India*

- As a nice speaker, Mr. Rolly Angeles was able to make good and provide a healthy environment during the training. He was able to provide a valuable example as per requirement and cracking good jokes and comments. *From Gourav Shrivastava, Executive Officer, Tata Power, India*

- The knowledge given was very different from our presumption. Very useful and good discreet knowledge. The importance is given to the change of culture which was the main solution for all the problems. Changing the Man from the Mirror was a very important message received. *From O.R. Kamble, Manager*

- Rolly covered all the salient and important points needed for conducting the machinery failure analysis. His vast field experience helped him connect with the participants and their viewpoints. *From Prakulla Kulede, Manager, Tata Motors, India*

- Concepts were explained in a practical manner which helped us understood the concept in better ways. The presentation was done excellently, including the humor part, which kept the audience always attentive. *From S. Dubery, Assistant Manager, Tata Power, India*

- The training was very good. It actually addressed the real issues faced by the maintenance team in any industry. The tools covered are also very vital tools to improve the reliability of the maintenance equipment. *From Tushar Mahark, Manager, Orissa Power Generation, India*

- This training is very important for me as part of maintenance engineering because we encounter many troubles without tracing the root cause of the breakdown. It may help us in troubleshooting analyzing the failure with the guidance of this training. I can apply this system at the production line, also additional information to understand and to eliminate hardship in repairing. *From Rowell Sarmiento, Continental Temic, Philippines*

- I believe that the training goal is achieved, open our management's perspective in the fast pacing industry. The training concept makes the delegates more adaptive. *From Laurence Talens, Continental Temic, Philippines*

- High level of expertise of the facilitator. Well-organized training course. Highly motivational to the participants. Training materials are prepared well. On-time training conduct. Clear delivery of terms and explanations and highly recommendable to other industries. *From Domel Tabuzo, Temic*

- The training is very helpful and informative. It may apply and be useful to the workplace. Visual aids, reading materials, and workshops are outstanding. I may recommend it to other colleagues and co-workers. The trainer is very intelligent and capable of handling a question from the trainees. The trainer has full knowledge and experience on the subject matter. *From Valentino Banayat, Continental Temic*

- Root Cause Failure Analysis is a very interesting course. I have learned that it's not only a problem-solving wherein when you hit the target, it stops, but it's all about change. *From Peejay Queddeng*

- Real events were used as a sample with complete details, data, video, and pictures. Actual presentation materials were provided to all participants free of charge. Workshops are based on real events with real results. The facilitator was lively, which created open discussions. The facilitator has a good sense of humor. *From Noel Tugonon, Continental Temic*

- The speaker is an expert on the subject matter. Preparation and materials are excellent. Manner of presentation is really great, especially the jokes that make the training more lively. *From Renato Corn Cruz, Continental Temic*

- Visual presentations and materials are adequate for the course. The facilitator/trainer is diligent enough to answer delegates' inquiries. Interactions between facilitators are greatly exercised during the training. Workshops had made the delegates think about working as a team and relating to the company's real/actual scenario. The facilitator is very active, alert, and very motivating in his own special way to encourage delegates to decide for a change for the betterment. *From Allan Dimailig, Temic*

- Really appreciate the small group arrangement as it gives more attention and clearer delivery. *From Zara Binti Mohamad Razal,*

- A different perspective on innovating maintenance. Nice course and resourceful instructor, and I enjoyed the music as well. *From Yap Wee Kiat, Southern Acid Industries, Malaysia*

- Gain new and good information and method on how to apply RCFA in my plant. Good materials for reference and handbook, notes, slides, and videos. This training is easy to understand. Interesting training session when mixed maintenance, music, and video clips. *From Mohamad Irfan Bin Mohamad*

- Learned the proper Root Cause Failure Analysis method. Understand the different types of wear, and identifying the physical cause, human cause, and latent cause of the problem. *From Roxan Mangliwan, Maintenance Supervisor, Hedcor Sibulan Inc.,*

- Strengthen and enhance the why-why we practice in finding the root cause of the problem. *From Michael Quidayan, Maintenance Manager, Hedcor Sibulan Inc.,*

- Good presentation. Able to present with audiovisual presentations. Able to deliver almost all the important points. *From Rafael go, Operation and Maintenance, Hedcor Sibulan Inc.,*

- It is based on actual events and can be applied to every aspect of problems. *From Roy Elmer Suanco, Hedcor Sibulan Inc.,*

- Very informative. Very applicable to my job by being able to know the causes of failure on our equipment. *From Jennifer Agton, Operations Supervisor, Aboitiz Power Corporation*

- The tone discussed is strongly relevant to the maintenance by knowing the right way to conduct an RCFA investigation and identify the root cause. *From Jefferson Tamano, Hedcor Sibulan Inc., Maintenance Supervisor*

- Methods of imparting knowledge and citing true to life events that relate to our problems. Mastery and the experience of the speaker making him a very effective one. *From Hanna bee Sordilla, Hedcor Inc.,*

- As always, great session. The concept, principles are clearly defined and explained clearly delineated the difference between RCFA and RCA. Speaker is an expert, honest, and really has the heart to teach and share. *From Leo Lungay, Operations Manager, Hedcor Sibulan Inc.,*

- Good mastery of the facilitator on the subject. Methods on imparting the knowledge are good. The topic is suitable for operation and maintenance, unlike the previous RCA training we had, which is too generic. *From Mirriam Saavedra, Asset Management Coordinator, Hedcor Inc.,*

- Practical experiences shared and discussed with the team. Detailed step-by-step procedures for analyzing equipment failures. *From Nuel, Hedcor Inc.,*

- Root Cause Failure Analysis is different from Root Cause Analysis. We are being taught how to perform root causes that can be applied in our equipment and plant. *From Marlon Torralba, Plant Operator, Hedcor Sibulan.*

- Lively. Provide us an understanding of why we need to perform Root Cause Failure Analysis. Learned a lot from this training. *From Pablito Mejia, Operation Supervisor, Hedcor Inc.,*

- Real-life examples. Covers a lot of subject on RCFA. Very educational and informative. *From Ariel Salipan, Hedcor Inc.,*

- That we have learned the true RCFA method by having an investigative team. Preserving the evidence and coming up with the physical cause on what caused the problem to find the human cause on what had flawed on the physical cause and stopping after finding the latency. *From Jayvee Almojuela, Maintenance Supervisor, Hedcor*

- I'm able to learn that RCFA is 100% evidence-based. Have three root causes which are the physical, human, and latent causes of the problem. *From Carlos Curiman, Maintenance. Hedcor*

- Learn how to address a certain failure using Root Cause Failure Analysis based on evidence and only evidence. From H. Alangdeo, Electrical Maintenance Engineer, Bakuh Grid, Hedcor

- It is my first time to know about RCFA and the causes, which are physical, human, and latent cause, and probing these causes as the root cause by identifying the probable cause or hypothesis by basing it on the evidence. *From Rocky Marquez, Maintenance Engineer, Hedcor*

Serving Maintenance Mankind Worldwide

This book is dedicated to all "Maintenance" out there in industries. My mission is to reach out to industries searching for ways to improve their maintenance human resources. I believe that the key to improving reliability is to provide our maintenance people the knowledge and education they need to maintain their equipment and assets. At my small firm, we serve maintenance mankind worldwide.

Figure E: We're Serving Maintenance Mankind Worldwide

Bibliography

- Anderson, T.L., **Fracture Mechanics, Fundamentals and Applications, Second Edition**, CRC Press Inc. Copyright 1995
- Angeles, Rolly, **Cutting Edge Maintenance Management Strategies,** Central books Printing, Philippines 2020
- Angeles, Rolly, **Decoding Reliability Centered Maintenance for Manufacturing Industries,** Central books Printing, Philippines 2021
- Angeles, Rolly, **Lubrication Tactics for Industries Made Simple,** Central books Printing, Philippines 2020
- Angeles, Rolly, **Maintenance Roadmap to Reliability**, Central books Printing, Philippines, 2016.
- Angeles, Rolly, **Problems, and Solutions on MRO Spare Parts and Storeroom,** Central books, Printing, Philippines 2020
- Angeles, Rolly, **Reliability, A Shared Responsibility for Operators and Maintenance,** Central books Printing, Philippines, 2018.
- Angeles, Rolly S. **World Class Maintenance Management – The 12 Disciplines**, Central books Printing, Philippines, 2009
- Callister Jr., William, **Material Science and Engineering, An Introduction, Second Edition,** John Wiley and Sons Incorporated., Canada. 1991
- Gotoh, Fumio. **Equipment Planning for TPM, Maintenance Prevention Design**, Portland, Oregon, Productivity Press, 1988.
- Mobley, R. Keith, **An Introduction to Predictive Maintenance, Second Edition**, Butterworth, Heinemann, 2002.
- Mobley, R. Keith, **Root Cause Failure Analysis, Second Edition**, Butterworth, Heinemann, 1999.
- Lim, Billi P.S. **Dare to Fail**. Selangor, Malaysia: Hard knocks Factory, 1996.
- Moubray, John, **Reliability Centered Maintenance II, Second Edition**, Great Britain: Butterworth, Heinemann, 1977.
- Nelms, C. Robert. **What You Can Learn From the Things That Go Wrong, A Guide Book to the Root Cause of Failure**, Virginia, USA: Failsafe, 2007.
- Pall Industries Hydraulics Co. **Contamination Control & Fundamentals**: New York, 1994.
- Rogers Commission Report, **Report to the President by the Presidential Commission on the Space Shuttle Challenger Accident**, (Washington DC, 1986)
- Schein, Edgar. **Coming To a New Awareness of Organizational Culture**, Sloan Management Review, Winter, 1984.
- Shirose, Kimura, and Kaneda. **P-M Analysis, An Advanced Step in TPM Implementation.** Portland Oregon: Productivity Press, 1995.

Glossary on Maintenance

This glossary is a collection and compilation of all my books on World Class Maintenance Management, the 12 Disciplines, Maintenance Roadmap to Reliability, Reliability a Shared Responsibility for Operators and Maintenance, Cutting-Edge Maintenance Management Strategies, Problems and Solutions on MRO Spare Parts and Storeroom, Lubrication Tactics for Industries Made Easy, Decoding Reliability-Centered Maintenance Process for Manufacturing Industries, and this latest book on Investigating Equipment Failures through Root Cause Failure Analysis.

8-Disciplines is an analytical problem-solving tool designed to determine the probable or most likely cause of the problem. This method can be applied to defects and equipment-related failures. This method establishes a permanent corrective action based on data and provides the probable cause of the problem.

Abnormality can be defined as any deficiency, disorder, slight irregularity, defect, bug, flaw, or any unwanted condition that could lead to other equipment problems.

Abrasive Wear is a type of wear that can be categorized by a single keyword, cutting. Abrasive wear occurs when hard particles are suspended in a fluid or projections from one surface roll or slide under pressure against another surface, thereby cutting the other surface. Abrasive wear can be either two-body or three-body abrasions.

Absolute Filtration refers to the smallest size of particle that will be removed during filtration. This means that if the filter's size is 5 microns and the rating is absolute, almost all the contaminants in the range of five microns and above should be removed by the oil filter.

Absolute Viscosity measures the resistance to flow when an external force is applied like a spindle driven by a motor. For example, when we buy water-based paint and open it, it takes some force to stir the paint. However, if we add water and mixed it with the paint, it is much easier to stir. This means that the viscosity decreases.

Additive is any blended chemical added to a lubricating oil or grease to improve its performance and properties to protect the base. It is also mixed with the oil to counter any negative effects of the oil from heat and contamination.

Adhesive Wear occurs when a peak or asperity from one surface comes in contact with the other surface's peak or asperity. There may be instantaneous micro-welding caused by the friction or heat involved due to the lubricant's loss of film.

Advance Discipline refers to the state-of-the-art non-destructive tools and diagnostics instruments and software such as CMMS or EAM for the maintenance requirements.

Age is the measure of a unit, item, or spare's total exposure to stress which can be expressed as the number of operating hours or other stress units since the asset started operating until their retirement or decommissioned.

Aging is the process where certain parts of the equipment deteriorate or wear out because of stress-induced over a given period. It is also termed as wear and tear or deterioration.

Age Exploration Method increases the interval for Preventive Maintenance overhauling and replacement by 10 % if aging or wear-out is not yet evident when it will be overhaul or replaced.

Analytical Ferrography (ASTM D7690) is an oil analysis test where solid debris suspended in a lubricant is separated and systematically deposited onto a glass slide called a ferrogram. The slide is examined under a microscope to distinguish particle size, concentration, composition, morphology, and surface condition of the ferrous and non-ferrous wear particles

Anthropometric Errors are a type of human error that occurs because the person simply cannot fit into the space provided, cannot reach something, or is not strong enough to move or lift something. Industries aiming for multiskilling should also consider anthropometric factors to avoid human errors.

Antifoaming or Defoamer is the process of removing entrapped air or bubbles in the lubricating oil. Almost all lubricating oil systems contain some air. Air is found in four phases: free air, dissolved air, entrained air, and foaming. A defoamer or an anti-foaming agent is a chemical additive in oil that reduces and hinders foam formation in industrial process liquids.

Anti-Wear additives (AW) provide the protective layer on moving parts that minimize metal contact effects. Anti-wear additives are used in many lubricating oils to reduce friction, wear, and scuffing under boundary lubrication conditions, where full-film lubrication cannot be fully maintained. Anti-wear additives are also called boundary lubrication additives.

Asperity, in tribology, in a micro-scale are topographical irregularities that are similar to peaks and valleys of a solid surface. This can only be seen if the surface is enhanced through a Scanning Electron Microscope (SEM). Once the two surfaces' asperities come into contact, lubrication is displaced, and there is metal-to-metal contact and fracture.

Asset Management is defined as executing that set of processes to realize value as the organization defines it from those assets.

Autonomous Maintenance is the activities in which operators perform their daily inspection, lubrication and parts replacement, minor repair and troubleshooting, accuracy checks, and so forth on their own equipment, aiming to keep the equipment in good running condition.

Availability is the proportion of time the equipment is available for use for its intended purpose, whether it is being utilized or not. The trend should be consistent and should not be lower than 98%. Calculation of availability can be reported on a daily, weekly, or monthly basis.

Available Time is given as the total time in a given period. 24 hours for one day, 168 hours for one week, and 720 hours for 30 days or 744 hours for 31 days.

Barcode is a machine-readable form of information on a surface that can be scanned by a barcode scanner. They are often known as UPC codes. The barcode is read by using a special scanner that reads the information directly. This information is transmitted into a database where it can be logged and tracked.

Basic Maintenance Discipline is the fundamental activity that should be performed on the equipment before proceeding with any other advanced or specialized disciplines. It also serves as the foundation of any maintenance management strategy. This includes training, KPI, basic equipment condition, Autonomous Maintenance, and Preventive Maintenance.

Bathtub Curve is a conditional probability curve representing the age-reliability relationship of certain items or characterized by early infant mortality region, a region of relatively constant reliability, and an identifiable wear-out age.

Bending Stress is the force an object encounters when subject to a load at a given point which causes the object to bend or warp. This type of stress usually occurs in objects when subjected to a tensile load.

Beta Filtration Rating is derived from the Multi-pass Test Method for Evaluating Filtration Performance of a Fine Filter Element (ISO 4572). The automatic particle counter measures the upstream particles and quantity per unit volume-of-fluid and the particle size and quantity downstream of the filter during the test.

Blending is a refinery operation that blends different component streams into various grades of gasoline. For gasoline, the fuel is blended to achieve a higher octane rating standard, creating different gasoline types. Most gasoline stations offer three octane levels, including regular, about 87, mid-grade, about 89, and premium, with an octane rating from 91 to 94.

Bottleneck is a constraint or congestion in a production operation that occurs when the products' inventory is build up quicker than the equipment can process. In manufacturing, bottleneck equipment is those assets that frequently fail in the processor have a lower UPH (units per hour) than other equipment in the process.

Boundary lubrication is affiliated with metal-to-metal contact between two sliding surfaces as the asperities come in direct contact with one another. This mostly happens during start-up and shutdown.

Breakdown Occurrences refer to the frequency of failure or breakdown of the system or component. The trend should be the lower, the better. This can be reported on a weekly or monthly basis.

Breakdown Loss sometimes referred to as equipment failure loss, is a loss when the machine stops since its function completely failed. The Japanese term for breakdown is kosho.

Brittle Fracture is a fracture that involves little or no permanent deformation on metals. Many non-metals lack ductility and are subject to brittle fractures. A brittle fracture occurs when a part is overloaded and breaks with no visible distortion and deformation.

C-level Executives are high-ranking executives of a company or any organization in charge of making company-wide decision-making. C stands for chief. This will include the chief executive officer (CEO), chief operating officer (COO), and chief information officer (CIO).

Capital Spares are the items within inventory that are purchase as spare parts for depreciable assets (e.g., capital equipment, backup engines, and redundancies). As such, these capital spares within the inventory can depreciate. In most cases, capital spares are not included as inventories since they are part of a company's property, plant, and equipment (PPE).

Carrying or Holding Costs are the accumulated cost on parts held in stock inside the storeroom. The holding or carrying cost begins to accumulate the day the items are put inside the storeroom. Usually, this will be around 10 % to 30% of the overall costs per year.

Causal Factor in root cause investigation is any unplanned, unintended contributor to an incident or failure. It can also be stated as the contributing cause to the incident. It can also be said that the causal factor is just one of the reasons the contributed to the incident.

Chronic Losses are usually made up of a wide variety of causes and frequently occur over time. The word chronic means repeating or recurring.

Circadian Rhythms which means around the day. It comes from the Latin words circa which means about and dies which means a day. These circadian variations are governed by a biological clock located in the brain.

Classification is used in FMEA/FMECA to highlight those failure modes with a severity rating of 9 to 10 or failures with the highest impact or consequences on the equipment or system being analyzed in the FMEA analysis.

Cleaning Materials Inventory is needed to sustain and maintain the plant's cleanliness and facilities, comfort rooms, toilets, and office spaces.

CMMS or Computerized Maintenance Management Software is a maintenance software used for the maintenance department to streamline and automate the maintenance process. CMMS should support these functions by capturing and automating administrative maintenance tasks and gathering relevant information to perform this process.

Complex Item refers to an item or asset whose functional failure can result from numerous failure modes.

Compression is when we applied a downward force on a vertical cylinder, the object will have an equal upward force, which is equivalent to the normal force. Once we increase the downward force, then we also increase the upward force.

Conditional Probability of Failure refers to the probability that an item will fail during a particular age interval, given that it survives to enter that interval.

Condition-Based Maintenance or Predictive Maintenance checks the equipment's actual condition using sophisticated measuring non-destructive instruments with precision accuracy. Predictive Maintenance instruments are just a higher form of the human senses. Performing inspection through the use of human senses is also included in On-Condition tasks for RCM.

Consequences of Failure result from a given functional failure at the equipment level and for the operating organization classified in RCM analysis: hidden failure consequences, safety consequences, environmental consequences, operational consequences, and non-operational consequences.

Consistency depends on the type and amount of thickener used and the viscosity of the base oil. Consistency is also known as the resistance to deform caused by an outside force. The measure of consistency is called penetration.

<u>**Glossary on Maintenance**</u>

Containment is a short-term action being initiated to keep the defects or failures from further damage. Containment is just a temporary solution and will not fix the problem.

Contamination in oil is anything that should not be present in the oil besides its base and additives. Contaminants may be solids, liquids, or even gases in bubbles that cause foaming.

Consumable spares are regularly consumed and used parts and items such as fasteners, seals, belts, oil filters, grease, lubricants, gloves, WD-40, rags, face masks, cleaning materials, and so on. These items are stored in the storeroom. They are used by maintenance regularly.

Corrective actions are actions taken to address the probable cause of identified non-conformances or incidents, manage their consequences, and prevent or reduce the likelihood of recurrence of failure. Identification and execution of corrective measures must be done for both the short term and the long term.

Corrective Maintenance is a maintenance task that has different meanings. In the majority of industries, this refers to repairing a failure after it happened. Other industries also refer to this as Predictive Maintenance tasks as the activities that are done when the P-F interval is nearing its functional failure. For those industries implementing TPM or Total Productive Maintenance, corrective maintenance means performing an improvement on the equipment.

Corrosion is a wear process of materials caused by chemical, electrochemical, or other reactions. The main difference between corrosion and rust is that corrosion occurs due to the chemical reaction on metal surfaces, while rust only affects iron metals exposed to air or moisture.

Corrosion inhibitor is a chemical compound substance that decreases the metal or alloy's corrosion rate when it comes in direct contact with the oil when added to the lubricating oil. Corrosion inhibitors are additives to the fluids that surround the metal or related object. The corrosive inhibitor's nature depends on the material being protected, which are most commonly metal objects, and on the corrosive agent that needs to be neutralized.

Corrosive wear is the deterioration of metal due to chemical or electrochemical reactions with its environment. Corrosion can occur when the substance is exposed to air or some chemicals while rusting mainly occurs when a metal is exposed to air or moisture.

Cortisol, a stress hormone that triggers your heart to work harder. Cortisol works with certain parts of our brain to control our mood, motivation, feelings, and fear.

Cracking Process is used to maximize the effectiveness of heavier oils. Heavier oil contains large strings of hydrogen and carbon molecules. Using a catalyst, these long strings of molecules are broken into small chains that transform heavier oil into lighter fluids such as gasoline and diesel fuels.

Criticality is a measure of how important an asset is to your process. The more critical the asset, the more impact it will have when failure happens. Criticality is based on what could happen if the failure occurs.

Critical Failure involves the loss of function or damage that could directly affect safety and environmental consequences. It can also have operational consequences with devastating impacts or aftermath.

Culture means how people perceived and do these things around their plants. It refers to their common values and beliefs, while others refer to them to shared thoughts and feelings. According to Schein, culture, is the pattern of basic assumptions that a given group had invented, discovered, or developed in learning to cope with its current problems of external adaptation and internal integration that had worked well enough to be considered valid. They have taught new members the correct way to perceive, think, and act concerning their day-to-day activities.

Defect and Rework Loss is a type of equipment loss caused when defects on products are found in which the product has to be reworked or scrapped.

Detection in FMEA/FMECA is an assessment of the design and machinery controls' ability to detect or capture that a failure mode is occurring independently. This is also termed as confidence.

Detergents are oil additives that keep the hot metal components free from deposits and neutralize acids, forming sludge and residues. The detergents used in lubricating oil can be organic soaps and salts of alkaline earth metals such as barium, calcium, and magnesium.

Design Speed Loss is a type of equipment loss on production caused by the difference between the design or theoretical speed and its actual operating speed. This is usually given in the PM manual document provided by the OEM.

Diffusional Interception is a filtration mechanism where the particles that strike through random moving gas molecules impact the medium and are held by adsorptive forces. The probability of removal is further increased by the effect of Direct Interception, Inertial Impaction, and Diffusional Interception combined together.

Direct Interception is where the contaminants are trapped in the pores between the medium fibers. It can also catch contaminants that are smaller than the pore size through bridging. In direct interception, the contaminants come in physical contact and are attached to the filter media.

Dispersants are added to the lubricant to prevent the accumulation or particle attraction of sludge and dirt in the oil. The dispersants' function is to suspend the contaminant in the oil rather than for the contaminant to settle at the bottom of the crankcase and form deposits so that it can be removed by the oil filter easily when the oil flows and circulates inside the engine system.

Dissolved water is when the water is dispersed in the oil in a homogeneous molecular solution. The gas or moisture cannot be removed from a conventional' nominal filtration; however, several absolute oil filtration on the market can remove a certain amount of moisture in the oil.

Distillation Process is a refinery process where crude oil will undergo a process called distilling or the distillation process. The oil is heated in an atmospheric distillation vessel until the crude oil reaches its boiling point, which will turn oil into vapor.

Downtime is the period when equipment or asset is not operating. This is the time that the equipment is unavailable for use. The equipment's unavailability may result from equipment failure, malfunction, or non-machine related such as PM shutdown. Downtime is classified as machine-related and non-machine-related. It refers to the amount of time spent on doing corrective maintenance on the equipment and assets.

Dropping Point is a test on grease that determines the temperature at which the grease drips off the testing unit in a non-decomposed condition. It indicates the grease's heat resistance or up to what temperature the grease can remain firm.

Economic Consequences refer to the consequences of the failure mode, which is evident and does not affect safety and the environment and has no operational consequences. The only consequence, in this case, is the direct cost of the repair.

Economic Order Quantity (EOQ) is one technique that can be used to optimize inventory levels by ordering the right quantity at a specific time interval to minimize inventory cost but still meet the users' demands. EOQ will answer how many orders need to be placed per year and how many items to place per order.

Electronic Data Interchange (EDI) can be best used by Purchasers in automating their transactions with their vendors and suppliers. It provides a direct link between the vendor and the buyer by automating the Purchase Request and Purchase Order system directly linked with an internal system.

Empowerment means power, control, authority, or dominion. The prefix "em" means to put on to or to cover with. Empowering is the passing of authority and responsibility to an individual. Empowerment occurs when the power goes to the operators, who then experience a sense of ownership and control over their jobs.

Emulsified Water can be in the form of microscopic droplets of water distributed in the form of an emulsion. Emulsified water is when the amount of dissolved water is greater than the saturation point. When the water is in an emulsified state in lubricating oil, its color turned into a white or milk state in layman's terms.

Enterprise Asset Management (EAM) software provides a holistic view of an organization's physical assets and infrastructure throughout their entire life cycle, starting from the design, procurement, installation, commission, operations, and finally, its retirement or disposal.

Environmental Consequences mean that a failure mode has environmental consequences if it causes a loss of function or other damage, leading to the breach of any known environmental standards or regulations.

Ergonomics originated from the Greek word ergon, meaning work, and nomoi, meaning natural laws. It is the science of refining the design of products to optimize them for human use. It is also sometimes known as human factors engineering. Ergonomics is a science that deals with designing and arranging things so that people can use them easily and safely.

Erosive Wear is a type of wear due to mechanical interaction between the surface and a fluid, a multi-component fluid, impinging liquid, or solid particles. Erosive wear occurs by the impact of particles upon a surface with a resultant material loss due to fracture.

Evidence is a piece of information that supports a conclusion. Evidence in its broadest sense includes anything used to determine or demonstrate the truth of an assertion. It is the lifeblood of any Root Cause Failure Analysis investigation.

Evident failures are failures that will become evident or visible to both the maintenance and operators when they occur independently.

External set-up is the activities, which can be performed while the machine is still running or is running the last production lot before the actual conversion will take place.

Extreme Pressure additives (EP) is an oil additive that is usually used in gearboxes to provide a layer of film on the metal's asperities so that when the two opposite asperities meet, instead of breaking up, they will simply slip or slide each other. It is not recommended to use EP additives on yellow metals.

Fact-Finding Group is a group of people that composed the Principal Investigator and Evidence Gathering Team. Their mission is to find out the cause of the failure, learn from it, and do something to prevent, mitigate or eliminate its occurrence.

Fail-Safe System is a system whose function is replicated o duplicated so that the function will still be available to the equipment after the failure of one of its sources, making failure tolerable and allowed to happen.

Failure is the inability of the equipment to perform its required function. The failure of a component is viewed as terminating its life. In general, failure refers to the state or condition of not meeting a desirable or intended objective. In life, failure may be viewed as the opposite of success.

Failure Finding Tasks are maintenance tasks design for hidden or unrevealed failures such as protective devices and redundant functions. These tasks are being performed to reduce the risks of multiple failures wherein both the protective device and protected function are both in a failed state. It is also termed as functionality inspection or Detective Maintenance.

Failure Effects describe, most likely, what happens when each failure mode occurs on its own. Writing the failure effect should compose of 20 to 60 words.

Failure Modes are the most likely, or probable cause of failure. These are not the root cause but are considered the possible or most likely cause of why the failure happens.

Failure Mode and Effects Analysis (FMEA) discipline were developed in the United States Military. The procedure for MIL-P-1629, titled Procedures for Performing a Failure Mode, Effects and Criticality Analysis, was written on November 9, 1949. It was used as a reliability evaluation technique to determine the effects of system and equipment failures.

Failure Rate is the expected rate of failure or the number of failures in a specified period. It is express in failures per million or billions of hours. It is also the probability of a failure in a stated unit of time and is the reciprocal of MTBF.

False Brinelling is caused by the continuous vibration of the duty equipment, thereby causing fretting damage on the bearing for standby equipment. The best way to address false brinelling is to rotate the shaft manually around 360 to 720 degrees + 60 degrees so that the ball's position changes from its position on the outer raceway of the bearing.

Fatigue Wear is a type of wear that results from a continuous cyclic slip under repetitive load applications for many thousands or millions of load cycles. Fatigue is the phenomenon leading to fracture under repeated or fluctuating stress having a maximum value less than the material's tensile strength. Fatigue fractures are progressive, which can start as micro cracks and propagates into larger ones that cause complete destruction of the components.

Fault-Tree Analysis was developed by Boeing Aerospace in the 1950s for use in the development stages of the design process. This is a mathematical tool that yields probabilities. Its primary intent is to predict the probability of a specific failure.

Filtration can be defined as removing contaminants from a fluid, liquid, or gas stream through some means of a porous medium. An oil filter's function is to remove contamination from a fluid, liquid, or gas medium through some porous medium to achieve a required fluid cleanliness level.

Finished Goods Inventory refers to the finished products of the plant available for customer purchase. These products will either be ship to their clients, picked up, or ready to sell.

Fire Point is defined as the lowest temperatures at which vapor of the material will catch fire while continue burning even after the ignition source is removed. The fire point is usually higher than the flashpoint, typically plus 10 to 24 °C or 50 to 75 °F beyond the flashpoint. The vapors produced at the flashpoint are not sufficient to ignite the fuel.

Flashpoint is the temperature at which the oil gives off vapors that can be ignited with a flame over the oil. The lower the flashpoint, the greater the oil's tendency to suffer vaporization loss at high temperatures and burn.

Flow rate is how much volume of a liquid can pass a certain point in a given time, measured in gallons per minute or liters per second.

Foaming is the formation of air and bubbles in a petroleum product, lubricant, or fuel oil that can reduce the product's effectiveness. Foaming can cause sluggish operation, air binding of oil pumps, and overflow of tanks or sumps. It can also result in excessive turbulence, improper fluid levels, air leaks, cavitation, or contamination with water or other foreign material.

Four-ball EP (Extreme Pressure) is a grease test called the 4-Ball EP, Four-Ball Extreme Pressure, 4-Ball Weld, or Load Wear Index. The Four-Ball EP test (ASTM D2596) measures the grease's ability to prevent wear during sliding contact under extreme pressure caused by heavy loads.

Fracture refers to the separation of a solid body into two or more pieces imposed upon by stress, which is usually static and at temperatures relative to the material's melting temperature. The applied stress can be considered as tensile, compressive, shear, or torsional.

Free Water can be any water or gas which is not dissolved in the lubricating oil. Free water is when the water and the oil separates. Since the water is denser and heavier than the water, it will settle at the bottom and easily be removed from the drain plug.

Friction is a force that is created whenever two surfaces move relatively across each other. As long as there is friction, heat is generated, and wear takes place. The area of contact will always be the point of wear.

Friction Modifier additives affect the frictional properties between two rubbing surfaces. These additives prevent scoring, reduce wear, noise, and prevent micro-pitting in industrial gear lubricants. Friction modifiers are commonly used in gasoline engine oils.

Function is the normal or characteristic actions of an item, sometimes defined in terms of performance capabilities. This is also what the users want the equipment or asset to do.

Function Loss Failure or failure of the primary function is when a failure occurs where the machine or equipment will totally stop and will definitely halt or stop operations completely.

Functional Failure is defined as the inability of an item to meet its specific performance standard. This was when the asset had failed completely.

Grease is defined as a solid to a semi-fluid product of dispersion of a thickening agent in a lubricant. It comes from the Latin word Crassus, which means fat. Grease is a thick, oily lubricant consisting of inedible lard, the fat of waste animal parts, mineral or petroleum-derived or synthetic oil, plus a thickening agent.

Grease Consistency depends on the type and amount of thickener used and the viscosity of the base oil. Consistency is also known as the resistance to deform caused by an outside force. The measure of consistency is called penetration.

Hershey Number is the dynamic viscosity (η) multiplied by the speed (N) divided by the normal load (P) per length of the tribological contact. The load also means the average pressure.

Heterodyning is a process in ultrasonic analysis that translates these frequencies into the audible range. Heterodyning is the mixing of two waves, which produces both the sum and difference of their original waves, which allows the shifting of a high-frequency sound to the audible or sonic range. It converts the ultrasonic range to an audible or sonic range through this process.

Hidden Failures are failures that will not become evident to the operator or user of the equipment when the failure occurs independently. To consider a hidden failure, a secondary failure should occur. Hidden failures will only be applicable for protective devices and redundant components.

Hidden Failures Consequences refer to the risks of multiple failures due to an undetected earlier failure of a hidden function item. This mostly refers to the aftermath when protective devices and redundancies fail.

Horizontal Replication, also called fan-out, is the process of repeating or doing the proposed tasks derived from the RCM analysis to similar equipment with the same operating context or condition. This may also apply to modification, redesign, or improvement as long as the equipment replicated also possessed the same problems.

Human Cause is the second part of Root Cause Failure Analysis, which refers to human errors, omissions, or commissions resulting from the physical roots. Either someone did something wrong, or the person did the wrong thing.

Human Nature is a concept that denotes the characteristics of human beings such as the way they think, their feelings, the way they feel and react naturally. Humans react based on their instinct.

Human Sensory Perception means humans are gifted with five senses: smell, touch, hearing, feel, and taste. Forget the sense of taste. Some failure modes will give some sort of warning or symptoms that they are on the verge of failing, such as smell, excessive vibration, noise, heat, or anything that can be detected by our human senses.

Hydraulics is the transmission and control of forces and motions through the medium of fluids. It is also the science of transmitting force or motion through the medium of a confined fluid. In a hydraulic system, the power is transmitted by compressing a confined fluid. This transfer of energy takes place because a quantity of liquid is subject to pressure.

Hydrocracking is a flexible catalytic refining process that can upgrade a large variety of petroleum products. Hydrocracking is commonly applied to upgrade the heavier fractions obtained from crude oils' distillation, including their excess.

Hydrodynamic Lubrication (HL) is also termed or called full-film or fluid film lubrication. Hydrodynamic lubrication is where the film is thick enough to separate interacting surfaces to minimize friction.

Hypothesis is a term used in root cause failure analysis which means the probable or most likely causes that need to be verified. In RCM, they are termed failure modes, while in a crime scene, they are referred to as the suspects.

Inertia Impaction is a process that helps remove contaminants smaller than the pore size of the filter medium. The contaminants are retained mechanically or through adsorption. This type of removing contaminants is more effective in gases than in fluids. This process is efficient for particles greater than 0.5 to 1 micron.

Infant Mortality Failures are failures, which occur at the beginning of life. Others refer to this as commissioning failures, start-up failures, or debugging failures right after Preventive Maintenance activities.

Infrared is a type of light in the electromagnetic spectrum that is not visible to our naked eyes. Our eyes can only see a small portion in the electromagnetic spectrum, the visible light spectrum, or the rainbow's colors in layman's terms. Infrared radiation lies between the visible and microwave portions of the entire electromagnetic spectrum.

Infrared Thermography is the science of actually allowing us to see the heat. An infrared inspection can detect heat that would normally be invisible to the naked eyes and represent them as an image, which we can see.

Inherent Reliability Level is the level of reliability of an item, equipment, or asset that is attainable by utilizing all the available maintenance tasks on RCM.

Initial Cleaning in Autonomous Maintenance Step 1 removes any form of unwanted abnormalities from the equipment. It consists of removing dirt, contaminants, grime, excess oil, grease, and other foreign objects that affect equipment parts and components that have accumulated over time on the equipment.

Initial task intervals also called proposed tasks, are maintenance task intervals assigned before the service maintenance program, subject to adjustments based on the actual operating experiences.

Inspection Tasks refer to scheduled tasks requiring testing, measurement, visual inspection, or human senses for detecting failure evidence done by both operators and maintenance. For inspecting hidden failures, this will default to Failure Finding tasks, while for evident failure inspection, it will default to On-Condition Tasks.

Instinct can be defined as an inborn impulse or motivation to action typically performed in response to specific external stimuli. It can be considered as an intuition, feeling, impulse, or gut feeling. This is how we react to certain situations.

Insurance Spares is a spare part that will replace a failed part in a piece of equipment whose penalty cost for downtime is very high. These parts do not become obsolete until the equipment is retired from service. In most cases, these parts are big and classified as non-moving parts.

Intangible Measurements cannot be quantified, measure, or contribute to the plant's bottom-line results. They can also be termed as qualitative measurements.

Intermediate Discipline is a discipline on world-class maintenance that refers to the different reliability and maintenance strategies that can be adopted once basic equipment conditions have been established in the asset. This includes Lubrication Management, MRO Spare Parts Management, Life Cycle Management, Root Cause Failure Analysis, Reliability, and Maintenance Strategies.

Internal Set-up are those activities that can be performed during conversion only when the machine had been totally shut down for conversion since a different product will be required to operate on the equipment.

Industrial Internet of Things (IIoT) uses wireless smart sensors and actuators to enhance manufacturing and industrial processes by hooking them on the equipment and providing data directly to your computer.

ISO 55000 is a standard described as the parent document of ISO 55001. It provides a general overview of asset management as a discipline and contains definitions of terms used in the ISO 55000 series of standards. This ISO standard aims to establish a clear and consistent understanding of the principles and requirements applied when developing and implementing an Asset Management System.

Karl Fisher Titration Test is an oil analysis instrument that measures the amount of dissolved moisture content in the oil. In this method, water reacts quantitatively with a Karl Fischer reagent. This reagent is a mixture of iodine, sulfur dioxide, pyridine, and methanol.

Kauro Ishikawa developed the Ishikawa or Fishbone Diagram in 1969. A fishbone is constructed by assigning the 4M's and 1E. 4M refers to man, machine, method, and materials, while 1E refers to the environment. The most common probable causes that relate to the problem are listed and grouped accordingly. The team brainstorms and focused on the most likely or probable causes of the failure. He also developed the Poke-Yoke process.

Kinematic Viscosity is a viscosity test, which takes the time for the oil to travel through a glass orifice of a capillary tube under the force of gravity.

Kosho is the Japanese term for breakdown. It is lost time due to equipment failure that may or may not cause downtime.

Knowledge-Based Mistakes are types of mistakes that occur when someone is, confronted with a situation that has not yet occurred before and had not been anticipated. In other words, there are no rules or procedures to follow. In situations like this, the person has to decide quickly about an appropriate course of action, and a mistake occurs due to the wrong decision.

Lagging Indicators are indicators and measurements that indicate the bottom-line results. Examples of this include productivity, repair, and maintenance costs.

Lapse occurs when someone misses out on a key step in a sequence of events or activities. For example, a mechanic leaves a tool behind after working on a machine or simply forgets to fit a key component while reassembling it. If a person will do step 1, step 2, step 3, step 4, and step 5, and the person missed out on step 4 in the process, then a lapse occurred. As humans age, our memory will not be as sharp compared to when we were young, just like in our high school days. Lapse is when people tend to forget things as they age.

Latent Cause is the last level on Root Cause Failure Analysis, which refers to hidden or concealed causes that need to be exposed so industries can learn from the things that go wrong. This is about looking in the mirror and admitting that each of us is also part of the problem.

Leading Indicators are the process of how the results were achieved. These indicators influence the final outcome of the results. These are also the indicators that lead to the lagging results.

Life Data Analysis is when a product, part, or component has operated successfully, or the time it operated before it failed, measured in hours, miles, cycles, strokes, minutes, or other measures.

Life Cycle Cost is the sum of the initial costs and the running costs of the equipment. It is the sum of the overall cost of equipment throughout its entire lifespan, including the costs incurred initially during the design stage up to the stage the equipment will be put out of service, disposed of, or decommissioned.

Lubricating oil, also called a lubricant or lube oil, is a class of oil used in equipment, engines, and machinery types to reduce the friction, heat, and wear between mechanical components to avoid metal-to-metal contact.

Lubrication Tasks refer to the scheduled tasks to assure the existence of completeness of lubrication films on the equipment. These tasks can either be Scheduled routine tasks or condition-based maintenance tasks.

Machine-Related Downtime, also known as unplanned downtime, is when the equipment is not operating due to machine-related downtimes such as breakdowns, set-up and conversion, cutting tool change, minor stoppages, and quality defects.

Maintenance Induced Failures are direct failures caused by maintenance itself. Examples of maintenance-induced failures are infant mortality failure or doing intrusive maintenance, incorrect overhauling practices, and inducing early failures right after endorsing the equipment back to operators. In this case, the maintenance reassembled the equipment incorrectly.

Maintenance Management is the art or science of managing maintenance resources. It is also the manner of managing to keep our physical assets in an existing state or condition.

Maintenance Prevention according to JIPM, Maintenance Prevention is defined as the use of the latest maintenance data and technology when planning or building new equipment to promote greater reliability, maintainability, economy, operability, and safety while minimizing maintenance cost and deterioration.

Marking is the process in a semiconductor industry of identifying, traceability, and distinguishing marks on the integrated circuit package at the end of line process.

Material Hardness is the measure of the material's resistance to plastic deformation. The most common method to determine the hardness of the material is the Rockwell Hardness Test, which can test the hardness of metals and alloys.

Maxi-Event or large-scale RCFA is performed on maxi events or failures with a large impact or consequences in the industry. It is recommended that an outsider do this event or an independent third party person act as the principal investigator to avoid bias. There will be three persons assigned as the evidence-gathering team for the maxi event, which will require a stakeholder meeting.

Mean Time between Assists (MTBA) is the average time the equipment performs its intended function between assists. It is also the productive or the operating time divided by the number of assists.

Mean Time Between Failure (MTBF) is defined as the average time between failures and therefore is a measure of trouble-free time. The common unit used is in hours. MTBF is derived from the US MIL-STD 217 for testing electronic parts. MTBF is a reliability engineering term that means the average amount of operating time between the occurrences of breakdowns that requires repair divided by the frequency of failures. The frequency of reporting MTBF should be either on a weekly or monthly basis. MTBF trend should be the higher, the better.

Mean Time to Fail (MTTF) is the expected time to fail of a system, part, or component. It is a basic measure of reliability for non-repairable systems. It is the meantime expected until the first failure of a piece of equipment.

Mean Time to Repair (MTTR) is the average time required to repair a component. Other terms used are Mean Time to Restore, Mean Time to recover, or Mean Time to React. It is also the average time required to perform corrective maintenance or repair on all removable items in equipment, product, or system.

Mercaptan was introduced by William Christopher Zeise in 1832. The Latin word mercurium captans means capturing mercury because the thiolate group bonds strongly with mercury compounds. These mercaptans are used as an odorant in sour gasoline into disulfides.

Micron is also known as a micrometer and is exhibited by the Greek symbol Mu or Mµ. A unit of one-micron length is equivalent to 39 millionths of an inch or 0.000039 or 0.0009906 millimeters.

Midi-Event or Medium Scale RCFA is performed on midi or medium-scale events. A Principal Investigator leads the investigation with one or two persons assigned as the evidence-gathering group.

Mini-Event or Small Scale RCFA is to encapsulate plastic material components to protect the IC or passive devices at the end of line process in semiconductor industries.

Moisture Vapor Transfer Rate (MVTR), or sometimes called Water Vapor Transmission Rate (WVTR), is a measure of water vapor passage through a substance. It is a measure of the permeability of vapor barriers. There are many industries where moisture control is critical such as semiconductor industries.

Moment represents the effectiveness of either a rotating, bending or twisting force upon an object. All materials have their respective hardness. The hardness of the material is the measure of the material's resistance to plastic deformation.

Motor oil, also called engine oil, is a lubricant used in internal combustion engines, power cars, motorcycles, diesel engines, engine-generators, mobile, mining equipment, etc. Inside the main engine are metal parts, which move relatively against each other, and the friction between these moving parts consumes more power by converting kinetic energy into heat.

MRO Spare Part is defined as a part of a machine ready to replace an identical part if it becomes faulty. It is also defined as those parts of the machine which are kept on standby to be substituted when a part of equipment fails, a repair is required, or the part simply becomes worn out and needs to be replaced

MRO Wear and Tear Spares are spare parts that must be replaced every time the equipment undergoes a Preventive Maintenance or Shutdown, and the equipment is disassembled and re-assembled for overhauls and parts replacement.

MSG-1 Document is a working paper prepared by the Boeing 747 Maintenance Steering Group published in July 1968 under the title: Handbook Maintenance Evaluation and Program Development (MSG1), which includes the first use of the decision diagram techniques to develop an initially scheduled maintenance program.

MSG-2 Document is a refinement of the decision diagram procedures in MSG1 published in March 1970 under the title: MSG-2 Airline/Manufacturer Maintenance Program Planning Document, which is the immediate precursor of the RCM document. This document was used to develop a scheduled maintenance program for Lockheed 1011 and Douglas DC10, military aircrafts such as Lockheed S3 and P3, McDonnell F4J.

MSG-3 Document refers to the Operator/Manufacturer Scheduled Maintenance Development developed by the Airlines for America (A4A), formerly the Air Transport Association or ATA. This document was developed in 1980, revised in 1988 and 1993, and to this day process is used for maintaining all types of civil aircraft.

Multipass Test is a controlled laboratory test where the effluent fluid is recirculated through the filter element while a new contaminant is continuously added. This test is used to determine the efficiency and the beta rating of the oil filter.

Multiple Failure is where the protected function had miserably failed because the protective device is said to be in a failed state.

Multi-purpose grease can be defined as grease combining the properties of two or more specialized greases that can be applied in more than one application. This permits the use of a single type of grease for a wide variety of applications.

Near Miss can be considered as minor accidents or close calls that have the potential for an injury, accident, property loss, or even death. A person was standing in front of a Wrecking Ball Crane. After a few minutes, the ball fell which nearly smashed the head of the person just a few feet away.

Noack Volatility test was named after Kurt Noack. It is measured by the principal European test called NOACK. It is the amount of oil lost (light molecules) over time at a given temperature and pressure. It directly impacts high engine temperature and oil effectiveness, especially on viscosity, emissions, and oil consumption. Today's oil has a NOACK volatility limit of 15 %. Oil's Volatility rate should be less than 15%.

Nominal Filtration refers to the average particle size of contaminants that will remain in the fluid after filtration. The efficiency of nominal filtration is 50%, and the beta rating is 2.

Non-Machine Related Downtime, also known as planned downtime, is when the equipment is not operating due to non-machine-related factors such as PM activities, operators break time, meetings, or any other downtime which is not caused by the machine.

<u>**Glossary on Maintenance**</u>

Non-Maintenance Induced Failures are failures on the equipment, which are not caused by the maintenance function, but other factors such as how the equipment was operated, design errors, commissioning errors, or cutting costs.

Non-Operational Consequences is one of the consequences of a failure mode, which does not directly affect safety, environmental, or operational consequences. The only consequences of this failure mode are the direct cost of repair and the chances of secondary damages to the equipment.

No-Scheduled Maintenance is a default task on RCM which means that there is no feasible scheduled maintenance. Failure is allowed to occur. Also similar to the terms run to fail, breakdown maintenance, reactive maintenance, or corrective maintenance. When the failure occurs, it will be subject to repair.

Occurrence in FMEA/FMECA is a rating according to the likelihood that a particular failure mode will occur within a specific period. Occurrence is an assessment of the likelihood that a particular failure mode will occur. This is also termed as the probability of failure to occur.

Octane Rating measures the fuel's ability to resist engine knocking caused by the air-fuel mixture. The higher the octane, the greater resistance the fuel has to knock during the combustion process. The octane ratings are a measure of the fuel's stability.

Office Supplies Inventory are inventory supplies used by the different plant offices ranging from printer ink, pencil, ball pen, bond paper, folders, envelopes, scissors, paper clips, memo pads, and so on in industries.

Oil Analysis is a maintenance management tool that allows users to monitor the oil condition for maximum equipment life and maximum lubricant drain interval. It tells us the actual condition of the oil in our equipment.

Oligomerization is a polymerization process in which a few, usually three to ten, of the basic building block of molecules are combined to form the finished product. Therefore, the product is formed with varying molecular weights and viscosities to meet a broad range of requirements.

On-Condition Tasks are maintenance tasks that entail checking the equipment's actual condition or using human senses, pressure, temperature inspections, gauges, and SPC charts.

One-Point Lesson is a learning tool for communicating standards, problems, and improvements in work processes and equipment. Workers and supervisors use one-point lessons to provide key information about everyday work and improvement opportunities.

Operational Consequences are a type of consequence where operations will be affected. The primary function of most equipment in any industry is connected to the need to earn revenue or support revenue-earning activities. A failure mode has operational consequences if it has a direct adverse effect on operational capability.

Overall Equipment Effectiveness (OEE) is the primary measurement used for plants and industries initiating Total Productive Maintenance. It is calculated by multiplying the equipment availability by its performance rate and quality rate expressed in percentage.

Oxidation is the breakdown of oil due to the extreme heat in the engine or equipment. This is the reaction between the oil and oxygen, causing acidic gases, sludge, and varnishes to form in the crankcase.

Oxidation inhibitors are added to the oil to extend its operating life. These additives are said to be sacrificial (perhaps like a suicide bomber), consumed while performing their duty of delaying oil oxidation, thus protecting the base oil. They are present in almost every lubricating oil and grease. This inhibitor aims to prevent oxygen from reacting with the oil, thus slowing its aging rate.

P-F interval is the interval between the emanation of potential failure and its decomposition to its final stage, a functional failure. At this stage, the equipment has already failed. The P-F interval is actually a warning period or the lead-time to failure or better known as the failure development period

P-M Analysis is a problem-solving tool design to analyze chronic losses according to the inherent principles and naturals laws that govern them. It was developed by Mitsugu Kaneda, Shirose Kunio, Yoshifumi Kimura. P stands for Phenomena and Physical. A phenomenon is a deviation from a normal to an abnormal state. M stands for Mechanism and the 4Ms.

Pareto's 80/20 Rule. Dr. Joseph Juran, a Quality Management pioneer, developed the use of the Pareto Principle. He worked in the US from 1930 to 1940. According to Vilfredo Pareto's observation, 80% of the land in Italy was owned by 20% of its population, where 20 percent of something is always responsible for 80 percent of the results. He recognized a universal principle he called the vital few and trivial many and placed it into writing.

Partial Functional Failure is when the asset is still functioning, but it is below the performance standards that the user wants the asset to function.

Penetration is the measure of the grease's consistency and depends on whether the consistency had been altered by handling. ASTM D217, and D1403 test, measures the penetration of unworked and worked greases.

Physical Cause is the first level of Root Cause Failure Analysis that refers to the physical cause of why the part or component failed. This is the technical explanation of why things broke or failed. This is said to be the metallurgical aspect as to why the failure occurred. It is also referred to as the Failure Analysis. The analysis will end on the component or part level.

Physiological Factors are a type of human error that refers to environmental stress affecting human performance. These stresses can include high or low temperatures, loud or irritating noise, excessive humidity, excessive or high vibration," exposure to toxic chemicals, radiation, or working too long without adequate break time, not to mention the day-to-day pressure from the boss.

Planned Maintenance is one of the TPM pillars, aiming to improve the current maintenance activities, whose goal is to zero out all unplanned breakdowns of the equipment or asset. They are also the mentors of Autonomous Maintenance.

<u>**Glossary on Maintenance**</u>

Planning is defined as the total process set up to ensure that the right resources and materials arrive at the right place, at the right time, and doing the right job in the right way. Planning also has a macro meaning when reference is made to an array of scheduled shutdowns for the year, the business plan, the marketing plan, budget, and others.

PLCC (Plastic Leaded Chip Carrier) is a plastic, square, surface mount chip package that contains leads on all four sides. The lead pins extend down and back under and into tiny indentations in the housing. PLCC is a type of integrated circuit package that can be used to enable us to be mounted on a printed circuit board directly soldered either to the board or within a socket.

Potential Failure is defined as an identifiable physical condition that indicates that a functional failure is about to occur or is in the process of occurring.

Poka-Yoke also referred to as Mistake Proofing, is mostly used by manufacturing industries, especially to help an equipment operator avoid errors and mistakes during operations. Its purpose is to eliminate product defects by preventing, correcting, or drawing attention to human errors. Shigeo Shingo, as part of the Toyota Production System, developed it. It was originally termed as Baka-Yoke, which was called foolproofing or idiot-proofing. Later on, Shigeo Shingo changed it to a milder name Poka-Yoke.

Pour Point is an oil analysis test that refers to the temperature at which oil will solidify due to cold temperature making the oil no longer capable of flowing. This oil analysis test only applies to countries with winter or extremely cold seasons. This refers to the W rating for engine oils. Note that W stands for winter.

Pour point Depressants are critical substances added to the additive compound to prevent the wax formation in the base oil from forming large crystal networks that can inhibit the flow of lubricating oil at cold temperatures, usually during the winter season.

Precision Maintenance involves performing maintenance work in a consistent, precise, and industry-accepted way. If properly implemented, maintenance should yield exactly the same results no matter who is performing the tasks whether the person is the most experienced or least experienced of the craft.

Predictive means to declare or indicate something in advance, especially foretell based on observation, experience, or scientific reason. Predictive comes from the Latin word "pre," meaning before, and "diction," also from the Latin word "dicare," meaning to proclaim.

Predictive Maintenance is a type of maintenance performed on the equipment based on the equipment's actual condition with specialized diagnostic monitoring tools or Predictive Maintenance instruments to determine potential failures. Predictive Maintenance is a maintenance activity geared to indicating where a piece of equipment is on the critical wear curve and predicting its remaining useful life.

Preventive Maintenance is a type of maintenance performed on a fixed schedule or time-based interval. Preventive Maintenance is used when the parts have a wear rate and are directly related to their age.

Primary function explains why the asset was purchased. It has something to do with the volume, output, speed, and quality

Proactive-Corrective-Maintenance, which I can define as any activity done on the equipment such as overhauling, replacement, or other means resulting from Predictive Maintenance tasks before the failure is about to happen.

Proactive Maintenance is about understanding the root cause of why a part keeps on failing and performing corrective measures to eliminate the problem completely and avoiding the recurrence of failure. In Proactive Maintenance, maintenance is ahead of failure.

Protective Device refers to those devices place in the equipment and assets to protect something. Samples of protective devices include led, sensors, alarms, emergency stops, lighting arresters, and so on. The protective device's function is to indicate that the protected function is still working and is functional.

Psychological Factors can be grouped according to those, which are unintended, and those errors which are intended. Unintended errors can be group into slips and lapses, while intended errors can be group according to mistakes and violations.

Pumpability is the ability of the grease to be pumped or pushed through a system. This indicates how easy or hard the grease can flow through lines, nozzles, and grease dispensing units.

Purchasing Cost is also called the Ordering Cost. This is the sum of the fixed cost that is incurred each time an item is ordered. This includes the physical activities required to process an order. Usually, the purchasing cost can include the purchaser's salary, telephone costs, receiving clerk salary, account payable costs, internet costs, electricity, and other costs.

Radio Frequency Identification (RFID) refers to a technology where the digital data encoded in RFID tags or smart labels are captured by a reader through radio waves. RFID systems consist of three components: an RFID tag, or smart label, an RFID reader, and an antenna.

Random Failures are failures that can occur at any given period. This means that the probability that an item will fail in any one period is the same as it is in any other period. This means that the conditional probability of failure is constant.

Raw Materials Inventory: These are inventory parts and items used by industries to produce their finished products. If you work in the automotive industry, this will refer to the different parts needed to assemble a car.

RCFA Logic Tree diagram is a tool that uses deductive logic to guide the thought process used to draw correct conclusions. Therefore, a logic tree is a disciplined methodology that prompts the user to answer questions that will eventually identify the root cause of a failed event.

Reactive-Corrective-Maintenance is the activity of repairing or troubleshooting the equipment after a failure occurs. This will be done after the failure happens, similar to the original definition of corrective maintenance.

<u>Glossary on Maintenance</u>

Reactive Maintenance is a type of maintenance strategy that simply means fixing it only when it fails. As the saying goes, when it ain't broke, don't fix it; when it fails, then we come and fix it. Other terms used to designate Reactive Maintenance are run to fail, run to destruction, corrective maintenance, fire-fighting mode, and stop the bleeding syndrome. RCM termed this as no-scheduled maintenance.

Redesign or Modification means that if the rest of the maintenance tasks are not feasible and worth addressing the failure mode, then the last option for maintenance will be to resort to redesign or modification, especially when the failure mode will have safety or environmental consequences. The goal here is to eliminate or reduce the consequences of failure from happening.

Redundancy or Standby means duplicating the system or component, which is not affected if failures and breakdowns occur. Failures are allowed or being tolerated through some forms of redundancy, standby, or when the asset has some form of duplicated function.

Reforming is a process designed to increase the amount of gasoline that can be produced from crude oil. Hydrocarbons in the naphtha stream have almost the same number of carbon atoms as gasoline, but their structure is generally more complex. Reforming rearranges naphtha hydrocarbons into gasoline molecules. The other products of reforming are light gases and a high-octane gasoline blending component called reformate.

Reliability is the probability that an item will operate without failure throughout a specified interval and that the item will perform its intended function under specified operational and environmental conditions under a given and specified time.

Reliability-Centered Maintenance is a process used to determine the most feasible equipment maintenance requirements in its present operating context or state that it is being operated. This strategy is used to define the correct maintenance tasks for equipment or asset with the aid of an RCM decision diagram or algorithm.

RCFA Logic Tree diagram is a tool that uses deductive logic to guide the thought process used to draw correct conclusions. Therefore, a logic tree is a disciplined methodology that prompts the user to answer questions that will eventually identify the root cause of a failed event.

RCM Decision Diagram is an algorithm that allows the users to make a decision-making process to identify the most appropriate and feasible tasks to address a particular failure mode

RCM Default Tasks are RCM tasks available when it is not feasible to perform a Preventive or an On-Condition task to address a particular failure mode. Default tasks on RCM include Failure Finding Tasks for hidden failures, Run to Fail, Switching intervals for redundant functions, and redesign or modification.

Recurrence Prevention lists things and actions that need to be done to indicate that the failure has not repeated itself. This will usually be done in the case of the physical, human, and system cause only.

Risk is something or someone that can cause a problem, harm, injury, damage, danger, hazard, or loss to the industry. Risk is the potential for uncontrolled loss that is of definite value to an organization. It is an intentional interaction with uncertainty.

Risks Priority Number or RPN in FMEA is the sum of severity multiplied by its detection and its occurrence. This is also the sum of the consequences multiplied by its criticality, confidence, and probability.

RNM Cost refers to the total repair and maintenance cost incurred on the equipment or system. The trend we want should be the lower the RNM cost, the better. RNM costs should be tracked and reported either weekly or every month.

Roll Stability of Grease: ASTM D1831 is a grease test that assesses how stable the grease is when it is subject to operating conditions. This test allows us to determine if the grease will handle the intended load and for how long before it begins to fail.

Root Cause Failure Analysis is a basic investigative tool that allows us to understand why things went wrong by identifying the problem's basic source or origin. Root Cause Failure Analysis is the investigation process or probing on a problem, which is entirely based on the evidence unfolded on equipment-related failures.

Rooticians are people who are knowledgeable in conducting a Root Cause Analysis or Root Cause Failure Analysis investigation. They are the fact-finding group deployed to find the cause of a problem or incident.

Rotable Spare Parts are reusable spares or components that can be reconditioned and reused, such as motors and engines. These can be a component or inventory item that can be repeatedly restored or refurbish to be used once again on the equipment into a fully serviceable condition after it failed. The value of rotable spare parts depends on the remaining useful life of the production equipment it supports.

Ruled-Based Mistakes usually occur when people believe that they follow the correct course of action when doing specific tasks based on a specific rule or procedure, but the course of action is inappropriate.

RULER (Remaining Useful Life Evaluation Routine ASTM D-6971-04 3 and ASTM D-6810-02) is an oil analysis test that measures the remaining useful life of the lubricant. This test can be used to determine whether the oil needs to be changed or can be extended. The RULER oil analysis test will measure the remaining level of antioxidant additive in the lubricating oils.

Run to fail is a maintenance strategy, which tells us that it is time to repair the failure when a machine fails. The failure will happen first, and then maintenance reacts by repairing it. This is the very essence of reactive maintenance. Maintenance is done at a point when there is repair or actual breakdown of the equipment. This occurs when repair action is taken on a problem only when the problem results in a machine failure or breakdown.

SAE JA1011 is also known as the Evaluation Criteria for Reliability-Centered Maintenance (RCM) Processes. This document describes the criteria to indicate that any process or analysis to derived the tasks is compliant with the RCM process requirements. This also separates the classical RCM from the Streamlined RCM.

Safety Consequences: A failure consequence where a failure mode can cause injury or kill someone else. Failure modes that can cause accidents or even near-misses, as a result, will be included in this category.

Safety Supplies Inventory refers to items regularly supplied to employees by the EHS, such as gloves, hard hats, goggles, safety shoes, ear protectors, overhaul outfits used as part of the plant's uniform in conducting their regular work routine provided to operators and maintenance people working in the plant.

Saponification is the process used to develop the grease thickeners where fatty acids are used to react with an alkali to form a chemical soap.

Schadenfreude means the pleasure we gained from another person's distress, especially if they perceived the person deserved it because they engaged themselves in a bad deed.

Scheduled Discard Tasks refer to removing an item or spare to be replaced at a specified time interval before the part wears out completely. This is included as a Preventive Maintenance task. Also termed as scheduled replacement of parts. The frequency of scheduled discard tasks is governed by the age at which the item or component shows a rapid increase in failure conditional probability.

Scheduled Restoration Tasks are Preventive Maintenance tasks that entail re-manufacturing a single component or overhauling an entire assembly on or before a specified age limit, regardless of its condition at the time, also termed as Scheduled Rework. The frequency of a scheduled restoration task is governed by the age at which the item or component shows a rapid increase in failure conditional probability.

Scheduled Maintenance Tasks, also called Preventive Maintenance tasks, are scheduled tasks that need to be done on the equipment or asset at scheduled intervals. In RCM, these refer to scheduled discard and scheduled restoration tasks.

Seal Swell is one of the functions of the oil in which the oil must be compatible with the seals and must not cause the seal to crack, shrink, or degrade; rather, it must cause the seals to expand slightly to ensure proper sealing.

Secondary function refers to the other functions of the asset besides the primary function. Most assets are expected to fulfill more than one function besides their primary functions.

Secondary Damage refers to the damages on the equipment that is caused by the primary failure mode. For example, a bearing seizes that it caused a fracture on the shaft or other parts due to excessive vibration.

Set-Up Loss, Conversion, or Changeover in several manufacturing industries, equipment is not dedicated, producing different product types with the same equipment. Hence, when it is time to change one product to another, it will require the time required to remove dies, jigs for one product, clean-up, prepare dies and jigs for the next product, reassemble the equipment, adjust the equipment, perform trial runs and make further adjustments until the product of acceptable quality is now obtained from the equipment.

Severity in FMEA/FMECA refers to the impact of the failure mode and its effects. Severity considers the worst scenario that can happen after a failure or breakdown takes place. It is a rating corresponding to the seriousness of the effects of a potential failure mode. This is termed as the consequences.

Shear force is the sum of the effect of shear stress over a surface that results in a shear strain. The important thing to consider is that the force acting on an object is parallel.

Shear Stress is also called tangential stress. Unlike both tensile and compressive stress, the force applied is not perpendicular but parallel to the area. Once a force is applied in parallel in opposite directions, it can cause the object to deform. Shear stress is a force acting parallel to a surface or to a planar cross-section of an object.

Shutdown Loss is the time when equipment is shut down for scheduled maintenance. However, shut down related work generally affects the operating time of the equipment. Shutdown-related work must be regarded as a loss, and reduction of shutdown work time which must be reduced.

Situation is a set of things happening and the conditions that exist at a particular time and place. Merriam Webster dictionary states that situation is how something is placed concerning its surroundings. The outcome is something that follows as a result or consequence of the situation. It is a result or effect of an action, situation, or event.

Skill is the product of personal motivation and thorough training. The end result is mastery, which is used to perform one's job correctly and efficiently, developed over time.

Soot is a by-product of the combustion process and can escape the piston rings through blow-by. This may be caused either by low compression, too much idling due to stop and start, such as traffic jams or air-fuel mixture. Soot can range from 3 microns and above,

Sour Crude Oil is a crude oil that contains a high amount of sulfur, which is one of the main impurities in the oil. It is very common to find crude oil containing sulfur impurities. When the crude oil's total sulfur content is more than 0.5%, the oil is called sour.

Spare Parts Management is the science of managing parts inside the storeroom when operations and maintenance required them in case of a breakdown that seeks replacement or Preventive Maintenance routine and replacement activities.

Squirrel Stores are maintenance secret hiding places for their own spare parts. Squirrel Stores are unofficial stores in which maintenance keeps their own spare parts for their own need and benefit.

Squirrelling is the act where maintenance keeps the spares parts themselves. Why does maintenance do this? It is because there are a lot of stock-outs, which means that the part reflects in the system, but physically it is out of stock.

<u>**Glossary on Maintenance**</u>

Slip occurs when somebody does something incorrectly or does something in the wrong sequence. If a person will perform something in sequences such as step 1, step 2, step 3, step 4, and step 5, and what the person did was step 1, step 2, step 4, step 3, and step 5, the person committed a slip. This is when a human action takes place that is not intended. The occurrence of a lack of one's trains of thought is derived mostly from unconscious behavior.

Snake Oil as a quack remedy or panacea. Snake oil is a euphemism for deceptive marketing that promises its customers a silver bullet solution to all their problems. As the word implies, these are considered too good to be true products. Examples of these are lubricant additives commercialized on Home TV shopping performing tests on oil beyond your wildest imagination but only to be recognized by the Federal Trade Commission as a fraud.

Spalling is a fracture of the running surface, and subsequent small, discrete material particles removal. It is the pitting or flaking away of bearing material. Spalling can occur in the inner ring, the outer ring of the bearing's rolling elements.

Spectroanalysis (ASTM D5185 and ASTM D4951) is the analysis of metal content and additive package. This test will check 19 elements and reports them in ppm. This oil analysis tests limit its test from 10 microns and below, and its disadvantage is detecting particles larger than 10 microns.

Sporadic Failures are failures that often indicate sudden large and abrupt deviations from the norm. This refers to catastrophic failures and is the opposite of chronic failures.

Stakeholders will include any of the following; the person being accused, the person whose behavior needs to change, the person who has to spend money, the person accusing, or anyone that the Principal Investigator and evidence gathering team think is involved in the incident.

Start-Up Loss is a type of loss that occurs during starting up the equipment. Start-up loss means that the material loss is caused at the initial stage of product launching, namely the loss caused during the period from a start-up of production to the stabilized production stage.

Strategy is a careful plan or method for achieving a particular goal, usually over a long period. It also includes the carrying out of plans to achieve the goal. It is particularly a long-term plan for success. The strategy also means to maintain and build a competitive advantage over the competition.

Static friction is the force that keeps a motionless object from being pushed or pulled across a surface. The only force that is acting on this block will be the weight of the object.

Strain is the change of shape of the object after compressive or tensile stress is applied, which can also be equal to the Modulus of Elasticity (E), which is equal to the stress divided by the strain of the object. For tensile and compressive stress, the force applied is perpendicular or at 90 degrees to the area. It is also the quantity that describes the amount of deformation that can occur within a material body whenever a load is applied.

Strength in the metallurgical sense is the metal part's property to resists the part's stress imposed upon the part.

Stress is defined as the force per unit area, often considered as the force acting through a small area within a plane.

Stock-Out is when maintenance requests a part in the storeroom where there is an actual inventory in the system, but the actual or physical inventory is zero.

Sweetening is replacing a certain amount of oil so that new additives can be added to the existing used oil inside the equipment. Unlike changing oil, where the oil's total volume is replaced, around 25 to 30% of the used oil will be replaced with new fresh oil during the sweetening process.

Synthetic oil will fall on Group IV and V Base Stocks for lubricating oil. Synthetic oil is the result of a chemical reaction called synthesis, in which the end result is a uniformly shaped molecules that are more resistant to heat and impossible to achieve through the crude oil refining process compared to mineral-based oil, whose molecules are uneven in size.

TAN / TBN Crossover is where the value of both TAN and TBN becomes equal. When fresh oil enters service, the TAN will be low, while TBN will be high. As the engine runs, TAN will increase, and TBN will fall. When the TAN value becomes higher than TBN, the TAN/TBN crossover had been reached which means that it is time to change the oil.

Tangible Measurements or quantitative measurements are those measurements that are easy to quantify and have a direct impact on the bottom line performance of the plant.

Tensile Stress is When you have an object in a vertical position such as a cylinder, and a downward force is applied to an object, there will always be an opposite force which will be equal to the downward force. Both the upward and downward forces will cause the object to be under tensile stress.

Tool Cutting Blade Change is an equipment loss in the equipment or machine incurred on swapping or changing any consumable tooling item when it has become worn-out, ineffective, or severely damaged.

Torsional Stress is the twisting of an object caused by force acting on the object's longitudinal axis. The twisting effect is known as the torsion. Torsional stress can lead to deformation. The twisting is called torque which is also called the twisting moment.

Total Acid Number (TAN) is a measure of the total acid concentration present in a lubricant. Occasionally, the decrease of an additive package may cause an initial increase in the TAN of fresh oil.

Total Base Number (TBN) is a measure of alkaline concentration present in a lubricant. Engine oil is formulated with alkaline additives to combat the build-up of acids in a lubricant as it breaks down. Once alkaline additives are depleted, the lubricant no longer performs its function, and the engine is at risk of oxidation, corrosion, sludge, varnish, and other problems.

Total Functional Failure occurs when there is a total loss of function on the equipment. This can happen both on the primary and secondary functions of the equipment.

Glossary on Maintenance

Total Productive Maintenance is a plant improvement methodology system that enables continuous and rapid improvement of the manufacturing process through employee involvement, employee empowerment, and closed-loop measurement of results. TPM is a production-driven improvement methodology designed to optimize equipment reliability and ensure efficient management of its assets. TPM aims to build up a corporate culture that thoroughly pursues production system efficiency, improvement, and Overall Equipment Effectiveness, originating from Japan.

Treating in oil refineries has several options for their treating processes, but the primary purpose of the majority of them is the elimination of unwanted sulfur compounds. A variety of intermediate and finished products, including gasoline, kerosene, jet fuel, and gases, are dried and sweetened. Sweetening is a major refinery treatment of gasoline that treats sulfur compounds to improve color, odor, and oxidation stability.

Tribology was coined by Peter Jost in 1966, derived from the Greek word tribos, meaning rubbing, and translating the word literally on the rubbing science. The word tribology was not widely used until the Oxford English Dictionary defines tribology as the branch of science and technology concerned with interacting surfaces in relative motion and associated matters such as friction, wear, lubrication, and design bearings.

Trim is the cutting of the dambar that shortens the leads together. The form is the forming and bending of the leads into the correct shape and position. Singulation is the cutting of the tie bars attached in the individual units to the lead frame resulting in the individual separation of each unit from the lead frame at the end of the line process in the semiconductor industry.

Uptime is the time in which our equipment and asset are operating. It is also the time during which our equipment and asset are working without failure. Other terms to designate uptime includes operating time, productive time, running time, or time the equipment is utilized.

Utilization is the proportion of time that the equipment is utilized for its intended function and purpose. It is also the time the equipment is operating and running.

Varnish is an oil contaminant made of oxidized or carbonaceous adhesive material covering the engine's internal surfaces. As the oil is oxidized, these compounds form a consistent, hard, and shiny substance. Varnish occurs because of many factors, such as oxidization, friction, and heat.

Velocity is the average speed at which the fluid can move from a given point. The velocity is measured in either mile per hour, feet per second, or in SI or International System of Units. This is an important factor in sizing the hydraulic lines.

Vibration can be defined by these common terms, such as swinging back and forth, oscillating, unbalance, and shaking. The vibration occurs when a machine or machine component moves from its neutral or normal position to a lower and upper extreme travel limit. This happens because the force is applied to the rotating equipment and components. Vibration can also be a cyclic or pulsating motion of a machine or machine components from its point of rest.

Vibration analysis is one of the most common types of Predictive Maintenance techniques that analyzes these plotted vibration signals to diagnose abnormal vibration. This is the most used technique for analyzing the condition of rotating machinery.

Violation occurs when someone knowingly and deliberately commits an error. Violations fall under three categories: routine violation, exceptional violation, and acts of sabotage.

Viscosity is the measurement of a fluid's resistance to flow. Viscosity indicates the oil's capacity to lubricate by providing a thin film to separate metal-to-metal contact. An increase or decrease in viscosity can indicate many things in the oil and the equipment. Viscosity is not a perfect Oil Analysis test as we still need to perform other tests to diagnose the exact problem of why the viscosity of the oil changes.

Viscosity Index (VI) of a lubricant describes how the oil's viscosity will change depending on the temperature change. This means that if the temperatures increase, the viscosity will decrease, and vice versa. The viscosity index is a characteristic used to indicate variations in the viscosity of the lubricating oil concerning temperature changes.

Visual controls are any means or devices used to provide ease in inspection and expose early problems and deviations to the operators. It easily alerts operators of problems in the equipment. It is a business management technique that originated in Japan, where information is communicated using visual signals instead of texts or other written instructions. The design is deliberate in allowing quick recognition of the information being communicated to increase efficiency and clarity. This is also used to spot the anomaly or abnormality on the equipment more easily.

Volatility is the property that describes the degree and rate at which the oil will vaporize under a given pressure and temperature. This means that the higher the volatility rate, the higher the oil lost due to evaporation.

Waddington Effect means the less invasive the Preventive Maintenance is, the better the maintenance's outcome can be as more maintenance can lead to early failures. This can be said as the original concept of infant mortality failures.

Water Vapor Transmission Rate (WVTR) is a measure of water vapor passage through a substance. It is a measure of the permeability of vapor barriers.

Water Washout is a test on grease that refers to ASTM D 1264, which covers evaluating the resistance of lubricating grease to water washout from a bearing.

Wear is defined as damage to a solid surface caused by the removal or displacement of materials by any means of mechanical action of a contacting solid, liquid, or gas. It may cause significant surface damage, and the process is usually gradual. The damage is usually thought of as gradual deterioration, correlated with the equipment's operating age.

Wear debris can be defined as particles produced from the breakdown of surfaces within a machine. These particles can range from a submicron size to chunks of metal as large as those seen by the naked eye.

Wear Metal Debris Analysis Tests is an oil analysis test that indicates the number of metal contaminants present in the oil. By knowing this information, maintenance can understand what parts inside the equipment are on the verge of wearing out. Each of these elements will have its own specific limits, as specified in the oil analysis tables.

Wear-Out Failures or Age-Related Failures are parts that will eventually survive and reach a specific age before they fail. Age specified may be in running hours, time, number of strokes, calendar days, number of revolutions, number of stress applied, or any other form.

Weibull Distribution is often used to describe the lifetime of parts. It is used to analyze and predict failure rates and describe the failure of parts and equipment.

Work-in-process (WIP) Inventory comprises of all the materials, components, parts, assemblies, and subassemblies that are being processed in production or are still waiting to be processed within the system.

World Class is the ability to compete anywhere globally to meet and beat any competitor anywhere in the world in terms of product, price, quality, and on-time delivery. (Extracted from Terry Wireman's book on World Class Maintenance)

World Class Maintenance Management is the art and science of managing maintenance resources performed by best-in-class industries worldwide.

<u>RSA Maintenance Books Collection in Series</u>

This is not only about a technical book on maintenance; it is a book that makes everyone in maintenance feel proud that they belong to the maintenance function. If you have been living through the day-to-day pressures of doing maintenance, then this is your story. *Rolly Angeles*

I have written these books inspired by the millions of maintenance mankind who want only the best equipment and assets to function the way we want. These books are written in series, explaining each discipline's on maintenance in detail based on its original concept on World Class Maintenance, The 12 Disciplines.

• Volume 1: World Class Maintenance Management – The 12 Disciplines
• Volume 2: Maintenance – Roadmap to Reliability
• Volume 3: Reliability – A Shared Responsibility for Both Operators and Maintenance
• Volume 4: Cutting – Edge Maintenance Management Strategies
• Volume 5: Problems and Solutions on MRO Spare Parts and Storeroom
• Volume 6: Lubrication Tactics for Industries Made Simple
• Volume 7: Decoding Reliability-Centered Maintenance Process for Manufacturing Industries
• Volume 8: RSA Reliability and Maintenance Newsletter Vault Collection, Subscribers Edition
• Volume 9: Investigating Equipment Failures through Root Cause Failure Analysis

Figure F: Rolly Angeles Reliability and Maintenance Books

Available at https://www.amazon.com/Rolly-Angeles/e/B07B3T1TXC/ref=ntt_dp_epwbk_0

Index

Glossary on Maintenance

Trustworthy, 219
twisting moment, 130, 335
two-body abrasion, 123

U

Ultimate Weapon, 285, 288
Unbiased, 218
Union Carbide, 6, 116, 207, 208, 209, 210
University of Hard Knocks, 40, 169
Usain Bolt, 4, 21, 30, 31, 32, 33, 38

V

varnish residue, 134
Vilfredo Pareto, 83, 326

violations, 147, 148, 323, 328
Visual Communication, 174, 243
Visual controls, 6, 173, 174, 242, 336

W

wear and tear, 117, 121, 309
White Star Liner, 158, 162, 207
William Murdock, 160
William Rogers, 195
Witch-Hunting, 15

Y

Yoshifumi Kimura, 81, 326

344

www.ingramcontent.com/pod-product-compliance
Lightning Source LLC
Chambersburg PA
CBHW060111120726
48003CB00009B/2590